International Workshop on Flavour Changing and Conserving Processes (FCCP 2017)

EPJ Web of Conferences Volume 179 (2018)

Anacapri, Italy
7 - 9 September 2017

Editors:

G. D'Ambrosio
M. Iacovacci
S. Mastroianni
M. Passera
G. Venanzoni

ISBN: 978-1-5108-6390-3

Printed from e-media with permission by:

Curran Associates, Inc.
57 Morehouse Lane
Red Hook, NY 12571

Some format issues inherent in the e-media version may also appear in this print version.

Printed by Curran Associates, Inc. (2018)

For additional information, please contact EDP Sciences – Web of Conferences at the address below.

EDP Sciences – Web of Conferences
17, Avenue du Hoggar
Parc d'Activité de Courtabœuf
BP 112
F-91944 Les Ulis Cedex A
France

Phone: +33 (0) 1 69 18 75 75
Fax: +33 (0) 1 69 28 84 91

contact-edps@webofconferences.org

Additional copies of this publication are available from:

Curran Associates, Inc.
57 Morehouse Lane
Red Hook, NY 12571 USA
Phone: 845-758-0400
Fax: 845-758-2633
Email: curran@proceedings.com
Web: www.proceedings.com

TABLE OF CONTENTS

Workshop "Flavour changing and conserving processes", FCCP2017

Web page: http://fccp2017.na.infn.it
E-mail: loc_fccp2017@lists.na.infn.it
Tel. +39 081 18891222/223
Fax +39 081 18891224

FCCP2017

"International Workshop on Flavour changing and conserving processes"

Villa Orlandi, Anacapri – Capri Island, Italy

7 - 9 September 2017

BULLETIN

1. Place and Date.

The "International Workshop on Flavour changing and conserving processes" (FCCP2017), is organized by the Physics Department of the University of Naples "Federico II" and by INFN (Istituto Nazionale di Fisica Nucleare).
It will take place in Anacapri, Capri Island, from Thursday, September 7 to Saturday, September 9, 2017
This is the second workshop of a series which aims to discuss the status and prospects of the physics of "Flavour changing and conserving processes". It address the scientific community involved in experiments like g-2, mu2e, meg, meee, COMET, EDM and more generally to the physics of K and B mesons.

2. Organizing Committee

G. D'Ambrosio (INFN, Naples)
M. Iacovacci (University of Naples)
M. Passera (INFN, Padova)
G. Venanzoni (INFN, Frascati)

3. Local Organizing Committee

G. D'Ambrosio (INFN, Naples)
G. Carlino (INFN, Naples)
M. Iacovacci (University of Naples)
S. Mastroianni (INFN, Naples)
M. Passera (INFN, Padova)
G. Venanzoni (INFN, Frascati)

4. Secretariat

effe erre congressi srl
Area Industria della Conoscenza
Via Diocleziano 107/A
80124 Napoli
Ph: + 39 081 2302417
Fax: + 39 081 2429572
www.frcongressi.it

5. Preliminary schedule of the workshop

Thursday 7th		
08:00-09:00	Registration	
Morning session	**Chairman:**	
09:00 -09:05	*Welcome*	
09:05 -09:15	*Introduction*	
09:15 -09:40	*General discussion on g-2*	Marc Knecht
09:40 -10:05	*Status of the Muon g-2 experiment at Fermilab*	Tim Gorringe
10:05 - 10:30	*Status of the Muong g-2/EDM experiment at J-Parc*	Glen Marshall
10:30 – 10:55	*QED contribution to the electron g-2*	Stefano Laporta

11:00-11:20	**Coffee break**	
11:20 -11:45	*Hadronic leading order contribution to the muon g-2*	Daisuke Nomura
11:45 -12:10	*Lattice progress on hadronic contributions to g-2 of the muon*	Marina Marinkovic
12:10 -12:35	*Hadronic light-by-light contribution to the muon g-2 on the lattice (1)*	Christoph Lehner
12:35 -13:00	*Hadronic light-by-light scattering contribution to the muon g-2 on the lattice (2)*	Andreas Nyffeler
13:00 -15:00	**Lunch time**	
Afternoon session	**Chairman:**	
15:00 -15:25	*Dispersion relations for hadronic light-by-light scattering*	Gilberto Colangelo
15:25 -15:50	*On hadronic light-by-light contribution to the muon g-2*	Johan Bijnens
15:50 -16:15	*Review on gamma-gamma physics at KLOE*	Dario Moricciani
16:15 -16:40	*Status of Padme experiment and review on dark photon searches*	Mauro Raggi
16:40-17:00	**Coffee break**	
17:00 -17:25	*The muon magnetic moment in the 2HDM and supersymmetric extensions*	Dominik Stockinger
17:25 -17:50	*Muon pair production with hadronic vacuum polarization*	Vladimir Sauli
17:50 -18:15	*Radiative corrections for a precision determination of the fine structure constant*	Luca Trentadue
18:15 -18:40	*Measurements of alpha and proposal for amu^HLO spacelike*	Umberto Marconi
20:00	**Welcome party**	

Friday 8th		
9:00 -15:00	***Boat excursion around Capri Island***	
Afternoon session	**Chairman:**	
15:05 -15:30	*Report from the Muon g-2 Theory Initiative at fermilab*	Christoph Lehner
15:30 -15:55	*Status of R measurements (SCAN)*	Fedor Ignatov
15:55 -16:20	*ISR measurements of hadronic cross sections*	Achim Denig
16:20 -16:45	*Muon electron scattering at NNLO*	Pierpaolo Mastrolia
16:45 -17:10	*Monte Carlo generators for muon electron scattering*	Fulvio Piccinini
17:10-17:30	**Coffee break**	
17:30 -17:55	*B anomalies at LHCB*	Marta Calvi
17:55 -18:20	*LHCB and Kaon physics: present status and perspectives*	Diego Martinez
18:20 -18:45	*Standard Model and New physics for epsilon*	Teppei Kitahara
18:45-19:10	*Next-to-leading order prediction for the decay of mu to 3e nu nubar*	Matteo Fael
21:00	**Social dinner**	

Saturday 9th		
Morning session	**Chairman:**	
09:00 -09:25	*Lepton Flavor Universality in B Decays*	Ferruccio Feruglio
09:25 -09:50	*Status of Mu2e experiment*	Simona Giovannella
09:50 -10:15	*Status of Mu3e and MEG experiments*	Angela Papa
10:15 - 10:40	*NA62 results on heavy leptons and K+ to pi+ nu nu*	Gianluca Lamanna
10:40 – 11:05	*Status of COMET experiment*	Phill Litchfield
11:05-11:20	**Coffee break**	
11:20 -11:45	*Charged lepton flavour violation: precise background calculation and effective field theoretical interpretation*	Marco Pruna
11:45-12:10	*Flavour Physics and extra dimensions*	Iyer Abhishek
12:10 -12:35	*Leptoquarks in Flavor Physics*	Dario Muller
12:35 -13:00	*The SHIP experiment and tau neutrino magnetic moment*	Giovanni De Lellis
13:00 -14:00	**Lunch break**	
14:00 -14:30	*Summary talk*	TBD

6. Social Events

- Thursday, September 7: Welcome cocktail at 20:00;
- Friday, September 8: Boat tour of the island (starting at 9:00)
- Friday, September 9: Social Dinner (21:00 h) at Restaurant " Materita"

Via Giuseppe Orlandi, 140, – Anacapri (http://www.ristorantematerita.it/)

7. Workshop Site

The Workshop will be held in Villa Orlandi, Anacapri, Capri Island.
All information can be accessed online through the web page:

http://fccp2017.na.infn.it/

A valid passport is required for entry in Italy from outside of the european Schengen area. Participants from some countries will also need a visa. Please see http://www.esteri.it/visti/index_eng.asp for details.

8. Registration and Accommodation

People willing to participate should carefully fill in the Registration Form on the web page of the workshop (http://fccp2017.na.infn.it/) not later than August 31, 2017. All participants are kindly requested to reserve their accommodation on their own.

9. Fee

The registration fee is 250 euros for participant and 80 euros for accompanying person. It includes proceedings, lunches, coffee breaks and social events. Accompanying people will have to pay 80 euros for social program, namely boat tour on Friday 8 and social dinner.

For fee payment, please refer to http://www.frcongressi.net/pay/index.php?token=4c4ca945c0375ec05037b27b40941da8

For any questions on fee payment and Hotel reservation contact:

effe erre congressi srl
Dr.ssa Alessandra Saioni,
Area Industria della Conoscenza
Via Diocleziano 107/A
80124 Napoli
Ph: + 39 081 2302417
Fax: + 39 081 2429572
www.frcongressi.it

10. Travel and information

You can get a fast boat to Capri from Naples' main port *Molo Beverello*. There are also some slower boats (car ferries) leaving from *Porta di Massa*, which is 500m from Molo Beverello (picture). Information about timetables can be found at the following link http://www.capri.net/salsa/lang/en/page/transport.html Please note that the timetables usually change end of March.

Arriving by plane: From Napoli Capodichino-Airport it takes 45 min. to reach the Port of Naples. The company *Alibus* runs a bus that leaves in front of terminal 1 every 20 minutes (approximately at .00, .20, .40). The bus makes two stops, one at the central train station, the second at Piazza Municipio. The port (Molo Beverello) is at the bottom of the square.

The ride costs 3 euro and the ticket can be bought on the bus. Taxis are also available (ride to Molo Beverello about 25 euro, verify price BEFORE boarding the car!)

Arriving by train: Stazione Centrale (15 minutes from the Molo Beverello) is connected with the port by bus and tram.

Anacapri is one of the two main villages in the island of Capri (the other one is called "Capri", like the whole island). The hotels where we are staying are located in Anacapri. To reach them, you can choose one of these ways:

1. From Marina Grande (the port where the boat leaves you) take the direct bus to Anacapri (runs on average every hour; cost=1,80 Euro; it takes 15 min.; buy a ticket at the ticket-desk). Tell the driver the name of your hotel.
2. At Marina Grande (the port where the boat leaves you) buy at the ticket-desk a ticket for the "Funicolare"=cable-car to Capri village (runs every 15 min.; it takes 4 min.); then, from Capri take the bus to Anacapri (runs every 10 minutes; it takes 15 min.). You need a second ticket for this ride. Tell the driver the name of your hotel.
3. From Marina Grande (the port where the boat leaves you) take a taxi (it costs about 25 Euros; it takes 15 min.; verify the price before boarding the taxi).

How to reach Villa Orlandi: from the Piazza Vittoria bus stop in Anacapri, follow the suggested route to Villa Orlandi.

Useful links:

www.capritourism.com
www.capri.net

11. Proceedings

The Proceedings of the Workshop will be published on-line by the Open Access Journal "EPJ Web of Conferences" (http://www.epj-conferences.org).
All contributing authors will be asked to prepare a paper to be included in the proceedings and submit it within the deadline of November 30, 2017. More information can be found on the web page of the Workshop.

12. Internet Connection

A WiFi connection will be available for participants.

13. Contact

If you have any questions or comments, please send an e-mail message to the following address: **loc_fccp2017@lists.na.infn.it**

14. Important dates

Deadline of the registration:	August 31, 2017
Arrival:	September 6, 2017
Beginning of the meeting:	September 7, 2017
End of the meeting:	September 9, 2017
Deadline for the proceedings:	November 30, 2017

PREFACE

The second edition of the International Workshop on "Flavour changing and conserving processes" 2017 (FCCP2017) took place on September 7-9, 2017 in Anacapri, Capri Island, Italy. The Workshop was organized by the Physics Department of the University of Naples "Federico II" and by INFN (Istituto Nazionale di Fisica Nucleare) – Sezione di Napoli. The focus has still been continued on the physics of flavour changing and conserving processes, the involved scientific community has been that of experiments like g−2, Mu2e, MEG, Mu3e, COMET, but also that of the K and B meson physics. More than thirty talks were presented covering the theory status as well as the latest results of experiments in various Laboratories of the world. We wish to thank everybody who contributed to the success of the Workshop. The speakers did an excellent job in presenting the latest results and all participants contributed to the success of the Workshop.

We look forward to the next edition FCCP2019 when new results are expected, from many sides, especially on the g-2 front.

Giancarlo D'Ambrosio
Michele Iacovacci
Massimo Passera
Graziano Venanzoni
Stefano Mastroianni

Editors
FCCP2017

edp sciences

EPJ Web of Conferences

Statement of Peer review

In submitting conference proceedings to EPJ Web of Conferences, I certify to the Publisher that I adhere to the Policy on Publishing Integrity of the journal in order to safeguard good scientific practice in publishing.

1. All articles have been subjected to peer review administered by the proceedings editors.
2. Reviews have been conducted by expert referees, who have been requested to provide unbiased and constructive comments aimed, whenever possible, at improving the work.
3. Proceedings editors have taken all reasonable steps to ensure the quality of the materials they publish and their decision to accept or reject a paper for publication has been based only on the merits of the work and the relevance to the journal.

Title, date and place of the conference FCCP2017, 7-9 September, Anacapri, Capri (IT)

Proceedings editor(s): G. D'Ambrosio, M. Iacovacci, S. Mastroianni, M. Passera, G. Venanzoni

Date and editor's signature May, 3rd - 2018

17, avenue du Hoggar - PA de Courtabœuf – BP 112 - 91944 Les Ulis Cedex A (France)
Tél. : 33 (0)1 69 18 75 75 - Fax : 33(0)1 69 07 45 17 – www.edpsciences.org

On the Hadronic light-by-light contribution to the muon $g-2$

Johan Bijnens[1,a]

[1] Department of Astronomy and Theoretical Physics, Lund University, Sölvegatan 14A, SE22362 Lund, Sweden

Abstract. This talk is about the hadronic light-by-light contribution to the muon anomalous magnetic moment, mainly our old work but including some newer results as well. It concentrates on the model calculations. Most attention is paid to pseudo-scalar exchange and the pion loop contribution. Scalar, a_1-exchange and other contributions are shortly discussed as well. For the π^0-exchange a possible large cancellation between connected and disconnected diagrams is expected.

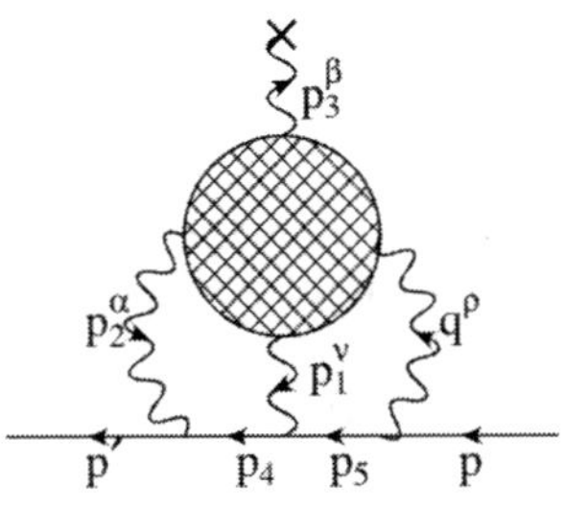

Figure 1. HLbL contribution to the muon anomaly a_μ. The crossed circle indicates the strong interaction part.

1 Introduction

This talk is mainly an update of my talk from two years ago [1] and has thus a very large overlap with it. In addition, this writeup should be read together with a number of other contributions to this and the previous workshop [2]. A more general introduction to the muon anomaly $a_\mu = (g_\mu - 2)/2$ was given in the talk by and Knecht [3]. An alternative method to obtain the hadronic light-by-light-contribution (HLbL) shown in Fig. 1 was discussed in the talk by Colangelo [4]. The present status of the lattice calculations of the same quantity were discussed by Lehner [5] and Nyffeler [6].

The main reason for this sessions is the measurement of the muon anomalous magnetic moment of [7] and the discrepancy with the standard model prediction and the new measurements in progress at Fermilab and under development at J-PARC. Reviews of the theory can be found in [8–10]. More references can be found in the remainder of this talk and the talks mentioned above. The present best estimate of the HLbL is $(11 \pm 4) \times 10^{-10}$ [8, 10] or $(10.5 \pm 2.6) \times 10^{-10}$ [9]. The main difference is an estimate of the errors which is always somewhat subjective. A new report with the aim of getting a consensus is under way as discussed in [11].

In this talk I will concentrate on the work done a long time ago [12–14] as well as some newer work on the pion loop [15]. I will also discuss more recent contributions about the pseudo-scalar exchange and quark-loop. I do not present a new final overall number but will argue that a good estimate for the pion-loop contribution is $-(2.0 \pm 0.5) \times 10^{-10}$.

An often asked question is why one cannot simply calculate the hadronic parts in Chiral Perturbation Theory (ChPT). ChPT is an effective field theory approximation to QCD valid at low energies. Since the muon has a low mass, at first sight a_μ should be a perfect quantity to calculate in ChPT. This is not true both for HLbL and the hadronic vacuum polarization contribution (HVP) there are integrals over all photon momenta present. The hadronic part is thus not only at low-energies. The lowest-order prediction for both HVP and HLbL is the same as for scalar QED and is finite. However, higher orders require a higher dimensional counterterm that is precisely the same as the muon Pauli term. We are thus left without a prediction beyond lowest order in ChPT. However, ChPT can (and should be) used to put as many constraints as possible on the underlying hadronic quantities.

We thus need to go beyond ChPT since we need high energies and beyond perturbative QCD since we need low energies for the hadronic quantities. The main options are experiment, dispersion relations, lattice QCD and models. For the HVP contributions models only play a role in understanding the results from the other approaches. For HLbL we have not quite reached that stage but important progress is being made as discussed by Colangelo, Lehner and Nyffeler. In the future, the main roles for models will be estimating the contributions that are not included in the systematic approaches.

The requirement for a model calculation is simple to formulate "do as well as you can." That means constraining your model as much as possible from experiment via measured states, form-factors and scattering processes and from theory by including as many long-distance constraints from ChPT and short-distance constraints from

[a] e-mail: bijnens@thep.lu.se

perturbative QCD. One should also use "common sense" in varying model parameters, making sure your model is general enough to describe what you need to describe and if different regions are treated differently consistency between them should be checked.

An overview of general properties of the underlying four-point functions and the early calculations is in Sect. 2. Sect. 3 discusses the numerically largest contribution, pseudo-scalar meson exchange. Next I discuss the pion-loop contribution in some detail since here I have new results [15]. The quark-loop, which has rather large theoretical errors, is discussed in Sect. 5. The remaining leading large N_c exchanges are scalar, discussed in Sect. 6, and a_1-exchange, Sect. 7. The π-loop contribution is treated in more detail since there is where I have some new results to present. Details are in Sect. 4. Conclusions and some possible future directions are given in the last section.

2 General properties and early work

The underlying object is the four-point function $\Pi^{\rho\nu\alpha\beta}(p_1, p_2, p_3)$ of four electromagnetic vector currents. We really need only the derivative w.r.t. p_3 at $p_3 = 0$,

$$\left.\frac{\delta\Pi^{\rho\nu\alpha\beta}(p_1, p_2, p_3)}{\delta p_{3\lambda}}\right|_{p_3=0} . \tag{1}$$

$\Pi^{\rho\nu\alpha\beta}(p_1, p_2, p_3)$ has in general 138 Lorentz structures which reduces to 43 gauge-invariant structures. Note that in four dimensions there really are 2 less, 136 and 41 [16]. Of the 138 more general structures 28 [15] actually contribute (improving the 32 estimate of [13]). Each of these functions depends on p_1^2, p_2^2, q^2 and before the derivative also on $p_3^2, p_1.p_3, p_2.p_3$. This should be compared with the lowest order hadronic vacuum polarization where there is one function of one variable. An alternative split is using the helicity amplitudes for off-shell photon-photon scattering as used in the dispersive work by Colangelo and collaborators [4]. The choice of basis is definitely not unique and different choices are appropriate for the different approaches.

After setting $p_3 \to 0$ the loop integrals over the photon momenta is 8 dimensional. Three of these integrations are trivial and using Gegenbauer polynomial methods two more can be done [10, 15, 17]. So, after having a model or a computation of $\Pi^{\rho\nu\alpha\beta}(p_1, p_2, p_3)$ there is a triple integral over p_1^2, p_2^2, q^2 left. The components and their derivatives become multiplied with functions of p_1^2, p_2^2, q^2 examples of which are in [10, 17] and the full results can be found in [15]. In the work I have been involved in we did the relevant integrations in Euclidean space, i.e. with $P_1^2, P_2^2, Q^2 = -p_1^2, -p_2^2, -q^2$ always positive.

How models actually contribute to the muon anomaly a_μ can be studied by rewriting the integral over P_1^2, P_2^2, Q^2 in the form [8]

$$a_\mu = \int dl_{P_1} dl_{P_2}\, a_\mu^{\mathrm{LL}} = \int dl_{P_1} dl_{P_2} dl_Q\, a_\mu^{\mathrm{LLQ}} \tag{2}$$

with $l_P = (1/2)\ln\left(P^2/\mathrm{GeV}^2\right)$. The reason for choosing the logarithm is that this way it is easiest to see which momentum region contributes. Alternatively one can integrate each momentum up to a cut-off Λ.

One should remember that the different contributions are usually defined within a given model or approach. What is included under the same name can therefore differ and one should be careful when drawing conclusions from comparing calculations.

The underlying problem is that the integration over photon momenta p_1, p_2 in the diagram in Fig. 1 contains both low and high momenta and mixed cases. Double counting is thus a serious issue when using both quark and hadron contributions. In Ref. [18] a partial solution was found by using chiral p and large N_c counting to distinguish different contributions. This does not fully solve the double counting issue but it is a good start. This suggestion was followed by two groups doing a more or less full evaluation of the HLbL. Kinoshita and collaborators [19–21] (HKS) used meson models, did the pion-loop using the hidden local symmetry model for vector mesons and the quark loop with simple vector meson dominance (VMD). Calculations were performed in Minkowski space. The one I was involved in [12–14] (BPP) tried to use a consistent model, the extended Nambu-Jona-Lasinio (ENJL) model as in [22, 23], as much as possible but adjusted using measured form-factors and QCD constraints. The calculations were done in Euclidean space. In fact, these two are still the only existing full calculations, but many parts have been evaluated using other approaches since then.

The main observations were:

- The largest contribution is π^0 (and η, η') exchange/pole. Be aware that exchange and pole or not precisely the same. Most estimates of this part are in reasonable agreement as discussed in Sect. 3.
- The pion loop can be sizable, with a large difference between the two evaluations and even larger numbers have been proposed. Further discussion is in Sect. 4.
- The other contributions are smaller but there are many and cancellations are present.
- Final numbers:
  BPP: $(8.3 \pm 3.1)\ 10^{-10}$, HKS: $(8.96 \pm 1.54)\ 10^{-10}$.

3 π^0-exchange

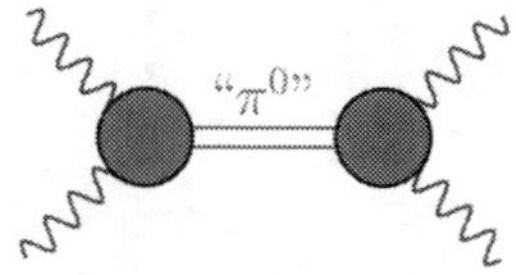

Figure 2. The π^0 exchange contribution. The blobs and the propagator need modeling.

The single largest numerical contribution is given by "π^0" exchange, depicted in Fig. 2. The blobs need modeling and the propagator in the ENJL model also has corrections to the $1/(p^2 - m_{\pi^0}^2)$. The pointlike vertex has a logarithmic divergence which is uniquely predicted [24, 25]. The VMD form-factor in the $\pi^0\gamma^*\gamma^*$ form-factor, the blobs, were modeled in [13] with a variety of form-factors and as a function of the cut-off Λ (corrected for the overall

Table 1. The π^0 exchange results of [13].

	$a_\mu \times 10^{10}$				
Λ GeV	Point-like	ENJL-VMD	Point-VMD	Transv. VMD	CELLO-VMD
0.5	4.92(2)	3.29(2)	3.46(2)	3.60(3)	3.53(2)
0.7	7.68(4)	4.24(4)	4.49(3)	4.73(4)	4.57(4)
1.0	11.15(7)	4.90(5)	5.18(3)	5.61(6)	5.29(5)
2.0	21.3(2)	5.63(8)	5.62(5)	6.39(9)	5.89(8)
4.0	32.7(5)	6.22(17)	5.58(5)	6.59(16)	6.02(10)

sign error discovered by [17]). We took the form-factor that was made to fit the then existing data integrated up to 2 GeV as our main result with a guesstimate of the error. This result was in quite good agreement with [20] which used the pointlike-VMD approach. This contribution has since been reevaluated many times using different models and approaches. A partial list is:

BPP [13]: $5.9(0.9) \times 10^{-10}$
Nonlocal quark model [26]: 6.27×10^{-10}
DSE (Dyson-Schwinger modeling)[27]: 5.75×10^{-10}
LMD+V [17]: $(5.8 - 6.3) \times 10^{-10}$
Formfactor inspired by AdS/QCD [28]: 6.54×10^{-10}
Chiral Quark Model [29]: 6.8×10^{-10}
Constraint via magnetic susceptibility [30]: 7.2×10^{-10}
$VV'P$ model [31]: 6.66×10^{-10}

All of these are in reasonable agreement, within the errors. Future improvements will come when more experimental results or the direct lattice calculations of the underlying formfactors are included. One way to define precisely the π^0 pole contribution is the dispersive approach [32] and the talk [4], but no numerical results have been published so far.

Two more comments are needed. The above numbers are for the π^0. One needs to take into account the η and η' exchange as well. The latter is enhanced due to the charge combinations in the $\eta'\gamma^*\gamma^*$ vertex. In large N_c models like the ENJL model, the pseudoscalar spectrum is not like QCD, one has a π^0, a $\tilde\pi$ ($\bar uu + \bar dd$ quark content) and a π_s ($\bar ss$). The $\tilde\pi$ has the same mass as the π^0 and due to the quark charges is contributes 25/9 times the π^0 contribution. Lattice QCD calculations with only connected diagrams included will have the $\tilde\pi$ contribution as well so there will be an unphysical enhancement compared to the QCD result for the pseudoscalar exchange part. This is discussed in more detail in [15]. In [13] we used pointlike-VMD to estimate the ratio of π^0, η, η' contributions as 5.58, 1.38, 1.04. Models that include large N_c-breaking effects and fit the mixings to data typically end up with very similar numbers. The total pseudoscalar exchange contribution I thus estimate to be

$$a_\mu^{PS} = (8 - 10) \times 10^{-10} \tag{3}$$

An example of a specific calculation is the AdS/QCD result of $a_\mu^{PS} = 10.7 \times 10^{-10}$ [33] which also includes excited pseudoscalars.

The other comment is that the short-distance behaviour of the four-point function is known in several limits. In particular when $P_1^2 \approx P_2^2 \gg Q^2$ the four point function is related to the axial-vector-vector-vector three-point function [34]. This three point function has a number of exact properties in QCD and we thus know how it behaves. The above models for π^0-exchange do not exhibit this behaviour. It can be implemented via making one of the blobs in Fig. 2 pointlike [34] and one then obtains 7.7×10^{-10} for the π^0-exchange contribution. Plots how this affects the contribution of different momentum regions are in [8]. The above behaviour of the four-point function must be obeyed in a full calculation, however whether one implements it via π^0-exchange is a choice. Models incorporating a short-distance quark-loop contribution have the short-distance part of this included [8, 26]. One can see this when comparing quark-loop plus pseudo-scalar exchange of [13] with pseudo-scalar exchange of [34].

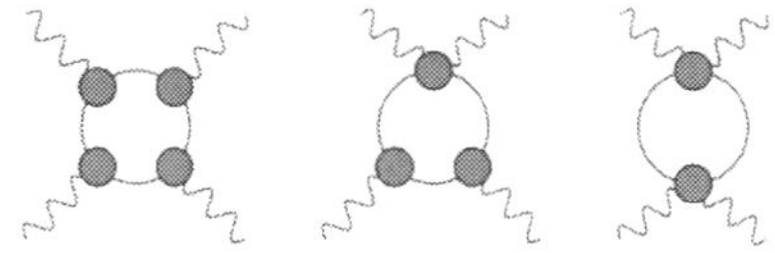

Figure 3. The charged pion loop contribution.

4 π-loop

The π-loop contribution to the four-point function is depicted in Fig. 3. The leftmost diagram is the naive one, the other two are required by gauge-invariance. In more general models also a diagram with three photons in one vertex and one with all four in the same vertex might be needed. These have been included in the calculations mentioned below when needed.

The simplest model is a point-like pion or scalar QED (sQED). This gives a contribution of about -4×10^{-10}.

The single photon vertex is in all determinations used as including the pion form-factor. For this one can use either the VMD expression or a more model/experimental inspired version. For the $\pi\pi\gamma^*\gamma^*$ vertex there were originally two main approaches used, full VMD (BPP) and the hidden local symmetry model with vector mesons (HKS). The former is essentially using sQED and putting a VMD-like form-factor in all the photon legs. This was proven to be a consistent procedure in [13]. We obtained there a result of -1.9×10^{-10} using an ENJL inspired pion form-factor. Using a simple VMD typically gives about -1.6×10^{-10}. This version is exactly what is called the model-independent part of the two-pion contribution in [4, 32, 35]. The reason for the lower number compared to the point-like pion loop is obvious in Fig. 4 where we show a_μ^{LLQ} of (2) as a function of $P_1 = P_2$ and Q.

HKS [19, 20] used a different approach. Due to the then existing arguments against full VMD they used the hidden local symmetry model with only vector mesons (HLS) and obtained -0.45×10^{-10}. The difference between this and the previous numbers was the reason for the large error quoted on the pion-loop. This difference was rather

 https://doi.org/10.1051/epjconf/201817901001

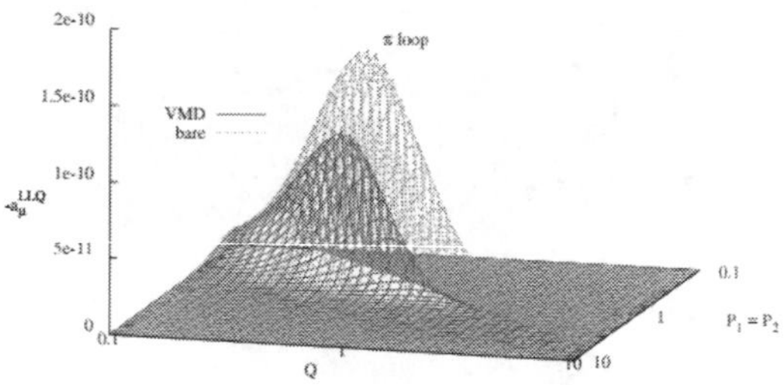

Figure 4. The momentum dependence of the pion loop contribution. Plotted is a_μ^{LLQ} of (2) as a function of $P_1 = P_2$ and Q. Top surface: sQED, bottom surface:full VMD.

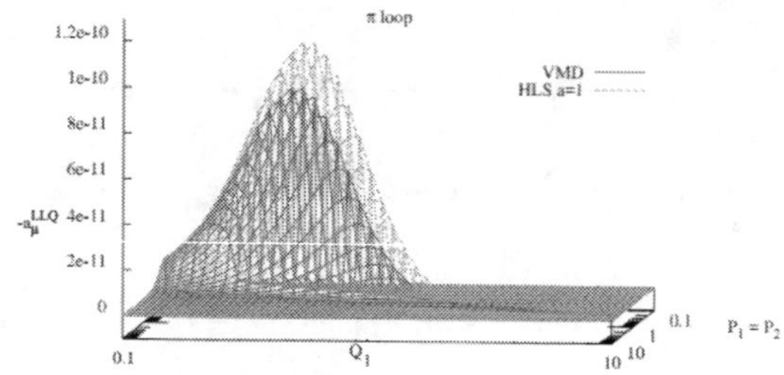

Figure 6. The momentum dependence of the pion loop contribution. $-a_\mu^{LLQ}$ of (2) as a function of $P_1 = P_2$ and Q. Top surface: HLS a=1, bottom surface: full VMD.

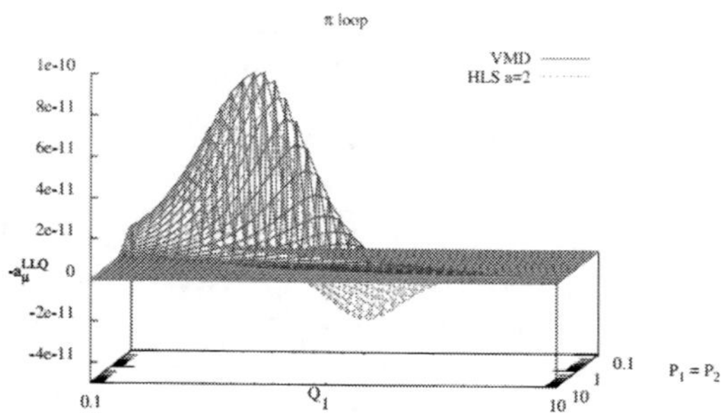

Figure 5. $-a_\mu^{LLQ}$ of (2) as a function of $P_1 = P_2$ and Q. Top surface: full VMD, bottom surface: HLS.

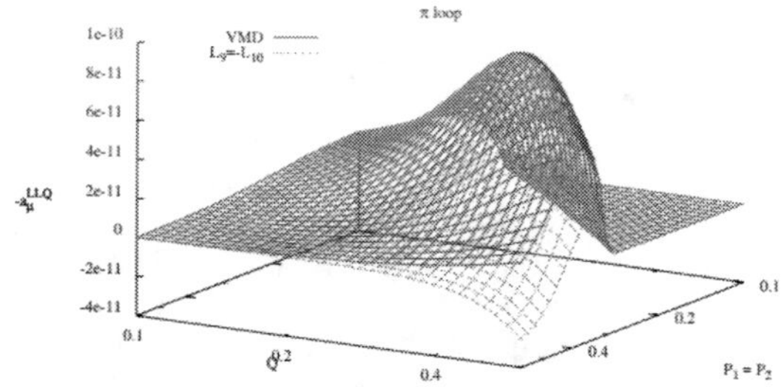

Figure 7. $-a_\mu^{LLQ}$ of (2) as a function of $P_1 = P_2$ and Q. Top surface: full VMD, bottom surface: ChPT with $L_9 = -L_{10}$ so the charge radius is included but no polarizability.

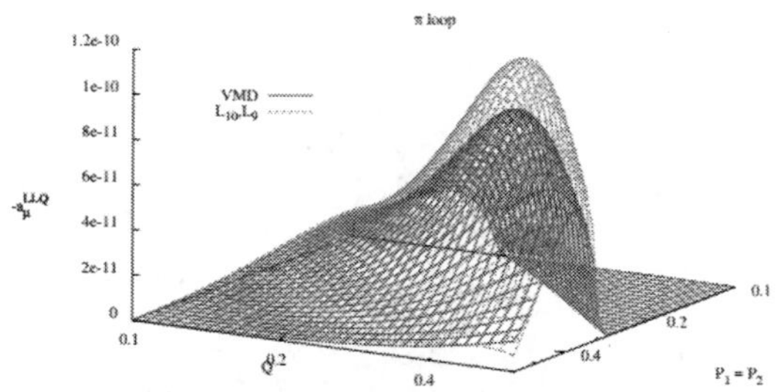

Figure 8. $-a_\mu^{LLQ}$ of (2) as a function of $P_1 = P_2$ and Q. Bottom surface: full VMD, top surface: ChPT with $L_9 \neq -L_{10}$ so the charge radius and the polarizability are included.

puzzling, one reason could be that the HLS model does not have the correct QCD short distance constraint when looking at the two-photon vertex with the same and large virtuality for both photons, the full VMD model has the correct behaviour. This version of the HLS model also does not give a finite prediction for the π^+-π^0 mass difference. The reason for the large numerical difference is indeed the short distance behaviour. The low momentum behaviour is very close but the negative contribution above 1 GeV, clearly visible in Fig. 5, is the main reason for the difference [15, 36]. A comparison as a function of the cut-off can be found in [37]. In fact, using the HLS with an unphysical value of the parameter $a = 1$, which then satisfies the abovementioned short-distance constraint gives very similar numbers as full VMD. This is shown in Fig. 6 From this we conclude that a number in the range $-(1.5$-$1.9)\times 10^{-10}$ is more appropriate with an error of half to 1/3 that.

More recently, it was pointed out that the effect of pion polarizability was neglected in these calculations and a first estimate of this effect given using the Euler-Heisenberg four photon effective vertex produced by pions [38] within Chiral Perturbation Theory. This approximation is only valid below the pion mass. In order to check the size of the pion radius effect and the polarizability we have implemented the low energy part of the four-point function and computed a_μ^{LLQ} for these cases. Partial results are in [36, 37] and the full results in [15]. The effect of the charge radius is shown in Fig. 7 compared to the VMD, notice the different momentum scales compared to the earlier figures. As expected, the charge radius effect is included in the VMD result since the latter gives a good description of the pion form-factor. Including the effect of the polarizability can be done in ChPT by using experimentally determined values for L_9 and L_{10}. The latter can be determined from $\pi^+ \to e\nu\gamma$ or the hadronic vector two-point functions. Both are in good agreement and lead to a prediction of the pion polarizability confirmed by the COMPASS experiment [39]. The effect of including this in ChPT on a_μ^{LLQ} is shown in Fig. 8 [15, 36, 37]. An increase of 10-15% over the VMD estimate can be seen.

ChPT at lowest order or p^4 for a_μ is just the pointlike pion loop or sQED. At NLO pion exchange with pointlike vertices and the pion loop calculated at NLO in ChPT are

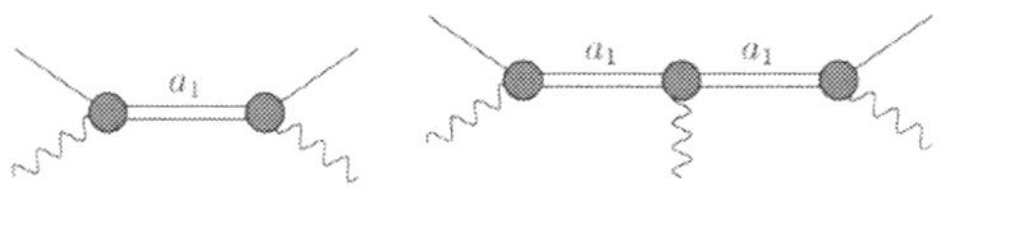

Figure 9. Left: the a_1-exchange that produces the pion polarizability. Right: an example of a diagram that is required by gauge invariance.

needed. Both gives divergent contributions to a_μ, so pure ChPT is of little use in predicting a_μ. If we want to see the full effect of the polarizability we need to include a model that can be extended all the way, or at least to a cut-off of about 1 GeV. For the approach of [38] this was done in [40] by including a propagator description of a_1 and choosing it such that the full contribution of the pion-loop to a_μ is finite. They obtained a range of $-(1.1\text{-}7.1)\times 10^{-10}$ for the pion-loop contribution. I find this range much too broad. One reason is that the range of polarizabilities used in [40] is simply not compatible with ChPT. The pion polarizability is an observable where ChPT should work and indeed the convergence is excellent. The ChPT prediction has also recently been confirmed by experiment. Our work discussed below indicates that $-(2.0\pm0.5)\times10^{-10}$ is a more appropriate range for the pion-loop contribution.

The work described below is published in [15]. Preliminary results have been reported at several conferences, see e.g. [41, 42]. The polarizability comes from $L_9 + L_{10}$ in ChPT. Using [43], we notice that the polarizability is produced by a_1-exchange depicted in Fig. 9. This is depicted pictorially in the left diagram of Fig. 9. However, once such an exchange is there, diagrams like the right one in Fig. 9 lead to effective $\pi\pi\gamma\gamma\gamma$ vertices and are required by electromagnetic gauge invariance. This was done in [40] via the propagator modifications. We deal with them via effective Lagrangians incorporating vector and axial-vector mesons.

If one looks at Fig. 9 one could raise the question "Is including a π-loop but no a_1-loop consistent?" The answer is yes with the following argument. We can first look at a tree level Lagrangian including pions ρ and a_1. We then integrate out the ρ and a_1 and calculate the one-loop pion diagrams with the resulting Lagrangian. In the diagrams of the original Lagrangian this corresponds to only including loops with at least one pion propagator present. Numerical results for cases including full a_1 loops are presented as well below [15]. As a technicality, we use anti-symmetric vector fields for the vector and axial-vector mesons. This avoids complications due to π-a_1 mixing. We add vector $V_{\mu\nu}$ and axial-vector $A_{\mu\nu}$ nonet fields. The kinetic terms are given by [43]

$$-\frac{1}{2}\left\langle \nabla^\lambda V_{\lambda\mu}\nabla_\nu V^{\nu\mu} - \frac{M_V^2}{2}V_{\mu\nu}V^{\mu\nu}\right\rangle + V \leftrightarrow A\,. \qquad (4)$$

First we add the terms that contribute to the L_i [43]

$$\frac{F_V}{2\sqrt{2}}\left\langle f_{+\mu\nu}V^{\mu\nu}\right\rangle + \frac{iG_V}{\sqrt{2}}\left\langle V^{\mu\nu}u_\mu u_\nu\right\rangle + \frac{F_A}{2\sqrt{2}}\left\langle f_{-\mu\nu}A^{\mu\nu}\right\rangle \quad (5)$$

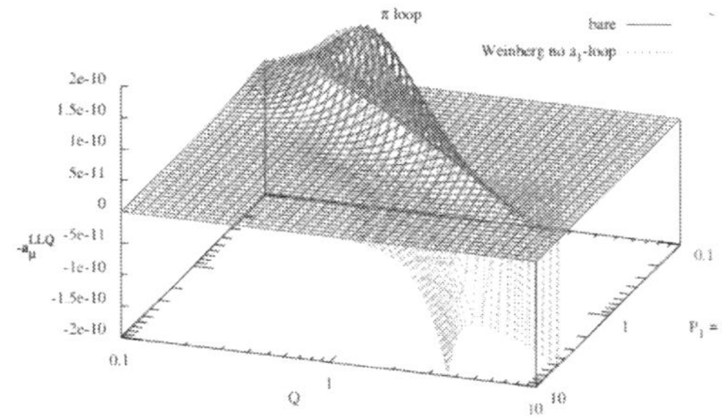

Figure 10. $-a_\mu LLQ$ as defined in (2) as a function of $P_1 = P_2$ and Q with a_1 but no full a_1-loop. Parameters determined by the Weinberg sum rules.

with $L_9 = \frac{F_V G_V}{2M_V^2}$, $L_{10} = -\frac{F_V^2}{4M_V^2} + \frac{F_A^2}{4M_A^2}$. The Weinberg sum rules imply in the chiral limit $F_V^2 = F_A^2 + F_\pi^2$, $F_V^2 M_V^2 = F_A^2 M_A^2$ and requiring VMD behaviour for the pion formfactor $F_V G_V = F_\pi^2$.

First, look at the model with only π and ρ. The one-loop contributions to $\Pi^{\rho\nu\alpha\beta}$ are not finite. They were also not finite for the HLS model of HKS, but the relevant $\delta\Pi^{\rho\nu\alpha\beta}/\delta p_{3\lambda}$ was. However, in the present model it is only finite for $G_V = F_V/2$ and then the result for a_μ is identical to the HLS model. The same comments as made for the HLS model thus also apply.

Next we do add the a_1 and require $F_A \neq 0$. After a lot of work we find that $\delta\Pi^{\rho\nu\alpha\beta}/\delta p_{3\lambda}|_{p_3=0}$ is finite only for $G_V = F_V = 0$ and $F_A^2 = -2F_\pi^2$ or, if including a full a_1-loop $F_A^2 = -F_\pi^2$. These solutions are clearly unphysical. We then add all $\rho a_1 \pi$ vertices given by

$$\begin{aligned}
&\lambda_1\left\langle\left[V^{\mu\nu}, A_{\mu\nu}\right]\chi_-\right\rangle + \lambda_2\left\langle\left[V^{\mu\nu}, A_{\nu\alpha}\right]h_\mu{}^\nu\right\rangle \\
&+ \lambda_3\left\langle i\left[\nabla^\mu V_{\mu\nu}, A_{\nu\alpha}\right]u_\alpha\right\rangle + \lambda_4\left\langle i\left[\nabla_\alpha V_{\mu\nu}, A_{\alpha\nu}\right]u^\mu\right\rangle \\
&+ \lambda_5\left\langle i\left[\nabla^\alpha V_{\mu\nu}, A_{\mu\nu}\right]u_\alpha\right\rangle + \lambda_6\left\langle i\left[V^{\mu\nu}, A_{\mu\nu}\right]f_-{}^\alpha{}_\nu\right\rangle \\
&+ \lambda_7\left\langle iV_{\mu\nu}A^{\mu\rho}A^\nu{}_\rho\right\rangle\,.
\end{aligned} \qquad (6)$$

These are not all independent due to the constraints on $V_{\mu\nu}$ and $A_{\mu\nu}$ [44], there are three relations. After a lot of work [15] we found that no solutions with $\delta\Pi^{\rho\nu\alpha\beta}/\delta p_{3\lambda}|_{p_3=0}$ exists except those already obtained without Λ_i terms. The same conclusions holds if we look at the combination that shows up in the integral over P_1^2, P_2^2, Q^2. We thus find no reasonable model that has a finite prediction for a_μ for the pion-loop including a_1. If we choose the parameters as fixed by the Weinberg sum rules and the VMD behaviour of the pion-form factor we obtain $-a_\mu^{LLQ}$ as shown in Fig. 10. Adding a full a_1-loop changes the plot only marginally. As long as we require the correct polarizability and a VMD-like form-factor behaviour, the plots look quite similar for all cases below 1 GeV. The integrated value up to Λ for a number of cases is how in Fig. 11. We see that all models end up with a value of $a_\mu = -(2.0 \pm 0.5) \times 10^{-10}$ when integrated up-to a cut-off of order 1-2 GeV. We conclude that that is a reasonable estimate for the pion-loop contribution. The main missing part is the π-π rescattering.

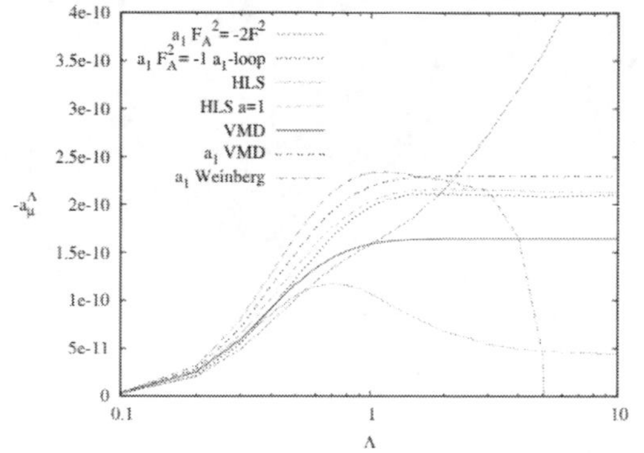

Figure 11. $-a_\mu$ using a variety of models for the pion loop as a function of Λ, the cut-off on the photon momenta. Figure from [15].

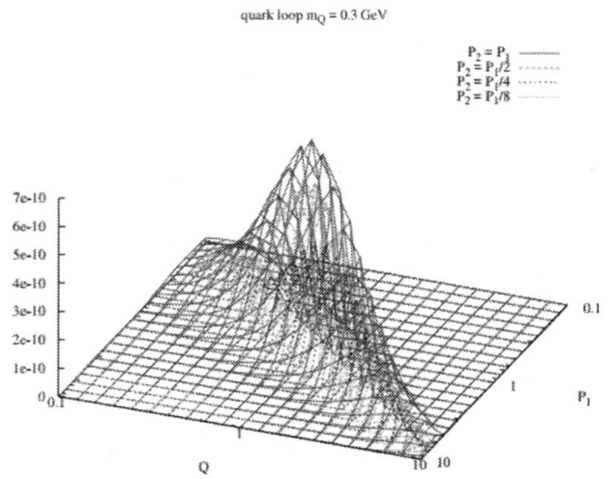

Figure 12. The quantity a_μ^{LLQ} defined in (2) as a function of P_1 and Q for several ratios P_1/P_1. The lines indicate the surface.

The dispersive approach has numbers that are compatible with the above and the inclusion of scalar exchange, $a_\pi^{\pi-loop} = (-2.4 \pm 0.1)\ 10^{-10}$. See the discussion in [4, 45, 46].

5 Quark-loop

The pure quark-loop contribution with a constant mass is known analytically. One of the surprises is that it converges rather slowly. A significant portion is from high momentum regions. With a constituent quark mass of 300 MeV and a cut-off of 1(2) GeV 50(25)% of the full contribution is still missing. A more visual illustration of this is the plot of a_μ^{LLQ} defined in (2) of this contribution. The contribution is plotted in Fig. 12 as a function of P_1 and Q for several ratios of P_2/P_1. The volume under the curve is proportional to a_μ. The contribution peaks for $P_1 \approx P_2 \approx Q$ and around 1 GeV. In [13] we used the ENJL up to a cut-off Λ and added a short-distance quark-loop where we used the quark-mass $M_H = \Lambda$ as a lower cut-off. The estimate used by HKS was a quark-loop damped by VMD factors in the photon legs. The results are given in Tab. 2. Notice especially the stability when we add the ENJL and the short-distance contribution in the region Λ = 1-8 GeV. The conclusion is that the quark-loop is about 2×10^{-10}. In the ENJL model the quark-loop and scalar exchange are needed together to have correct chiral symmetry. The sum of both is very similar to the quark-loop estimate of HKS.

Table 2. The quark-loop contribution with VMD damping, the ENJL model and with a heavy quark mass as cut-off. The numbers are $a_\mu \times 10^{10}$.

Cut-off Λ GeV	VMD	ENJL	mass-cut	sum ENJL masscut
0.5	0.48	0.78	2.46	3.2
0.7	0.72	1.14	1.13	2.3
1.0	0.87	1.44	0.59	2.0
2.0	0.98	1.78	0.13	1.9
4.0	0.98	1.98	0.03	2.0
8.0	0.98	2.00	.005	2.0

There are a number of estimates of the quark-loop that lead to much larger numbers. These have all in common that there is a momentum region with a fairly small (constituent) quark mass that is not shielded by a VMD-like mechanism. The most prominent example of this is the DSE estimate of [47] $10.7(0.2) \times 10^{-10}$. The present status of this calculation is given in [16]. It not yet a full calculation but includes an estimate of some of the missing parts. This DSE model describes a lot of low-energy phenomenology in a way very similar to the ENJL model. I am quite puzzled by the difference in results.

Similar size numbers are obtained in models with a low constituent quark mass where no VMD-like dynamical effects are included. Examples are the nonlocal chiral quark model [48] with $11.0(0.9) \times 10^{-10}$ and a number of estimates within the chiral quark model $(7.6-8.9) \times 10^{-10}$ [29], $(11.8-14.8) \times 10^{-10}$ [49] and $(7.6-12.5) \times 10^{-10}$[50]. The interpretation varies from an estimate to the full HLbL or just a part that needs to be added to other contributions.

6 Scalar exchange

The estimate of the scalar exchange contribution in the ENJL model is -0.7×10^{-10}. Similar size estimates have been obtained when exchanging a sigma-like particle. It should be pointed out that the scalar in the ENJL model has a phenomenology similar to the sigma but is quite a different underlying object.

A problem here is to distinguish scalar exchange from two-pion or pion-loop contributions. This is one of the areas where the method of [4, 32] as used in [45, 46] allowed for major progress.

7 a_1-exchange

The exchange of axial vectors in the ENJL model was estimated in [13] to be about 0.6×10^{-10}, but due to the high mass involved, even with a cut-off of 2 GeV only half the contribution was there. The ENJL part also includes some pseudo-scalar meson exchange due to the structure of the calculation.

Axial-vector meson exchange in a more phenomenological way was done using two multiplets in [34] who obtained 2.2×10^{-10}. It was later found that when correct antisymmetrization is included, this becomes smaller by a significant factor and is again in the ballpark of the ENJL result. This was noticed by F. Jegerlehner. He obtains about $(0.76 \pm 0.27) \times 10^{-10}$ for the axial-vector exchange [51, 52]. The evaluation of [53] is also in reasonable agreement with the ENJL estimate.

8 Conclusions

The present number for the HLbL contribution to the muon anomaly, $a_\mu = (g_\mu - 2)/2$, is (11 ± 4) or $(10.5 \pm 2.6) \times 10^{-10}$ [8–10] depending somewhat on which error estimates and which contributions are taken into account. In this talk I have given an overview of a number of model estimates with the emphasis on my old work [12–14] as well as a number of newer developments. For the latter I have spent quite some time on our reevaluation of the pion loop contribution [15, 36, 37, 41, 42], as well as given a number of arguments why the HLS number of [19, 20] should be considered obsolete. The conclusion is that the pion-loop contributes with $-(2.0 \pm 0.5) \times 10^{-10}$.

One of the remaining problems in the model approach is that the class of models with an "unshielded" quark-loop at relatively low-energies for the photons tend to obtain larger numbers. Whether this is a real phenomenon or not is a question which needs to be settled. My own opinion there is that I see no counterpart of it in $\gamma\gamma \to$ hadrons at low to intermediate energies beyond the already included single meson and two-pion exchanges.

For contributions of different mechanisms, progress can be expected both from the dispersive approaches mentioned and experiment restricting the couplings of off-shell or virtual photons to meson that go into the modeling. Alternatively, a full new model calculation that includes phenomenology beyond what the ENJL does, is very desirable as well as more work on the short-distance aspects.

Acknowledgements

I thank the organizers for the very nice atmosphere and the many opportunities to discuss the issues regarding HLbL. This work is supported in part by the Swedish Research Council grants 621-2013-4287, 2015-04089 and 2016-05996 and by the European Research Council (ERC) under the European Union's Horizon 2020 research and innovation programme (grant agreement No 668679).

References

[1] J. Bijnens, EPJ Web Conf. **118**, 01002 (2016) [arXiv:1510.05796 [hep-ph]].

[2] G. D'Ambrosio, M. Iacovacci, M. Passera, G. Venanzoni, P. Massarotti and S. Mastroianni, EPJ Web Conf. **118** (2016).

[3] M. Knecht, *General discussion on $g-2$*

[4] G. Colangelo, *Dispersion relations for hadronic light -by -light scattering*

[5] C. Lehner, *Hadronic light-by-light contribution to the muon $g-2$ on the lattice*

[6] A. Nyffeler, *Hadronic light-by-light scattering contribution to the muon $g-2$ on the lattice*

[7] G. W. Bennett *et al.* [Muon g-2 Collaboration], Phys. Rev. D **73**, 072003 (2006) [hep-ex/0602035].

[8] J. Bijnens and J. Prades, Mod. Phys. Lett. A **22**, 767 (2007) [hep-ph/0702170].

[9] J. Prades, E. de Rafael and A. Vainshtein, "Hadronic Light-by-Light Scattering Contribution to the Muon Anomalous Magnetic Moment," (Advanced series on directions in high energy physics. 20) [arXiv:0901.0306 [hep-ph]].

[10] F. Jegerlehner and A. Nyffeler, Phys. Rept. **477**, 1 (2009) [arXiv:0902.3360 [hep-ph]].

[11] C. Lehner, *Report from the Muon g-2 Theory Initiative at Fermilab*, see also https://indico.fnal.gov/event/13795/page/13

[12] J. Bijnens, E. Pallante and J. Prades, Phys. Rev. Lett. **75**, 1447 (1995) [Phys. Rev. Lett. **75**, 3781 (1995)] [hep-ph/9505251].

[13] J. Bijnens, E. Pallante and J. Prades, Nucl. Phys. B **474**, 379 (1996) [hep-ph/9511388].

[14] J. Bijnens, E. Pallante and J. Prades, Nucl. Phys. B **626**, 410 (2002) [hep-ph/0112255].

[15] J. Bijnens and J. Relefors, JHEP **1609**, 113 (2016) [arXiv:1608.01454 [hep-ph]].

[16] G. Eichmann, C. S. Fischer, W. Heupel and R. Williams, arXiv:1411.7876 [hep-ph].

[17] M. Knecht and A. Nyffeler, Phys. Rev. D **65**, 073034 (2002) [hep-ph/0111058].

[18] E. de Rafael, Phys. Lett. B **322**, 239 (1994) [hep-ph/9311316].

[19] M. Hayakawa, T. Kinoshita and A. I. Sanda, Phys. Rev. Lett. **75**, 790 (1995) [hep-ph/9503463].

[20] M. Hayakawa, T. Kinoshita and A. I. Sanda, Phys. Rev. D **54**, 3137 (1996) [hep-ph/9601310].

[21] M. Hayakawa and T. Kinoshita, Phys. Rev. D **57**, 465 (1998) [Phys. Rev. D **66**, 019902 (2002)] [hep-ph/9708227].

[22] J. Bijnens, C. Bruno and E. de Rafael, Nucl. Phys. B **390**, 501 (1993) [hep-ph/9206236].

[23] J. Bijnens, Phys. Rept. **265**, 369 (1996) [hep-ph/9502335].

[24] M. Knecht, A. Nyffeler, M. Perrottet and E. de Rafael, Phys. Rev. Lett. **88**, 071802 (2002) [hep-ph/0111059].

[25] M. J. Ramsey-Musolf and M. B. Wise, Phys. Rev. Lett. **89**, 041601 (2002) [hep-ph/0201297].

[26] A. E. Dorokhov and W. Broniowski, Phys. Rev. D **78**, 073011 (2008) [arXiv:0805.0760 [hep-ph]].

[27] T. Goecke, C. S. Fischer and R. Williams, Phys. Rev. D **83**, 094006 (2011) [Phys. Rev. D **86**, 099901 (2012)] [arXiv:1012.3886 [hep-ph]].

[28] L. Cappiello, O. Cata and G. D'Ambrosio, Phys. Rev. D **83**, 093006 (2011) [arXiv:1009.1161 [hep-ph]].

[29] D. Greynat and E. de Rafael, JHEP **1207**, 020 (2012) [arXiv:1204.3029 [hep-ph]].

[30] A. Nyffeler, Phys. Rev. D **79**, 073012 (2009) [arXiv:0901.1172 [hep-ph]].

[31] P. Roig, A. Guevara and G. López Castro, Phys. Rev. D **89**, no. 7, 073016 (2014) [arXiv:1401.4099 [hep-ph]].

[32] G. Colangelo, M. Hoferichter, M. Procura and P. Stoffer, JHEP **1409**, 091 (2014) [arXiv:1402.7081 [hep-ph]].

[33] D. K. Hong and D. Kim, Phys. Lett. B **680**, 480 (2009) [arXiv:0904.4042 [hep-ph]].

[34] K. Melnikov and A. Vainshtein, Phys. Rev. D **70**, 113006 (2004) [hep-ph/0312226].

[35] G. Colangelo, M. Hoferichter, B. Kubis, M. Procura and P. Stoffer, Phys. Lett. B **738**, 6 (2014) [arXiv:1408.2517 [hep-ph]].

[36] J. Bijnens and M. Z. Abyaneh, EPJ Web Conf. **37**, 01007 (2012) [arXiv:1208.3548 [hep-ph]].

[37] M. Z. Abyaneh, arXiv:1208.2554 [hep-ph].

[38] K. T. Engel, H. H. Patel and M. J. Ramsey-Musolf, Phys. Rev. D **86**, 037502 (2012) [arXiv:1201.0809 [hep-ph]].

[39] C. Adolph *et al.* [COMPASS Collaboration], Phys. Rev. Lett. **114**, 062002 (2015) [arXiv:1405.6377 [hep-ex]].

[40] K. T. Engel and M. J. Ramsey-Musolf, Phys. Lett. B **738**, 123 (2014) [arXiv:1309.2225 [hep-ph]].

[41] K. Kampf *et al.*, arXiv:1308.2575 [hep-ph].

[42] M. Benayoun *et al.*, arXiv:1407.4021 [hep-ph].

[43] G. Ecker, J. Gasser, A. Pich and E. de Rafael, Nucl. Phys. B **321**, 311 (1989).

[44] S. Leupold, private communication.

[45] G. Colangelo, M. Hoferichter, M. Procura and P. Stoffer, Phys. Rev. Lett. **118**, no. 23, 232001 (2017) doi:10.1103/PhysRevLett.118.232001 [arXiv:1701.06554 [hep-ph]].

[46] G. Colangelo, M. Hoferichter, M. Procura and P. Stoffer, JHEP **1704**, 161 (2017) doi:10.1007/JHEP04(2017)161 [arXiv:1702.07347 [hep-ph]].

[47] T. Goecke, C. S. Fischer and R. Williams, Phys. Rev. D **87**, no. 3, 034013 (2013) [arXiv:1210.1759 [hep-ph]].

[48] A. E. Dorokhov, A. E. Radzhabov and A. S. Zhevlakov, Eur. Phys. J. C **75**, no. 9, 417 (2015) [arXiv:1502.04487 [hep-ph]].

[49] R. Boughezal and K. Melnikov, Phys. Lett. B **704**, 193 (2014) [arXiv:1104.4510 [hep-ph]].

[50] P. Masjuan and M. Vanderhaeghen, arXiv:1212.0357 [hep-ph].

[51] F. Jegerlehner, talk presented at the workshop on Hadronic contributions to the muon anomalous magnetic moment: strategies for improvements of the accuracy of the theoretical prediction, 1-5 April 2014, Waldthausen Castle near Mainz.

[52] F. Jegerlehner, EPJ Web Conf. **118**, 01016 (2016) [arXiv:1511.04473 [hep-ph]].

[53] V. Pauk and M. Vanderhaeghen, Eur. Phys. J. C **74**, no. 8, 3008 (2014) [arXiv:1401.0832 [hep-ph]].

The SHiP physics program

Giovanni De Lellis[1,a]

[1] Dipartimento di Fisica "E. Pancini", Università "Federico II" and INFN, Naples, Italy

Abstract. The discovery of the Higgs boson has fully confirmed the Standard Model of particles and fields. Nevertheless, there are still fundamental phenomena, like the existence of dark matter and the baryon asymmetry of the Universe, which deserve an explanation that could come from the discovery of new particles. The SHiP experiment at CERN meant to search for very weakly coupled particles in the few GeV mass domain has been recently proposed. The existence of such particles, foreseen in different theoretical models beyond the Standard Model, is largely unexplored. A beam dump facility using high intensity 400 GeV protons is a copious source of such unknown particles in the GeV mass range. The beam dump is also a copious source of neutrinos and in particular it is an ideal source of tau neutrinos, the less known particle in the Standard Model. Indeed, tau anti-neutrinos have not been directly observed so far. We report the physics potential of such an experiment including the tau neutrino magnetic moment.

1 The SHiP experiment

The discovery of the Higgs boson is certainly a big triumph of the Standard Model. In particular, given its mass, it could well be that the Standard Model is an effective theory working all the way up to the Planck scale. Nevertheless, there are several phenomena deserving an explanation that the Standard Model is unable to provide: the existence of dark matter and its nature, the baryonic asymmetry of the Universe and neutrino masses. It is therefore clear the new physics is there and presumably several new particles have still to be discovered.

Searches for new physics with accelerators are being carried out at the LHC, especially suited to look for high mass particles with ordinary couplings to matter. Complementary searches for very weakly coupled and therefore long-lived particles require a beam dump facility. Such a facility is made of a high density proton target, followed by a hadron stopper and a muon shield. Apart from residual muons, the only remaining particles are electron, muon and tau neutrinos on top of hidden, long-lived particles produced either in proton interactions or in secondary particle decays.

A new experiment, Search for Hidden Particles (SHiP), has been proposed [1], designed to operate at a beam dump facility to be built at CERN and to search for weakly coupled particles in the few GeV mass range. Since a high intensity tau neutrino flux is produced by such a facility from Ds decays, the experimental apparatus foresees a neutrino detector to study the tau neutrino cross-section and discover the tau anti-neutrino. This detector would also be suited to detect dark matter through the scattering off the atoms of its target. The physics case for such an experiment is widely discussed in Ref. [2]. We will review here the physics potential of this experiment for a selection of physics channels, including the tau neutrino magnetic moment.

In five years, the facility will integrate 2×10^{20} 400 GeV protons, produced by the SPS accelerator complex, impinging on a 12 interaction length (λ_{int}) target made of Molybdenum and Tungsten, followed by a 30 λ_{int} iron hadron absorber. Hidden particles in the GeV mass range would be produced mostly by the decay of charmed hadrons produced in proton interactions. D_s mesons, copiously produced among charmed hadrons, are a source of tau neutrinos through their fully leptonic decay. Therefore, the SHiP facility is ideal also to study the physics of tau neutrinos, the less known particle in the Standard Model.

Figure 1 shows the SHiP facility to be placed in the North Area: downstream of the target, the hadron absorber filters out all hadrons, therefore only muons and neutrinos are left. An active muon shield is designed with two sections with opposite polarity to maximize the muon flux reduction: it reduces the muon flux from 10^{10} down to 10^5 muons per spill. 4×10^{13} protons are extracted in each spill, designed to be 1s long to reduce the detector occupancy [3]. A first successful test of the SPS cycle with a 1s long spill was performed in April 2015.

The neutrino detector is located downstream of the muon shield, followed by the decay vessel and the detector for hidden particles. The Collaboration has been requested by the SPSC Committee to prepare a Comprehensive Design Report by 2018, in the framework of the Physics Beyond Colliders working group, launched in 2016 at CERN. This working group is meant to prepare proposals as input

[a] e-mail: giovanni.de.lellis@cern.ch, on behalf of the SHiP Collaboration

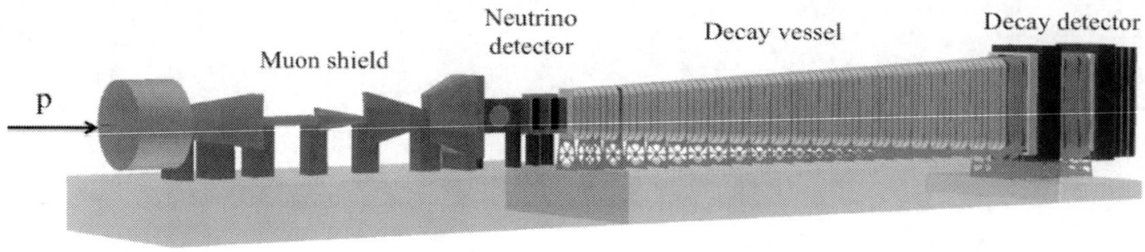

Figure 1. The beam dump facility and the SHiP detector.

for the High Energy panel of the European Strategy to be held in 2020, that will recommend about the approval of the experiment. The construction and installation will last until the third long shutdown of the LHC and the data taking is assumed to start in 2026.

The neutrino detector is made of a magnetised target region, followed by a muon spectrometer, as shown in Figure 2. The neutrino target is based on the emulsion cloud chamber technology employed by the OPERA experiment [4], with a compact emulsion spectrometer, made of a sequence of very low density material and emulsion films to measure the charge and momentum of hadrons in magnetic field. Indeed, this feature would allow to discriminate between tau neutrinos and anti-neutrinos also in the hadronic decay channels of the tau lepton. The emulsion target is complemented by high resolution tracking chambers to provide the time stamp to the event and connect muon tracks from the target to the muon spectrometer. The muon spectrometer is based on the concept developed for the OPERA apparatus: a dipolar iron magnet where drift tubes provide the momentum and resistive plate chambers provide the tracking within the iron slabs.

The emulsion target will also act as the target of dark matter as well as of any very weakly in- teracting particle produced at the accelerator, when its mass is in the GeV range. The detector is being optimised to search for interactions with the atoms (both electrons and nuclei) of very weakly interacting particles. The optimisation concerns e.g. the target material, the sampling frequency of the emulsion cloud chamber and the timing performances of the target tracker that would enable the separation between neutrinos and heavy particles based on the time of fligth measurement.

The detector for hidden particles is located in the downstream part of a 60 m long evacuated decay vessel, with an elliptical transverse section of 5×10 m^2, with the longer axis vertically. The hidden particles are supposed to decay within the vessel. The requirement to have less than 1 neutrino interaction in the vessel over five years sets the pressure to about 10^{-3} mbar. A magnetic spectrometer is located downstream of the vessel: it is made of straw tubes with a material budget of 0.5% X_0 per station, achieving a position resolution of 120 μm per straw, with 8 hits per station on average. This gives a momentum resolution of about 1%. The challenge is to build 5m long straw tubes and the large vacuum vessel. The vessel would be sorrounded by a liquid scintillator layer to tag particles coming from outside. Downstream of the spectrometer, an hadronic and electromagnetic calorimeter and a muon filter are used to identify particles.

2 Search for hidden particles

Extensions of the Standard Model in the low mass region foresee the existence of particles as singlets with respect to the Standard Model gauge group. These particles couple to different singlet compos- ite operators (so-called Portals) of the Standard Model. The neutrino can be considered as the first example of a hidden particle: it was postulated in 1930 to explain the puzzle of the beta decay, it is very light and extremely weakly interacting. The SHiP detector located immediately downstream of a beam dump has the potentiality to discover all of them, since it is sensitive to very weakly interacting and long lived particles in a wide unexplored range of their masses and couplings. We describe in the following its sensitivity to heavy neutral leptons, as an example. For the sensitivity to the other portals, we refer to Ref. [1].

2.1 Search for heavy neutral leptons

Neutrino masses are usually explained by the see-saw mechanism with a lagrangian term given by

$$L = i\bar{N}_I\partial_\mu\gamma^\mu N_I - Y_{I\alpha}\bar{N}_I^C\tilde{H}L_\alpha - M_I\bar{N}_I^C N_I + h.c. \quad (1)$$

where 3 right-handed heavy neutrinos N_I are introduced. The mass of active neutrinos is given by $m_\nu \simeq m_D^2/M_I$ where $m_D \simeq Y_{I\alpha}v$ and $v \simeq 246$ GeV is the Higgs vacuum expectation value. A massive neutrino with M_I of the order of 1 GeV and Yukawa couplings $Y_{I\alpha}$ of the order of 10^{-7} would account for an active neutrino mass $m_\nu \simeq 0.05$ eV, as measured by neutrino oscillation experiments.

If several conditions are satisfied, among which CP is not conserved, these right-handed neutrinos could also explain the leptogenesis and therein the baryogenesis [5, 6]. Such models do not pretend to explain also dark matter: a dark matter candidate could be provided by a fourth right-handed neutrino or an axion. Moreover they do not require

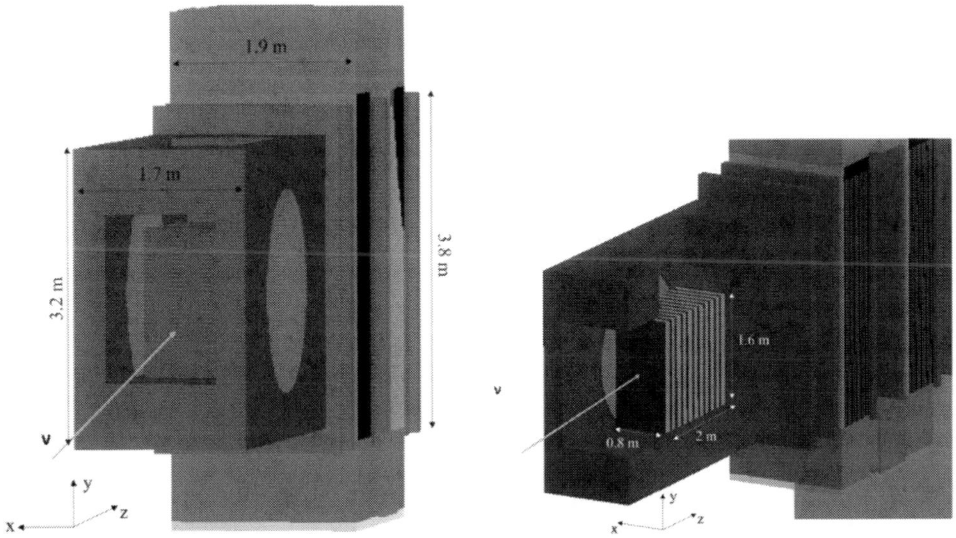

Figure 2. The neutrino detector upstream of the decay vessel in different views.

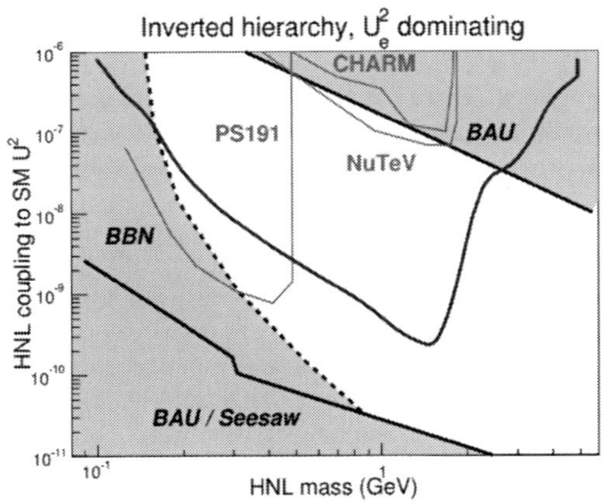

Figure 3. Exclusion limits at 90% C.L. set for inverted hierarchy with U_e^2 dominating, according to Ref. [10].

two of the three heavy neutrinos to be quasi-degenerate because CP violation does not need to be enhanced by mass degeneracy.

In the attempt of explaining also dark matter with three heavy neutrinos, it was first introduced in 2005 the so-called νMSM [7, 8], later followed by refined versions [9]. The role of the three right-handed neutrinos is different as well as their mass: M_1 could be a dark matter candidate with the mass of several keV while M_2 and M_3 should be states with a mass of about 1 GeV and quasi-degenerate to provide the baryon asymmetry of the Universe.

The states N_2 and N_3 could be produced in the decay of sufficiently massive particles like charmed hadrons. These states would be long lived particles that in turn could decay, for example, into $\pi\,\mu$.

Figure 3 shows the experimental and cosmological bounds on the search for heavy neutral leptons. It also reports superimposed the exclusion limits at 90% C.L. SHiP could set in case no signal is found, assuming the model reported in Ref. [10] with inverted hierarchy and a dominant electron coupling, $U_e^2 : U_\mu^2 : U_\tau^2 = 48 : 1 : 1$. The parameter space explored by SHiP extends in the cosmologically relevant region that is experimentally unexplored.

3 Physics with the neutrino detector

The tau neutrino is the less known particle in the Standard Model. Four candidates were firstly reported in 2001 by the DONUT experiment [11] and the observation of this particle was finally confirmed in 2008 when 9 candidates events were reported with an estimated background of 1.5 [12]. In the same paper they reported, for the first time, the tau neutrino cross-section where the constant term was measured to be $\sigma_{\nu_\tau}^{const} = (0.39 \pm 0.13 \pm 0.13) \times 10^{-38} cm^2 GeV^{-1}$. The large uncertainty is due to the poor statistical sample and to the rather scarce knowledge of the incoming flux. On top of the large uncertainty, DONUT could not separate tau neutrinos from anti-neutrinos. The OPERA experiment [4] has detected five tau neutrinos [13–17], leading to the discovery of tau neutrino appearance from muon neutrino oscillations [17]. The only leptonic decay observed by OPERA [15] shows negative charge as expected from a ν_τ interaction. Therefore, so far there is no direct evidence for tau anti-neutrinos.

The number of ν_τ and $\bar{\nu}_\tau$ emerging from the molybdenum target can be estimated as:

$$N_{\nu_\tau+\bar{\nu}_\tau} = 4N_p \frac{\sigma_{c\bar{c}}}{\sigma_{pN}} f_{D_s} Br(D_s \to \tau) = 6.6 \times 10^{15} \quad (2)$$

where N_p is the number of interacting protons (all incoming ones); $\sigma_{c\bar{c}} = 18.1 \pm 1.7$ μbarn [18] the associated charm production per nucleon; $\sigma_{pN} = 10.7$ mbarn the hadronic cross-section per nucleon in a Mo target. The inelastic cross-section pA shows the $A^{0.71}$ dependence [19]; $f_{D_s} = (8.8 \pm 0.6^{+0.5}_{-0.9})\%$ [20] is the fraction of D_s mesons produced; $Br(D_s \to \tau) = (5.54 \pm 0.24)\%$ [21] is the D_s branching ratio into τ; the factor 4 accounts for the charm pair production and for the two ν_τ produced per D_s decay. The SHiP facility is therefore a ν_τ factory, with 6.6×10^{15} tau neutrinos produced, equally divided in neutrinos and anti-neutrinos. Given the neutrino target mass of about 10 tons, one expects more than 10000 interactions of tau neutrinos and anti-neutrinos.

The prediction of the tau neutrino yield is affected by a large uncertainty: indeed, simulation studies of proton interactions in heavy and thick targets show that a sizeable contribution (about a factor of 2) to the charmed hadron yield comes from the cascade production [22]. Charmed hadrons are produced either directly from interactions of the primary protons or from subsequent interactions of the particles produced in the hadronic cascade showers. The SHiP Collaboration has proposed the SHiP-charm project [23], aiming at measuring the associated charm production by employing the SPS 400 GeV/c proton beam. This proposal includes a study of the cascade effect to be carried out using ECC techniques, i.e. slabs consisting of a replica of the SHiP experiment target [1] interleaved with emulsion films. The detector is hybrid, combining the emulsion technique with electronic detectors to provide the charge and momentum measurement of charmed hadron decay daughters and the muon identification. This allows a full kinematical reconstruction required by the double-differential cross- section measurement. According to the simulation performed, the delivery of 2×10^7 protons on target would allow the detection of about 1000 fully reconstructed charmed hadron pairs. An optimisation run is scheduled for 2018 and the full measurement is planned after the long shutdown LS2 of the CERN accelerator complex, with 5×10^7 protons on target and a charm yield of about 2500 fully reconstructed interactions.

The charged-current ν_τ ($\bar{\nu}_\tau$) differential cross-section is given by five structure functions. The contribution to the cross-section of F_4 and F_5 structure functions, introduced by Albright and Jarlskog [24], is negligible in muon and electron neutrino interactions due to the light charged lepton mass. On the contrary, given the non-negligible mass of the τ lepton, tau neutrino scattering is sensitive to their contribution. Indeed, the hypothesis of $F_4 = F_5 = 0$ would significantly increase the ν_τ and $\bar{\nu}_\tau$ charged-current deep-inelastic cross sections and the corresponding number of expected ν_τ and $\bar{\nu}_\tau$ interactions. The region of the energy spectrum below 40 GeV is highly sensitive to the non-zero values of these structure functions.

In the Standard Model, neutrinos may interact with a photon through second order diagrams giving a magnetic moment proportional to the neutrino mass and therefore totally negligible and undetectable. Beyond standard model theories predict a neutrino magnetic moment many orders of magnitude higher. Testing the neutrino magnetic moment is therefore a test of possible physics beyond the standard model. Limits on the magnetic moment of muon and electron neutrinos are rather stringent while the one of tau neutrinos is practically unconstrained. An anomalous magnetic moment of the tau neutrino would result in tau neutrino interactions with electrons. Quasi-elastic electron neutrino interactions where the final state proton is undetected constitute the main background. An upper limit $\mu_\nu < 1.3 \times 10^{-7} \mu_B$ can be achieved with the SHiP experiment in case no excess is observed.

3.1 Strange parton distribution

Charmed hadrons are produced in neutrino and anti-neutrino charged-current interactions at the level of about 5%. Experiments based on calorimetric technology identify charmed hadrons only in their muonic decay channel, when two opposite sign muons are produced in the final state. A cut of 5 GeV is applied to muons in order to suppress the background due to punch-through pions. The nuclear emulsion technology, instead, identifies topologically the charmed hadron by detecting its decay vertex. Energy cuts are therefore much looser, thus providing a better sensitivity to the charm quark mass. Moroever, a large statistical gain is provided by the use of hadronic decay modes [25]. Indeed, despite the fact that 1.280.000 ν_μ and 270.000 $\bar{\nu}_\mu$ charged-current interactions were collected by the NuTeV/CCFR Collaboration, only 5102 ν_μ and 1458 $\bar{\nu}_\mu$ events were identified as charm production induced by neutrino charged-current interactions. The largest number of events in an emulsion experiment was collected by CHORUS with 2013 charm candidates from ν_μ and only 32 from $\bar{\nu}_\mu$ [26]. SHiP will integrate about 10^5 charm candidates, more than one order of magnitude larger than the present statistics, with a large ($\sim 30\%$) contribution from anti-neutrinos. Charm production in neutrino scattering is extremely sensitive to the strange quark content of the nucleon, especially with anti-neutrinos where the s-quark is dominant. SHiP will improve significantly the uncertainty on the strange quark distribution in the nucleon as shown in the left plot of Figure 4 in terms of $s^+ = s(x) + \bar{s}(x)$ in the $0.02 < x < 0.35$ range.

3.2 Dark Matter search

The dark matter at an accelerator can be produced through the decay of a dark photon and it can thus be copiously produced at a beam dump facility. Given the proton energy, a dark matter in the GeV and sub-GeV mass range can be produced at SHiP. Direct searches performed in the underground laboratories rely on the observation of nuclear recoils induced by the scattering of galactic dark

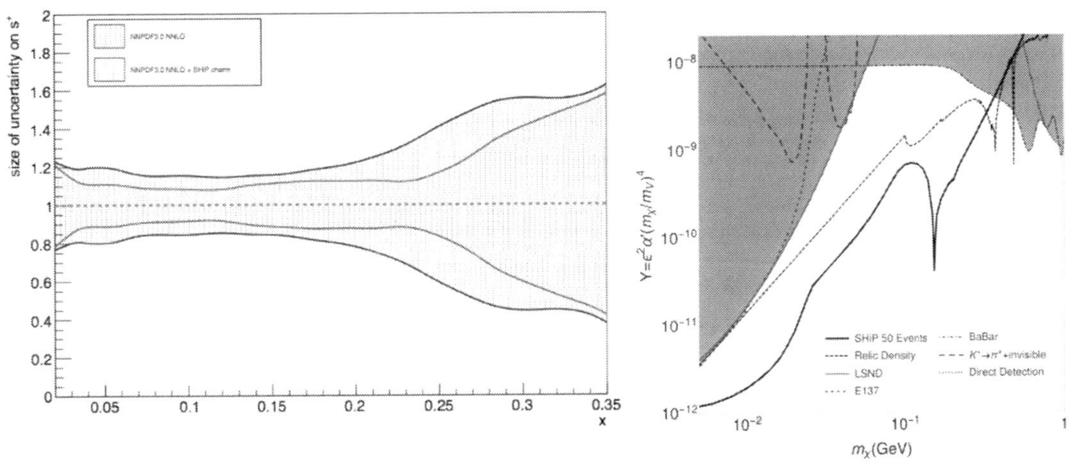

Figure 4. Left: Improvement of the accuracy on s^+ with SHiP (red) compared to the present status (blue) in the $0.02 < x < 0.35$ range. Right: SHiP sensitivity to light dark matter in the (m_χ, Y) plane.

matter. Given the non-relativistic nature of galactic dark matter, induced recoils show a kinetic energy of the keV order if the dark matter mass is around 1 GeV. This makes the search for dark matter at around 1 GeV or below extremely difficult. The sensitivity of current experiments in that range is therefore limited by the capability of observing such low energy recoils. At the accelerator, on the contrary, the dark matter would be ultra-relativistic and it could be observed through its scattering off the electrons of the lead-emulsion target of the neutrino detector. The elastic interaction of dark matter particles with electrons produces one electron in the final state, thus mimiking elastic interaction of neutrinos that constitute the main background for this search. In Ref. [1] we have estimated the total background. Following the approach described in Ref. [27], the right plot of Fig. 4 shows the SHiP sensitivity to light dark matter in the plane (m_χ, Y) where m_χ is the dark matter mass, $Y = \varepsilon^2 \alpha' \frac{m_\chi^4}{m_V^4}$, m_V the mass of the dark photon as dark matter parent, ε the coupling. This plot is drawn for the fixed ratio $\frac{m_V}{m_\chi} = 5$ and $\alpha' = 0.5$. The deep peak at around 150 MeV is due to the ρ/ω resonances in the dark photon production. The slope change around 27 MeV is due to π^0 production. The sensitivity to light dark matter turns out to be very competitive with all the planned experiments in the next decade.

The neutrino detector is in general sensitive to the search for very weakly interacting particles when they interact with the atoms of the emulsion target.

References

[1] M. Anelli et al., *A facility to Search for Hidden Particles (SHiP) at the CERN SPS*, arXiv:1504.04956.

[2] S. Alekhin et al., *A facility to search for hidden particles at the CERN SPS: the SHiP physics case*, arXiv:1504.04855.

[3] A. Akmete et al., *The active muon shield in the SHiP experiment*, JINST 12 (2017) no.05, P05011.

[4] N. Agafonova et al., *The OPERA experiment in the CERN to Gran Sasso neutrino beam*, JINST **4** (2009) P04018.

[5] M. Drewes, et al., *Leptogenesis from a GeV Seesaw without mass degeneracy*, JHEP **1303** (2013) 096.

[6] M. Drewes, *The phenomenology of right handed neutrinos*, Int. J. Mod. Phys. **E22** (2013) 1330019.

[7] T. Asaka and M. Shaposhnikov, *The nuMSM, dark matter and baryon asymmetry of the universe*, Phys. Lett. **B620** (2005) 17.

[8] M. Shaposhnikov, *A possible symmetry of the nuMSM*, Nucl. Phys. **B763** (2007) 49.

[9] L. Canetti et al., *Sterile neutrinos as the origin of dark and baryonic matter*, Phys. Rev. Lett. **110** (2013) 061801.

[10] L. Canetti and M. Shaposhnikov, *Baryon Asymmetry of the Universe in the NuMSM*, JCAP **1009** (2010) 001.

[11] K. Kodama et al., *Observation of tau neutrino interactions*, Phys. Lett. **B504** (2001) 218.

[12] K. Kodama et al., *Final tau-neutrino results from the DONuT experiment*, Phys. Rev. **D78** (2008) 052002.

[13] N. Agafonova et al., *Observation of a first ν_τ candidate in the OPERA experiment in the CNGS beam*, Phys. Lett. **B691** (2010) 138.

[14] N. Agafonova et al., *New results on $\nu_\mu \to \nu_\tau$ appearance with the OPERA experiment in the CNGS beam*, JHEP **1311** (2013) 036.

[15] N. Agafonova et al., *Evidence for $\nu_\mu \to \nu_\tau$ appearance in the CNGS neutrino beam with the OPERA experiment*, Phys. Rev. **D89** (2014) 051102.

[16] N. Agafonova et al., *Observation of tau neutrino appearance in the CNGS beam with the OPERA experiment*, PTEP **2014** (2014) 101C01.

[17] N. Agafonova et al., *Discovery of τ neutrino appearance in the CNGS Neutrino Beam with the OPERA experiment*, Phys. Rev. Lett. **115** (2015) 121802.

[18] C. Lourenco, H.K. Wohri, *Heavy flavour hadroproduction from fixed-target to collider energies*, Phys. Rept. **433** (2006) 127.

[19] J. Carvalho, *Compilation of cross sections for proton nucleus interactions at the HERA energy*, Nuclear Physics **A725** (2003) 269.

[20] H. Abramowicz et al., *Measurement of charm fragmentation fractions in photoproduction at HERA*, JHEP **1309** (2013) 058.

[21] K.A. Olive et al., *Review of Particle Physics*, Chin. Phys. **C38** (2014) 090001.

[22] H. Dijkstra and T. Ruf, http://cds.cern.ch/record/2115534/files/SHiP-NOTE-2015-009.pdf.

[23] A. Akmete et al. [SHiP Collaboration], CERN-SPSC-2017-033, SPSC-EOI-017 (2017).

[24] C. H. Albright and C. Jarlskog, *Neutrino Production of m^+ and e^+ Heavy Leptons.1*, Nucl. Phys. B **84** (1975) 467.

[25] G. De Lellis et al., *Charm physics with neutrinos*, Physics Reports **399** (2004) 227.

[26] A. Kayis-Topaksu et al., *Measurement of charm production in neutrino charged-current interactions*, New J. Phys. **13** (2011) 093002.

[27] P. deNiverville et al., *Light dark matter in neutrino beams: production modelling and scattering signatures at MiniBooNE, T2K and SHiP*, Phys. Rev. D95 (2017) no.3, 035006.

Status of the Mu2e experiment

S. Giovannella[1,a] on behalf of the Mu2e Collaboration

[1] Laboratori Nazionali di Frascati dell'INFN, via Enrico Fermi 40, 00044 Frascati, Italy

Abstract. The Mu2e experiment at Fermilab searches for the charged-lepton flavor violating neutrino-less conversion of a negative muon into an electron in the field of an aluminum nucleus. The dynamics of such a process is well modelled by a two-body decay, resulting in a mono-energetic electron with an energy slightly below the muon rest mass. If no events are observed, in three years of running Mu2e will improve the current limit by four orders of magnitude. Such a charged lepton flavor-violating reaction probes new physics at a scale inaccessible with direct searches at either present or planned high energy colliders. The experiment both complements and extends the current search for muon decay to electron-photon at MEG and searches for new physics at the LHC. This paper focuses on the physics motivation, the design and the status of the experiment.

1 Introduction

Differently from their hadronic counterpart, charged lepton flavor transitions are not allowed in the Standard Model (SM) with massless neutrinos. Even including neutrino mass, charged lepton flavor violation (CLFV) processes are extremely suppressed in the SM, with rates smaller than 10^{-50} [1]. On the other hand, a broad variety of new physics Beyond the Standard Model (SUSY, Leptoquarks, GUT, ...), predicts significantly larger rates, within the reach of next generation of CLFV experiments [2]. Because of the negligible background, any CLFV experimental detection would be a clear indication of new physics.

Searches in muon channels are of particular interest because of their high rates and the possibility of carrying out clean measurements free of hadronic corrections in the calculation. Comparison of current limit and expectations from next generation experiment with other channels are reported in Fig. 1. The experimental search for CLFV with muons ($\mu \to e\gamma$, $\mu \to 3e$, and $\mu N \to eN$, *i.e.* the muon to electron conversion in the field of a nucleus) is progressing extremely fast in the last decades. Current best limits are BR($\mu \to e\gamma$) $< 4.2 \times 10^{-13}$ at 90 % C.L. (MEG [3]) and $R_{\mu e} < 7 \times 10^{-13}$ (SINDRUM-II [4]). A solid international program exists with the MEG upgrade [5] underway, a proposed $\mu \to 3e$ experiment at PSI (Mu3e [6]) and with the approved programs on the muon to electron conversion at FNAL (Mu2e) [7] and J-PARC (COMET/DeeMe) [8, 9].

Process	Current Limit	Next Generation exp
$\tau \to \mu\eta$	BR < 6.5 E-8	
$\tau \to \mu\gamma$	BR < 6.8 E-8	10^{-9} - 10^{-10} (Belle II)
$\tau \to \mu\mu\mu$	BR < 3.2 E-8	
$\tau \to eee$	BR < 3.6 E-8	
$K_L \to e\mu$	BR < 4.7 E-12	
$K^+ \to \pi^+ e^- \mu^+$	BR < 1.3 E-11	
$B^0 \to e\mu$	BR < 7.8 E-8	
$B^+ \to K^+ e\mu$	BR < 9.1 E-8	
$\mu^+ \to e^+\gamma$	BR < 4.2 E-13	10^{-14} (MEG)
$\mu^+ \to e^+e^+e^-$	BR < 1.0 E-12	10^{-16} (PSI)
$\mu N \to eN$	$R_{\mu e}$ < 7.0 E-13	10^{-17} (Mu2e, COMET)

Figure 1. Current best limits on CLFV searches and expectations from next generation experiments.

2 The Mu2e Experiment

The goal of the Mu2e experiment is to improve by four orders of magnitude the best previous measurement and reach a single event sensitivity of 3×10^{-17} on $R_{\mu e}$, the rate of neutrino-less conversion of a muon into an electron in the field of a nucleus with respect to the dominant muon capture process. The experimental technique consists of a high intensity beam of low momentum muons stopped in an aluminum target and trapped in orbit around the nucleus, with a lifetime in the bound state of $\tau_\mu = 864$ ns. The distinctive signature of the conversion electron (CE) is a mono-energetic electron with momentum very close to the muon rest mass, $E_{CE} = 104.96$ MeV (Fig. 2). Muons stopped on aluminium have a 39% probability of undergoing a three-body decay when orbiting around the nucleus. The electron spectrum of this Decay-In-Orbit (DIO) process substantially differs from that of free decay, due to the presence of a large recoil tail that falls rapidly as the electron energy approaches the kinematical endpoint. The CE line has to be distinguished, with a high momentum res-

[a]e-mail: simona.giovannella@lnf.infn.it

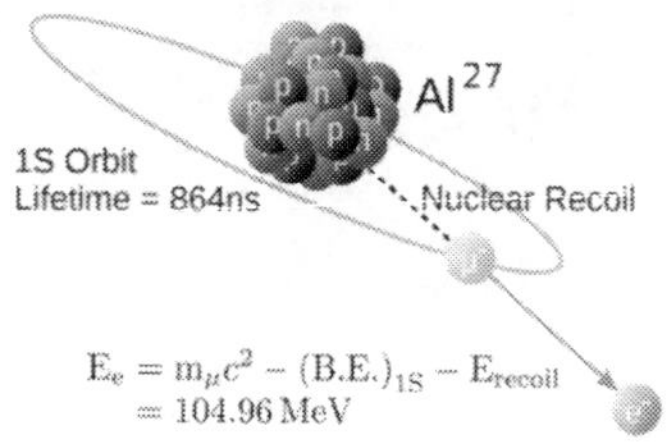

Figure 2. Drawing of the muon to electron conversion process.

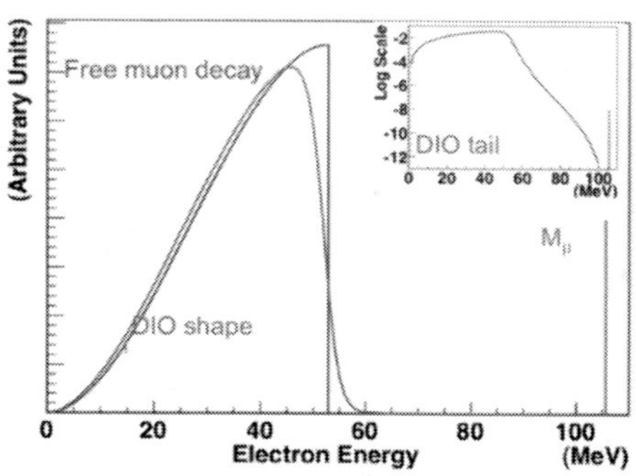

Figure 3. Energy spectrum for electrons produced from muon decays in orbit. An ideal resolution is assumed.

olution detector, from the DIO electron spectrum, Fig. 3.

Apart from the DIO contribution, an additional background source comes from the radiative pion capture (RPC), $\pi + N \rightarrow \gamma + N'$. Here, the electron positron pair, produced either by internal or external conversion, becomes a source of fake CE candidates when the e^- momentum is in the selection window.

In order to reach the required sensitivity, the experiment has to collect 10^{18} stopped muons with a number of background events less than 0.5. These considerations have driven the design strategy of Mu2e, based on four key elements:

1. A high intensity muon beam
   The goal is to increase the muon intensity by 10^4 w.r.t. previous experiments to reach 10^{11} muons/s on target. This is obtained combining a high rate particle production and a curved solenoidal system to create a transport channel that selects both charge and momentum.

2. A pulsed beam structure
   Mu2e has selected an aluminium target where the muon lifetime in the bound system (τ_μ = 864 ns) well matches the bunch period of the Fermilab accelerator (micro-bunch of 1694 ns period). The trick is to wait for the prompt backgrounds to decay and start the data acquisition ~ 700 ns after the bunch arrival time (Fig. 4).

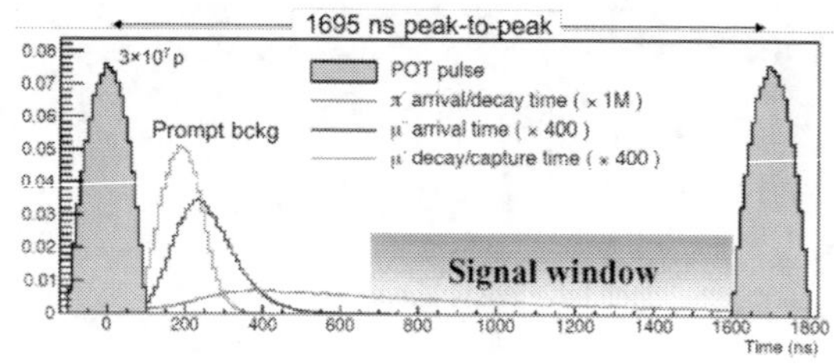

Figure 4. Mu2e bunch structure. The signal window is delayed by 700 ns to allow the decay of the prompt background.

3. A proton extinction better than 10^{-10}
   The number of protons traveling in the beam in the out of time window has to be reduced to the indicated level with respect to the in time protons.

4. A redundant high-precision detector
   This is needed to analyse the products from muon interaction on target to separate CE and DIO spectra and make the contribution from additional background sources negligible.

The layout of the Mu2e experiment is shown in Fig. 5. A series of superconducting solenoids forms a graded magnetic system composed of a Production Solenoid, PS, a Transport Solenoid, TS, and a Detector Solenoid, DS. The PS contains a tungsten target that is struck by an 8 GeV pulsed proton beam. A gradient field in the PS (from 2.5 to 4.6 Tesla) acts as a magnetic lens to focus the produced low energy particles (pions, muons and a small number of antiprotons) into the transport channel. The S-shaped Transport Solenoid efficiently transfers low energy, negatively charged particles to the end of the beamline while allowing a large fraction of pions to decay into muons. Positive and negatively charged particles drift in opposite directions while traveling through the curved solenoidal field, and a mid-section collimator removes nearly all the positively charged particles. The DS uses a graded field from 2 to 1 Tesla in the upstream region where the stopping target resides to increase acceptance for CE events. An uniform magnetic field of 1 Tesla occupies the region of the tracker and calorimeter systems. Approximately 50% of the muon beam, whose momentum is ~ 50 MeV, is stopped by the target; the surviving beam stops on the beam dump at the end of the cryostat. A muon stopping rate of 10 GHz allows the experiment to reach the final goal of 10^{18} stopped muons on target in three years of running. Muons that stop in the aluminium target are captured in an atomic excited state and promptly cascade to the 1S ground state with 39% decaying in orbit and the remaining 61% captured by the nucleus. Low energy photons, neutrons and protons are emitted in the nuclear capture process. These make up an irreducible

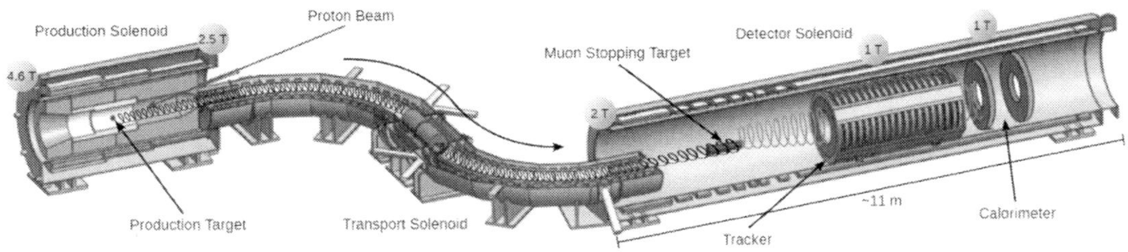

Figure 5. Layout of the Mu2e experiment. The Cosmic Ray Veto, the Extinction Monitor and the Stopping Target Monitor are not shown.

source of accidental activity that is the origin of a large neutron fluence on the detection systems. Together with the flash of particles accompanying the beam, the capture process produces the bulk of the ionizing dose observed in the detector system and its electronics.

The Mu2e tracker is the primary device to measure the momentum of the electron and separate it from background. The crystal calorimeter plays a crucial role in providing particle identification capabilities and a fast online trigger filter, while aiding the track reconstruction capabilities. An external veto for cosmic rays surrounds the solenoid. An extinction monitor detects scattered protons from the production target to evaluate the fraction of out-of-time beam and a stopping target monitor measures the rate and the number of negative muons that stop in the target.

3 Tracker

The Mu2e tracker system [10] has been designed to maximizing acceptance for conversion electrons (CE), minimizing the contamination from the muon Decay-In-Orbit (DIO) background. Nuclear modifications push the DIO spectrum towards the CE signal; energy loss and detector resolution produce overlap of the two processes. The selected design is based on nearly 20,000 low mass straw drift tubes of 5 mm in diameter, with 15 μm Mylar wall and 25 μm sense wire. Straws are oriented transversally to the solenoid axis and arranged in 18 stations for a total lenght of 3.2 metres. A central hole, 38 cm in diameter, makes the device blind to low momentum background particles. (Fig. 6).

Tracker performance has been studied by Monte Carlo using Mu2e full simulation. Results are reported in Fig. 7. The core momentum resolution of 115 keV/c, with a 3% high tail slope of 179 keV/c, is well within physics requirements and stable when incresing accidental hit rate. The total track efficiency of ~ 9% is fully dominated by geometric acceptance.

An eight channel tracker prototype has been built and tested with cosmics. In Fig. 8 the extracted position resolution is compared with Monte Carlo expectations. The shift observed in the transverse resolution is due to the

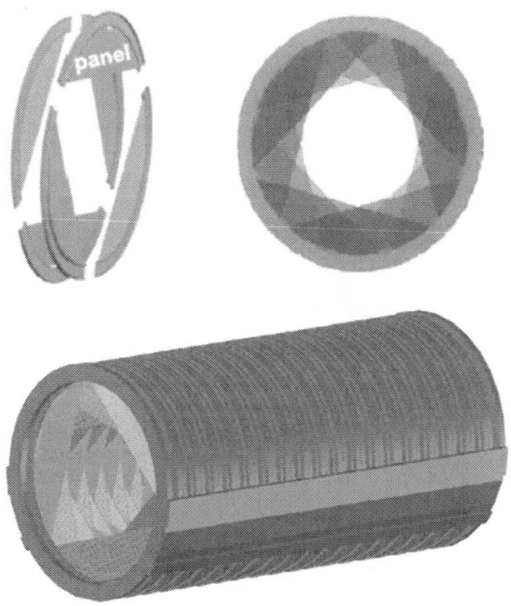

Figure 6. Sketch of the Mu2e straw tracker system. The basic element is the panel, where straws are organized in two staggered layers. Six panels arranged as shown in top-left form a plane; two planes rotated by 30° constitute a station, top-right. The tracker, containing 18 stations, is 3.2 meter long.

T_0 calibration differences. The transverse resolution extracted with a Gaussian fit is (0.133 ± 0.022) mm for data and (0.102 ± 0.001) mm for Monte Carlo simulation. The values extracted for the longitudinal resolution are $\sigma_{\rm data}$ = (42 ± 1) mm and $\sigma_{\rm MC}$ = (43 ± 1) mm.

A first pre-production prototype with final design was recently built and is being tested. A vertical slice test on fully instrumented panels with the entire FEE chain will follow.

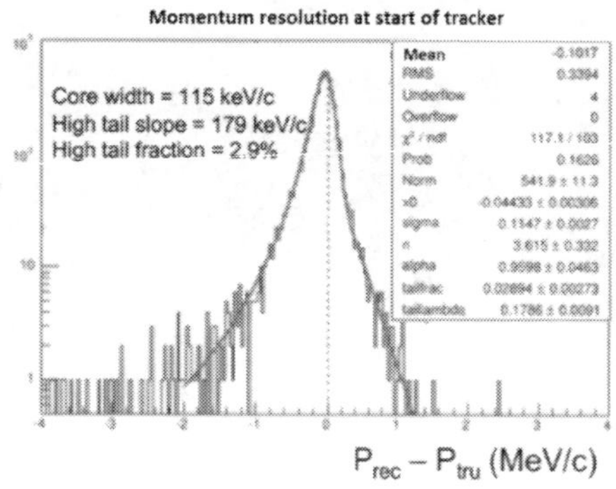

Figure 7. Momentum resolution for conversion electrons obtained with full simulation.

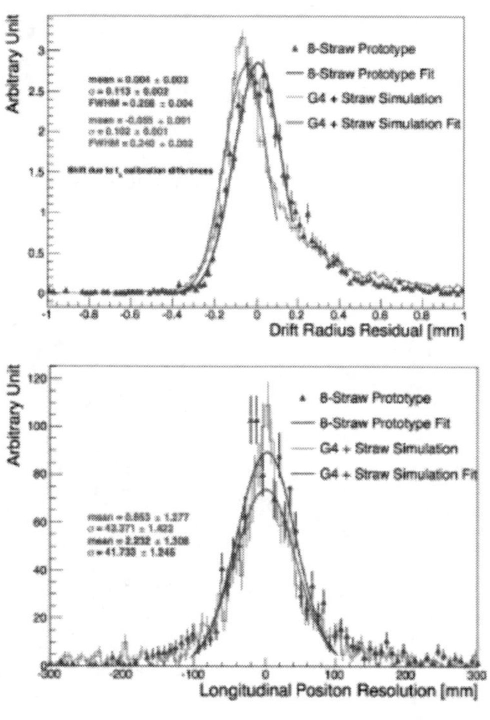

Figure 8. Transverse (top) and longitudinal (bottom) position resolution for an eight channel prototype of the tracker. Data from minimum ionizing particles (blue triangles) are compared with Monte Carlo simulation (red crosses). Resolution is extracted with Gaussian fits to the spectra.

4 Calorimeter

The Mu2e calorimeter [11] has to provide confirmation for CE signal events, a powerful e/μ separation - with a muon rejection factor of $\sim$ 200, a standalone trigger and seeding for track reconstruction. An energy resolution of $O(5\%)$ and a time resolution $<$ 500 ps for 100 MeV electrons

Figure 9. Sketch of the calorimeter system.

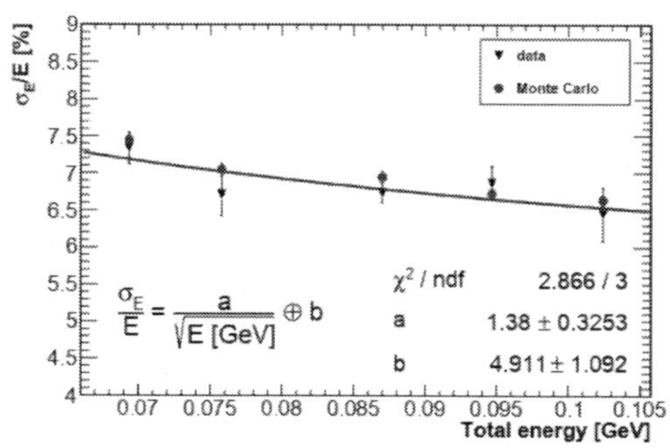

Figure 10. Energy resolution obtained with a (9×9) cm^2 calorimeter prototype.

are sufficient to fulfill these requirements. The calorimeter design consists of two disks made by 674 undoped CsI scintillating crystals with $(34 \times 34 \times 200)$ mm^3 dimension (Fig. 9). Each crystal is read-out by two custom array large area (2×3 of 6×6 mm^2 cells) UV-extended Silicon Photo-Multipliers (SiPMs). The crystals will receive an ionizing dose of 90 krad and a fluence of 3×10^{12} n/cm^2 in three years running. The photosensors, being shielded by the crystals, will get a three times smaller dose.

A small calorimeter protoype, a 3×3 array of CsI crystals coupled to a single multi-pixel photon counter, has been tested at the Frascati Beam Test Facility with electron beams of 80–120 MeV [12]. A time resolution of 100 ps and an energy resolution of 6.5% has been obtained for 100 MeV particles (Figs. 10, 11). For the latter, a significant leakage contribution is present, confirmed by simulation.

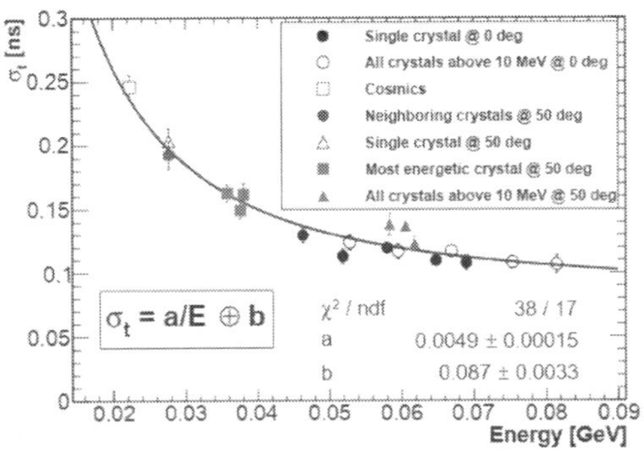

Figure 11. Time resolution obtained with a (9×9) cm^2 calorimeter prototype.

Figure 12. Sketch of the Cosmic Ray Veto.

Pre-production components both for crystals and SiPMs have been received from different vendors. They have been characterized and irradiation test have been carried out for a small subsamples. Pre-production components have been used to build a large calorimeter prototype, with 51 crystals and 102 SiPMs and front end boards, testing integration and assembly procedures.

5 Cosmic Ray Veto

The major background source in Mu2e is due to cosmic ray muons that produce fake CE candidates when interacting with the detector materials. These events occur at a rate of approximately one/day. In order to reduce their contributions in the experiment lifetime, the external area of the DS, and a part of the TS, are covered by a Cosmic Ray Veto (CRV) system [7], shown in Fig. 12. The requirement for the CRV system is to obtain a veto efficiency of at least 99.99% for cosmic ray tracks while withstanding an intense radiation environment. Comprised of four staggered layers of scintillation slabs (Fig. 13), the CRV counters are read out with two embedded wavelength shifting fibers, each one in optical contact with a (2×2) mm^2 Hamamatsu SiPM. Test beams on full size prototype have been carried out demonstrating that the needed light yield can be reached [13]: the measured number of photoeletrons obtained at 1 meter from the readout end provides a safety factor of $\sim$ 40% with respect to the requirement. Irradiation of SiPMs with neutrons have also been done to understand the maximum level of fluence acceptable for operations.

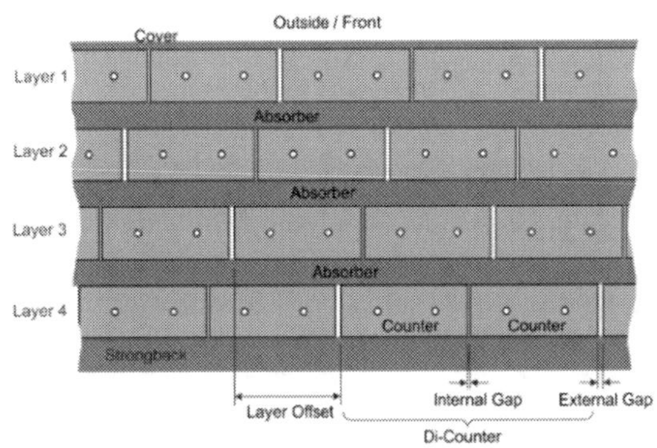

Figure 13. Internal structure of the Cosmic Ray Veto.

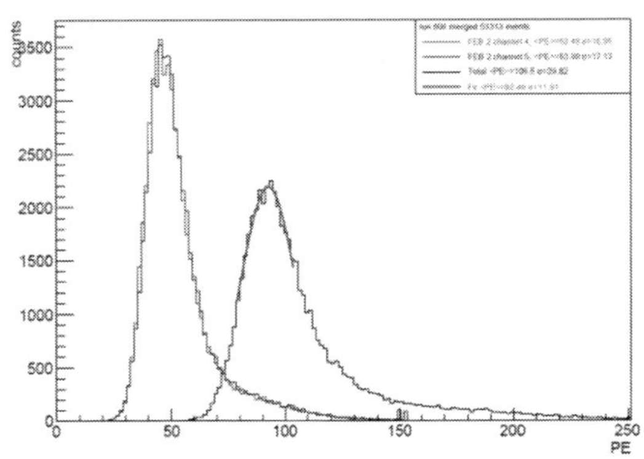

Figure 14. Number of photo-electrons obtained for 120 GeV protons impinging at 1 meter from the readout end. The two peaks are obtained with one (left) and two (right) fiber readout.

6 Expectations with full simulation

At 100 MeV, the momentum resolution is dominated by the fluctuations in the energy loss in the stopping target and bremsstrahlung in the tracker, with the trajectory altered by multiple scattering. The resolution for CE tracks is well-parametrised by a Crystal Ball function with a negative bremsstrahlung tail, a gaussian core of 116 keV and a long exponential positive resolution tail. The finite resolution has a large effect on the DIO falling spectrum that translates to a residual contamination in the signal region. Fig. 15 shows the signal and background distributions as

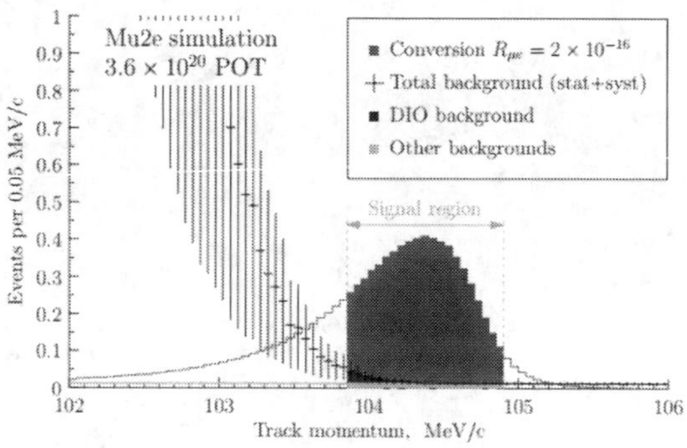

Figure 15. Distribution of the electron momentum spectrum for simulated background and signal events.

seen by a full simulation of the experiment, including pileup, in the following conditions:

- 3.6×10^{20} protons on target;
- a corresponding number of 6×10^{17} muon stops;
- $R_{\mu e}$ of 2×10^{-16}.

A momentum window is selected by maximising the signal over background. The estimate of the background contributions is presented in Table 1, for a total background of 0.46 events. Largest contribution comes from cosmic rays (0.24 events) and DIOs (0.14 events). The number of reconstructed signal events is 6.66. This counting corresponds to set a limit on $R_{\mu e}$ below 8×10^{-17} at 90 % C.L., in good agreement with the design goal of the experiment.

7 Status of the experiment

The Mu2e experiment has, as of this writing, successfully procured all superconducting cables, completed civil construction and obtained CD-3 (Critical Decision 3) from DOE. CD-3 grants permission to start the construction for the accelerator, the magnetic system, the muon beam line and all the detector components. The heart of the Mu2e apparatus is provided by the superconducting magnetic system whose design, fabrication, assembly and commissioning drives the schedule of the experiment. The status of the magnet as well as the construction and testing of the superconducting cables is satisfactory. An international bid for the DS and PS has been concluded and the construction phase for the large magnets is started at General Atomics, San Diego, USA. For the TS, after the construction of one module prototype by ASG Superconducting, in collaboration with the INFN group of Genova, the contract has been awarded to ASG. Construction of the 52 coils is progressing well, with the first 32 already being completed. The first TS module, assembled with two coils, is being delivered and then fully qualified at the HAB test facility in Fermilab. All systems are concluding the prototyping phase. Production of final components is starting, and it will move to full regime by the end of 2017. The schedule foresees a completion of the installation of detectors and commissioning with cosmic rays at the end of 2020.

8 Conclusions

The Mu2e experiment will exploit the world's highest intensity muon beams of the Fermilab Muon Campus to search for CLFV, improving current sensitivity by a factor 10^4 and with a discovery capability over a wide range of New Physics models. A low mass straw tube tracker, a pure CsI crystal calorimeter with SiPM readout and a high efficiency cosmic ray veto have been selected to satisfy the demanding requirements. The construction of the detectors is beginning at the moment of writing. Detector installation is foreseen in 2020, followed by Mu2e commissioning and data taking.

Acknowledgments

We are grateful for the vital contributions of the Fermilab staff and the technical staff of the participating institutions. This work was supported by the US Department of Energy; the Italian Istituto Nazionale di Fisica Nucleare; the US National Science Foundation; the Ministry of Education and Science of the Russian Federation; the Thousand Talents Plan of China; the Helmholtz Association of Germany; and the EU Horizon 2020 Research and Innovation Program under the Marie Sklodowska-Curie Grant Agreement N.690385. Fermilab is operated by Fermi Research Alliance, LLC under Contract No. De-AC02-07CH11359 with the US Department of Energy, Office of Science, Office of High Energy Physics.

References

[1] S. M. Bilenky *et al.*, Phys. Lett. B 67, 309 (1977) 309
[2] F. Grancagnolo *et al.*, Nucl. Phys. Proc. Supp. 248-250 (2014) 1-146.
[3] A. M. Baldini *et al.*, Eur. Phys. J. C 76 (2016) 434.
[4] The SINDRUM II Collaboration, Eur. Phys. J. C 47 (2006) 337-346
[5] A. M. Baldini *et al.*, arXiv:1301.7225 (2013).
[6] A. Blondel *et al.*, arXiv:1301.6113 (2013).
[7] L. Bartoszek *et al.*, "Mu2e Technical Design Report", ArXiv:1501.05241, Fermilab-TM-2594, Fermilab-Design-2014-01.
[8] R. Abramishvili *et al.*, http://comet.kek.jp/Documents_files/PAC-TDR-2016/COMET-TDR-2016_v2.pdf (2016).
[9] H. Natori *et al.*, Nucl. Phys. B Proc. Suppl. 248-250 (2014) 52-57
[10] M. J. Lee on behalf of the Mu2e collaboration, "The Straw-tube Tracker for the Mu2e Experiment", Nuclear and Particle Physics Proceedings 273-275 (2016) 2530.
[11] N.Atanov *et al.*, "Design and status of the Mu2e electromagnetic calorimeter", Nucl. Instr. and Meth. A 824 (2016) 695.

Category	Background process	Estimated Yield (events)
Intrinsic	Decay in orbit (DIO)	0.14 ± 0.11
	Muon Capture (RMC)	0.000 ± 0.004
Late Arriving	Pion capture (RPC)	0.025 ± 0.003
	Muon decay in flight	< 0.003
	Pion decay in flight	$0.000^{+0.001}_{-0.000}$
	beam electrons	$(2.5\pm 1.2) \times 10^{-4}$
Miscellaneous	pbar	0.047 ± 0.024
	Cosmic rays	0.24 ± 0.071
	Total	0.46 ± 0.13

Table 1. Expected background list as evaluated by full simulation.

[12] O. Atanova *et al.*, "Measurement of the energy and time resolution of a undoped CsI + MPPC array for the Mu2e experiment", JINST 12 (2017) P05007, ArXiv:1702.03720.

[13] A. Artikov *et al.*, FERMILAB-PUB-17-386-PPD, arXiv:1709.06587, submitted to Nucl. Instr. Meth. A.

[14] G. Blazey *et al.*, "Radiation Tests of Hamamatsu Multi-Pixel Photon Counters", submitted to Nucl. Instr. Meth. A.

 https://doi.org/10.1051/epjconf/201817901004

Fermilab muon g-2 experiment

Tim Gorringe[1,a] on behalf of the Fermilab muon g-2 collaboration.

[1] Department of Physics and Astronomy, University of Kentucky, Lexington, KY, 40502, USA

Abstract. The Fermilab muon g-2 experiment will measure the muon anomalous magnetic moment a_μ to 140 ppb – a four-fold improvement over the earlier Brookhaven experiment. The measurement of a_μ is well known as a unique test of the standard model with broad sensitivity to new interactions, particles and phenomena. The goal of 140 ppb is commensurate with ongoing improvements in the SM prediction of the anomalous moment and addresses the longstanding 3.5σ discrepancy between the BNL result and the SM prediction. In this article I discuss the physics motivation and experimental technique for measuring a_μ, and the current status and the future work for the project.

1 Introduction

These are interesting times in physics. With dark matter and dark energy being thought to comprise about 95% of the universe's mass, the standard model of particle physics seems astonishingly incomplete. Unfortunately – despite enormous efforts at high energy colliders, in deep underground searches, and elsewhere – no definitive evidence of new interactions, particles or phenomena has been found in any sub-atomic experiments.

It is nearly 20 years ago that the BNL 821 collaboration reported its now celebrated 540 ppb measurement [1] of the muon anomalous magnetic moment, a_μ. The BNL result and SM prediction were found to differ by roughly 3.5 σ. This discrepancy has survived experimental and theoretical scrutiny and today remains the most tantalizing and persistent hint from sub-atomic experiments of physics beyond the standard model.

In this article I will describe the new Fermilab muon g-2 experiment [2] – the direct descendant of BNL 821. The Fermilab experiment is designed to determine the muon anomaly a_μ to an unprecedented 140 parts-per-billion precision. Such precision will provide a unique test of the standard model and offer a broad sensitivity to new physics that complements other efforts at the energy, intensity and cosmic frontiers. Most importantly, it addresses the longstanding discrepancy between the SM prediction and the BNL experiment and – whether it confirms or refutes this discrepancy – will surely have broad impact and enduring value in nuclear, particle and astrophysics.

In what follows I offer an experimentalist's perspective on the physics interest, discuss the techniques exploited in the anomaly's measurement, and provide the current status and future work of the project.

Note an alternative measurement of the muon anomaly – employing a novel source of polarized muons and different approach to muon storage – is being pursued by the J-PARC E34 experiment [3]. It is described elsewhere in the conference proceedings.

2 Physics motivation

While collider experiments offer exceptional access to interactions and particles that lie within their energy range, the precision measurement of fundamental quantities offers opportunistic windows on new physics that's capable of reaching to much higher energy scales. In particular – through the subtle effects of quantum fluctuations – the muon magnetic moment is a unique portal to new phenomena at collider energies and beyond.

The muon anomaly a_μ is important because it's calculable and measurable with extraordinary precision; thereby offering both an exacting test of the standard model and a sensitive probe of any new physics (for details see Refs. [4, 5]). The BNL result [1] of

$$a_\mu^{exp} = (116\,592\,089 \pm 63) \times 10^{-11} \qquad (1)$$

has been one of the most-cited measurements of the past twenty years. Despite the intense theoretical and experimental scrutiny the longstanding discrepancy between the SM prediction and the BNL result has only strengthened over the years since the experiment.

Presumably all forces of nature – both known and unknown – give contributions to a_μ. Within the standard model the muon anomaly a_μ receives contributions from QED loops involving electrons and photons, electroweak loops involving W, Z and Higgs bosons, and hadronic loops involving quarks and gluons,

$$a_\mu^{SM} = a_\mu^{QED} + a_\mu^{EW} + a_\mu^{had}. \qquad (2)$$

By far the largest contribution is the famous Schwinger term $\alpha/2\pi$ involving a roughly part-per-thousand correction of the muon magnetic moment. However, the

[a] e-mail: gorringe@pa.uky.edu

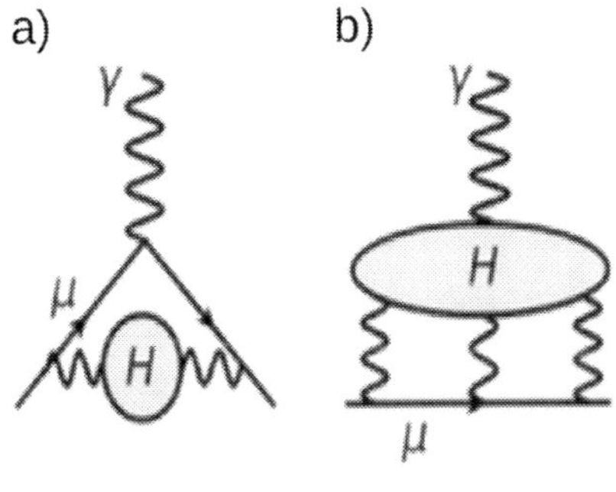

Figure 1. Feynman diagrams for a) the leading-order hadronic vacuum polarization contribution and b) the hadronic light-by-light scattering contribution.

overwhelming uncertainty in calculating a_μ^{SM} arises from hadronic vacuum fluctuations and how quarks and gluons dress the muon magnetic moment (see Fig. 1). As is discussed elsewhere in these conference proceeding, the most recent updates for the leading-order hadronic contributions $a_\mu^{had:LO}$ of 6922 ± 25, 6926 ± 33 and $6881 \pm 41 \times 10^{-11}$ from the Teubner *et al.*, Davier *et al.* and Jegerlehner [6–8] have lowered the uncertainties by roughly 25-50% over earlier estimates [9–11]. By comparison the uncertainties in the QED contribution, 0.08×10^{-11} [12], and the electroweak contribution, 1.0×10^{-11} [13], are completely negligible.

The largest uncertainty in the hadronic term arises from the leading-order hadronic vacuum polarization term $a_\mu^{had:LO}$. Since $a_\mu^{had:LO}$ isn't yet calculable with sufficient accuracy, its value is evaluated via dispersion relationships from an energy-weighted integral of the measured $e^+e^- \to$ hadrons cross section. The energy-weighting emphasizes the low-energy region that is dominated by the ρ and ω resonances and the $e^+e^- \to 2\pi$ channel. Recent estimates of $a_\mu^{had:LO}$ have benefited from new measurements and new analyses of e^+e^- data as well as more sophisticated treatments of the correlated systematic errors between the e^+e^- datasets. At present the uncertainty in $a_\mu^{had:LO}$ is dominated by the systematic errors in the experimental results for the various $e^-e^+ \to \pi\pi$ channel. The relevant data were obtained using initial-state radiation and energy-scan techniques and – while generally in agreement – there remains tension between BaBar data and other data at energies just above the ρ, ω resonances.

The other important contribution to the hadronic uncertainty is the hadronic light-by-light term a_μ^{hLBL}. For many years the estimation of a_μ^{hLBL} was limited to specific models of the hadronic loop in the four-photon vertex with the so-called "Glasgow consensus" $a_\mu^{hLBL} = (105 \pm 26) \times 10^{-11}$ emerging as the accepted value for the hadronic light-by-light contribution. One new approach, using lattice QCD, offers the prospect of a first-principles calculation of a_μ^{hLBL} at the 10% level. Another new approach, using dispersion relations, offers the prospect of a data-driven extraction of a_μ^{hLBL} by relating the quantity to pion form factor and $\gamma^*\gamma^* \to \pi\pi$ scattering data.

The effort to determine a_μ^{had} was discussed in great detail at this conference and represents a remarkable triumph of combined theoretical and experimental labor. Many talks spotlighted the continuing push to better quantify the hadronic corrections and associated uncertainties with theoretical tools including dispersion relations, lattice QCD, effective theories and hadronic models as well as ingenious new experimental ideas. The SM prediction for a_μ is currently known to about 400 ppb – to bring this uncertainty on par with experimental goals will need a roughly 0.2% determination of the hadronic vacuum polarization term and a roughly 10% determination of the hadronic light-by-light term.

Using the recent updates for the hadronic correction, the differences in a_μ between the BNL result and the SM prediction are $a_\mu^{exp} - a_\mu^{SM} = (288 \pm 80)$, (281 ± 73), and $(313 \pm 77) \times 10^{-11}$ [6–8]; a discrepancy of 3.6 to 4.1 σ. To set the scale – this discrepancy is roughly twice the entire contribution of electroweak effects – these effects mostly arising from one-loop diagrams involving virtual W and Z bosons.

New physics will typically make contributions to a_μ from one-loop diagrams at scales of $O(Cm_\mu^2/M^2)$ where C and M are the coupling constant and the mass scale of the new interaction. For couplings of order $O(1)$ the discrepancy would imply multi-TeV mass scales. By comparision, for electroweak-scale coupling $O(\alpha/4\pi)$ it would imply a sub-TeV mass scale and for milli-scale couplings $O(10^{-3})$ it would imply a GeV mass scale.

The anomalous magnetic moment represents a uniquely inclusive probe of new physics that nicely complements the specific sensitivities of collider experiments, underground searches, *etc.* Indeed the BNL result has major consequences for many SM extensions including extra Higgs, supersymmetric partners and dark photons, in some cases it pushing practitioners into small corners of model space. If new physics is discovered elsewhere the muon anomaly will surely play an important role in interpreting the signal. And - if new physics isn't discovered elsewhere – the anomaly is then the wide net to catch new physics.

3 Experimental technique

As mentioned, the Fermilab muon g-2 experiment [2] is designed to measure a_μ to 140 ppb precision, a goal commensurate with anticipated improvements in the present knowledge of the hadronic contributions from new e^+e^- data and theoretical advances.

The Fermilab experiment will use the same basic technique as the earlier BNL experiment. It involves injecting fills of polarized muons with magic momenta 3.094 GeV/c into the 1.54 T magnetic field of the 7.1 m radius storage ring. Within the ring the muons orbit at the cyclotron frequency ω_c and their spins rotate at the Lamor frequency ω_s. If a_μ was zero – *i.e.* muons were Dirac particles with g-factor $g \equiv 2$ – the two frequencies would be identical. However, for a non-zero anomalous magnetic moment the

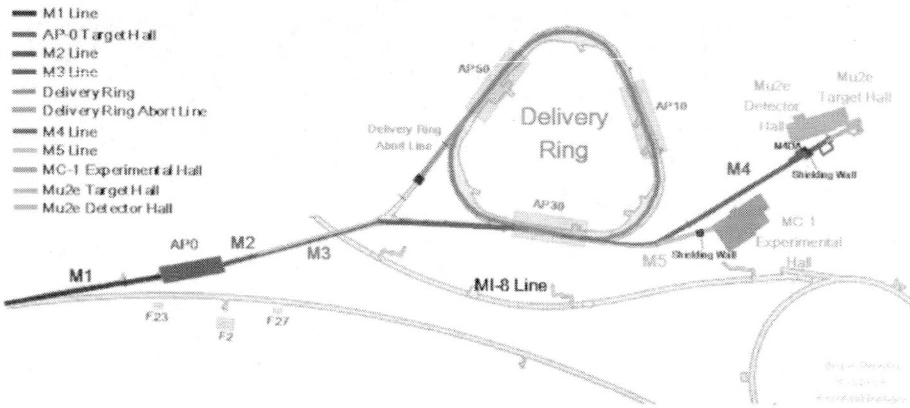

Figure 2. Schematic of Fermilab muon campus showing the M1 line for 8 GeV proton delivery, the AP0 production target for pion production, the M2 / M3 lines for 3.11 GeV/c pion collection and 3.09 GeV/c muon capture, the delivery ring for separating muons, pions and protons, the M5 line for beam delivery to the g-2 experiment, and the line M4 for beam delivery to the Mu2e experiment.

two frequencies differ by $\vec{\omega_a} \equiv \vec{\omega_s} - \vec{\omega_c}$ where

$$\vec{\omega_a} = \frac{q}{m}[a_\mu \vec{B} - (a_\mu - \frac{1}{\gamma^2 - 1})\vec{\beta} \times \vec{E}/c], \quad (3)$$

is the so-called anomalous precession frequency. At the magic momentum 3.094 GeV/c – where $\gamma = 29.3$ – the term in Eqn. 3 involving the electric field vanishes. Thus measurement of ω_a and B, and knowledge of q and m, together permit the determination of a_μ. One attractive feature of the experimental technique is the direct measurement of the anomalous part of the magnetic moment arising from vacuum fluctuations.

In practice, the experiment involves precisely measuring two frequencies; the muon anomalous precession frequency ω_a that was introduced above and a proton Lamor precession frequency ω_p for the B-field determination.

At the magic momentum the muons have a time-dilated lifetime of 64.4 μs and their spin advances about one extra turn for every thirty orbits of the storage ring. The frequency ω_a can be obtained from its sinusoidal modulation of the time distribution of the positrons emitted in the muon decays. The positrons are detected by calorimeters that are located on the inner circumference of the storage ring (the positrons have lower momenta than stored muons and spiral inwards). The origin of the ω_a-modulation is the self-analyzing nature of the muons, *i.e.* the high energy decay positrons are preferentially emitted along the muon spin direction. When the positron emission is directed along the muon orbit the Lorentz boost yields higher-energy positrons. When the positron emission is directed opposite the muon orbit the Lorentz boost yields lower-energy positrons. Consequently, the lab-frame energy of emitted positrons oscillates sinosoidally with frequency ω_a. Thus, by measuring either the integrated positron energy versus time or the high-energy positron counts versus time, the frequency ω_a is extractable.

The relevant B-field in Eqn. 3 is the magnetic field averaged over the muon beam distribution. A combination of fixed NMR probes and movable NMR probes are used to measure the proton Lamor frequency ω_p in the storage ring magnetic field. Nearly 400 fixed probes are distributed around the ring's circumference at locations either just above or just below the vacuum vessel. Additionally, 17 movable probes are mounted to a trolley that can circumnavigate the vacuum vessel interior in dedicated, beam-off, trolley runs. A special NMR probe is used for the 20 ppb absolute calibration of the entire NMR system. This system – together with accompanying measurements of the spatial distribution of the stored beam – thereby enables the determination of the beam-weighted, Lamor frequency $\overline{\omega}_p$.

From the determination of the muon's anomalous frequency, ω_a, and the proton's Lamor frequency, $\overline{\omega}_p$, the anomaly a_μ is extracted via

$$a_\mu = \frac{\omega_a/\overline{\omega}_p}{\mu_\mu/\mu_p - \omega_a/\overline{\omega}_p} \quad (4)$$

where μ_μ/μ_p is the muon-to-proton magnetic moment ratio. The quantity μ_μ/μ_p is needed as the frequency ω_a is proportional to the muon's anomalous magnetic moment whereas the frequency $\overline{\omega}_p$ is proportional to the proton's total magnetic moment. Fortunately, the ratio μ_μ/μ_p is precisely known from muonium spectroscopy – it having previously been measured at LAMPF to a 26 ppb precision and is currently being remeasured at J-PARC with a 10 ppb goal (see Ref. [5]).

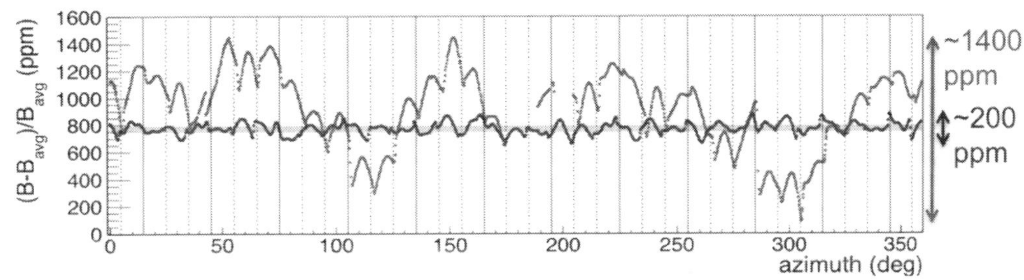

Figure 3. Azimuthal variation of the 1.45 T magnetic field of the storage ring during the shimming program. The red curve indicates the field variation in October 2015 and the blue curve indicates the field variations in June 2016. The shimming program untimately achieved a uniformity of ±50 ppm (the pink band).

4 Experiment status

The Fermilab muon campus is shown in Fig. 2. A number of existing components of Fermilab's accelerator complex are being reused to deliver fills of polarized, magic momentum muons to the g-2 storage ring. Protons bunches are first accelerated to 8 GeV using the booster ring. These bunches are then dispatched via the recycler ring to the AP0 target facility which is now repurposed as a pion source. Existing transfer lines are tuned to collect the 3.11 GeV/c pions from proton spallation and capture the 3.094 GeV muons from $\pi \rightarrow \mu\nu$ decay. Lastly the recycler ring – now renamed the delivery ring – is used to separate in time the muons from pions and protons with the muons then being directed to the g-2 storage ring via a newly-built transfer line. The design eliminates the prompt "flash" of pions and protons and – based on simulations and measurements – is expected to a yield an incident rate of 5-10×10^4 μ^+ per fill.

The Fermilab g-2 experiment is reusing the 1.45 T superconducting storage ring constructed for the earlier Brookhaven g-2 experiment – *i.e.* involving relocating the storage ring from Upton, NY to Batavia, IL. The move, which included transporting the 15 m-diameter superconducting coils and 650-ton steel yoke, was completed in 2013. The coils were shipped by barge – first south down the Atlantic coast and north up the Mississippi river – on a one-month, 3200-mile journey. The reassembly of the storage ring in the experimental hall of the newly constructed MC-1 building was completed in 2014. Two major milestones – cooling-down the superconducting coils and powering-up the main magnet – were achieved in 2015.

During the period 2015-16 the meticulous process of shimming the ring magnet to reach the required uniformity of magnetic field was carried out (see Fig. 3). This interative procedure progressed from the coarse adjustment of large pole pieces to the fine adjustment of tiny shimming wedges and greatly benefited from modern real-time measurement of field strengths versus shimming adjustments. Ultimately, the procedure led to a ±50 ppm field uniformity – about a four-fold improvement in the field uniformity over the BNL experiment.

In addition to the 5172 A main magnet for the 1.45 T magnetic field the storage ring incorporates a superconducting inflector, three pulsed kickers, and four electrostatic quadrupoles. The superconducting inflector and pulsed kickers are together responsible for channeling the injected muons onto the central orbit of the storage ring. The electrostatic quadrupoles provide the vertical confinement of stored muons.

Over the past year various new or refurbished detector packages have been installed. These packages include the twenty four electromagnetic calorimeters and the associated laser calibration system for the precision measurement of the positron time and energy distribution as well as in-vacuum straw trackers for positron tracking and in-vacuum scintillating fibers for muon tracking. A further detector package – comprising a plastic Cerenkov counter and scintillating fiber arrays – was installed to provide time and profile measurements of the beam upstream of the superconducting inflector. The full assembly of the muon storage ring and various detector packages are shown in Fig. 4.

Each calorimeter comprises a 6 × 9 segmented array of 2.5 × 2.5 × 14 cm^3 PbF_2 crystals [14, 15]. The Cerenkov light from each crystal is detected by commercial silicon photomultipliers (SiPMs), amplified by custom charge-sensitive amplifiers, and digitized by custom, 12-bit, 800 MSPS waveform digitizers. The continuously digitized traces from every crystal of every calorimeter are readout and processed by a 20 GB/s, 100 TFlop data readout and real-time processing system.

The first engineering run of the muon beam, storage ring and various detector systems was conducted in June, 2017. On May, 27 the first signals from a fill were recorded by the electromagnetic calorimeters and the straw trackers. On June, 7 the first evidence of "many-turn" storage of a fill was recorded by the electromagnetic calorimeters and the in-vacuum scintillating fibers. By June, 10 the first positron time distribution or so-called "wiggle plot" showing the time-dilated lifetime $\gamma\tau_\mu$ and anomalous precession frequency ω_a was reconstructed from the high energy positron hits in the electromagnetic calorimeters (see Fig. 5).

Figure 4. The 7.1 m radius, 1.45 T field, muon storage. The muons are injected at the top-left of the storage ring and the positrons are detected by the twenty four calorimeters abutting the ring's inner circumference.

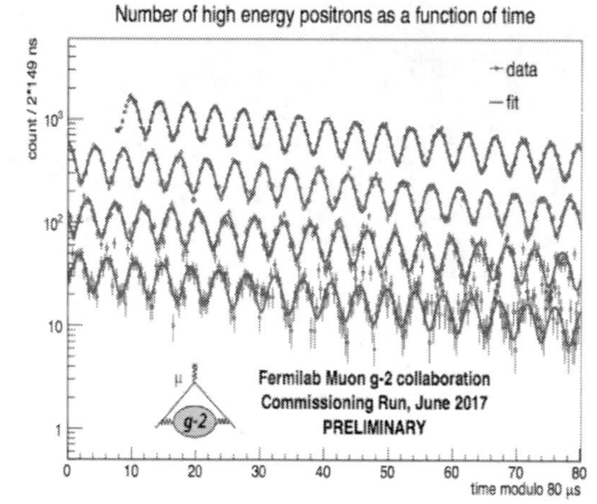

Figure 5. The wiggle plot of the positron time distribution from the June 2017 commissioning run showing the time-dilated muon lifetime $\gamma\tau_\mu$, anomalous precession frequency ω_a, and "best fit" curve.

The run permitted a number of studies of muon delivery and muon injection that currently are informing our efforts to optimize muon storage. Efficient storage requires careful focusing of the beam through the narrow aperture of the inflector, careful timing of the kicker pulsing that displaces particles onto the ring's orbit, and careful optimizing of the electrostatic quadrupoles to provide vertical confinement and avoid beam resonances. The run represented a major milestone in advancing g-2 to data-taking.

5 Experiment future

The goal of determining a_μ to 140 parts-per-billion precision requires improved measurements of the muon anomalous frequency ω_a and the proton Lamor frequency $\overline{\omega}_p$ in comparison to the Brookhaven experiment.

Concerning $\overline{\omega_p}$, we require a roughly three-fold reduction in systematic uncertainties from 170 ppb in the Brookhaven experiment to 70 ppb in the Fermilab experiment. This frequency is a weighted average over the muon azimuthal and transverse distributions in the storage ring. Better determinations of the stored muon distribution using the new positron tracking detectors and the magnetic field variation using the refurbished NMR trolley – along with aforementioned improvements in the field uniformity from the shimming program – are parts of accomplishing this goal.

Concerning ω_a, we require both a twenty-fold increase in detected positrons and a three-fold reduction in systematic uncertainties in comparision to BNL 821. The new Fermilab muon campus is the keystone in the twenty-fold increase in the positron statistics. The latter is achieved by increasing the fills per second of the storage ring, increasing the muons per fill in the storage ring, and extending running time. At time of writing, the push from the low fill rate, low muon rate, engineering run to high fill rate, high muon rate, production running is new underway.

To achieve the goals for ω_a systematics the collaboration built the new detector, electronics and acquisition system for the ω_a measurement. The system was designed to handle the increased rates at Fermilab and also minimize biases in the extraction of the anomalous frequency ω_a from the positron time distribution. The measurement requires extraordinary stability of the detectors and the electronics over the 700 μs measurement periods following the muon beam injection as the positron rates decrease by about four orders-of-magnitude. Serious concerns include pulse pileup, gain changes and pedestal drifts during the measurement periods.

One part of reducing systematics is the readout electronics and acquisition system that facilitates different approaches to determining ω_a. One category of recorded data – the T-method datset – consists of chopped islands of consecutive samples that are triggered by an above-threshold signal. This dataset is used to construct the time distribution of high energy positrons hits by fitting the individual crystal pulses and clustering into reconstructed positron hits. Another category of recorded data – the Q-method dataset – consists of periodic flushes of fill-summed histograms of the 700 μs contiguous traces of ADC samples. This dataset is used to construct the time distribution of the integrated positron energy without the fitting of individual crystal pulses and clustering into reconstructed positron hits. The two datasets incur different systematic dependencies from pulse pileup, gain changes and pedestal shifts and act as "multiple experiments" for measuring ω_a.

The collaboration is currently beginning a running period from November 2017 through June 2018 with similar periods planned for 2018-19 and 2019-20. Our goals are achieving 1-2× the BNL statistics by summer 2018, 5-

10× the BNL statistics by summer 2019, and 20× the BNL statistics by summer 2020.

6 Acknowledgments

The author would like to thank the organizers of the FCCP2017 workshop for a wonderful meeting. The author would also like to thank his Fermilab muon g-2 collaborators and the finanicial support of the U.S. National Science Foundation grant PHY-1503552.

References

[1] G. W. Bennett *et al.* [Muon G-2 Collaboration], "Final Report of the Muon E821 Anomalous Magnetic Moment Measurement at BNL," Phys. Rev. **D73**, 072003 (2006). [hep-ex/0602035].

[2] J. Grange *et al.* [Muon g-2 Collaboration], "Muon (g-2) Technical Design Report," arXiv:1501.06858 [physics.ins-det].

[3] M. Otani [E34 Collaboration], "Status of the Muon g-2/EDM Experiment at J-PARC (E34)," JPS Conf. Proc. **8**, 025008 (2015).

[4] J. P. Miller, E. de Rafael and B. L. Roberts, "Muon g-2: Review of Theory and Experiment," Rept. Prog. Phys. **70**, 795 (2007).

[5] T. P. Gorringe and D. W. Hertzog, "Precision Muon Physics," Prog. Part. Nucl. Phys. **84**, 73 (2015) [arXiv:1506.01465 [hep-ex]].

[6] K. Hagiwara, A. Keshavarzi, A. D. Martin, D. Nomura and T. Teubner, "g-2 of the muon: status report," Nucl. Part. Phys. Proc. **287-288**, 33 (2017).

[7] M. Davier, A. Hoecker, B. Malaescu and Z. Zhang, "Reevaluation of the hadronic vacuum polarisation contributions to the Standard Model predictions of the muon g-2 and alpha(mZ) using newest hadronic cross-section data," arXiv:1706.09436 [hep-ph].

[8] F. Jegerlehner, "Muon g-2 Theory: the Hadronic Part," arXiv:1705.00263.

[9] M. Davier, A. Hoecker, B. Malaescu and Z. Zhang, "Reevaluation of the Hadronic Contributions to the Muon g-2 and to alpha(MZ)," Eur. Phys. J. **C71**, 1515 (2011). Erratum-ibid. **C72**, 1874 (2012).

[10] K. Hagiwara, R. Liao, A.D. Martin, D. Nomura and T. Teubner, "$(g\text{-}2)_\mu$ and alpha(M_Z^2) re-evaluated using new precise data," J. Phys. **G38**, 085003 (2011).

[11] F. Jegerlehner, R. Szafron, "$\rho^0-\gamma$ mixing in the neutral channel pion form factor F_π^e and its role in comparing e^+e^- with τ spectral functions," Eur. Phys. J. **C71**, 1632 (2011). [arXiv:1101.2872 [hep-ph]].

[12] T. Aoyama, M. Hayakawa, T. Kinoshita and M. Nio, "Tenth-Order QED Contribution to the Electron g-2 and an Improved Value of the Fine Structure Constant," Phys. Rev. Lett. 109 (2012).

[13] C. Gnendiger, D. Stockinger and H. Stockinger-Kim, "The electroweak contributions to $(g-2)_\mu$ after the Higgs boson mass measurement," Phys. Rev. **D88** (2013) 053005.

[14] A. T. Fienberg *et al.*, "Studies of an array of PbF_2 Cherenkov crystals with large-area SiPM readout," Nucl. Instrum. Meth. A **783**, 12 (2015). arXiv:1412.5525 [physics.ins-det]].

[15] J. Kaspar *et al.*, "Design and performance of SiPM-based readout of PbF_2 crystals for high-rate, precision timing applications," JINST **12**, no. 01, P01009 (2017). [arXiv:1611.03180 [physics.ins-det]].

https://doi.org/10.1051/epjconf/201817901005

Status of R(s) measurements by energy scan method

Fedor Ignatov[1,a]

[1] Budker Institute of Nuclear Physics, SB RAS, Novosibirsk, 630090, Russia

Abstract. The precise low-energy $e^+e^- \to hadrons$ cross section are used in SM prediction of various fundamental quantities, like the anomalous magnetic moment of the muon. Direct energy scan experiments were always one of the main tool to provide this information. Status of $R(s)$ measurements using this method is presented, some of the most recent results are discussed in details.

1 Introduction

The $R(s)$ function is defined as the total hadron production from electron-positrion annihilation normalized to the two muon cross section:

$$R(s) = \frac{\sigma^0(e^+e^- \to \gamma^* \to hadrons)}{\sigma^0(e^+e^- \to \gamma^* \to \mu^+\mu^-)}$$

This is one of the fundamental quantities in high energy physics, as it reflects number of quarks and colors ($R(s) \sim 3\sum_q e_q^2(1+\delta_{QCD}(s))$). Measurements of $R(s)$ can be utilized to test perturbative QCD and measure α_s [1]. QCD sum rules provide a method of extracting from the values of R such important parameters as quark masses, quark and gluon condensates, the value of Λ_{QCD} [2]. The total $e^+e^- \to hadrons$ cross section gives an input through dispersion integral to the calculation of various fundamental quantities: the running fine structure constant $\alpha_{QED}(M_Z)$ [3], hyperfine splitting in muonium [4], the anomalous magnetic moment of the muon $a_\mu = (g_\mu - 2)/2$[5]. The $(g_\mu - 2)/2$ value provides a powerful test of the Standard Model with intriguing $3 \div 4$ sigma deviation between the most recent experimental measurement and the theoretical prediction. The main theoretical uncertainty comes from hadronic part of the anomaly, where the value and error are dominated by low energy $R(s)$ below $\sqrt{s} < 2$ GeV and, particularly, the $e^+e^- \to \pi^+\pi^-$ channel gives the biggest contribution (73% of a_μ^{had}).

The measurements of hadron production at colliders has already celebrated 50 years, starting from first published measurement of $e^+e^- \to \rho \to \pi^+\pi^-$ cross-section (Fig. 1), which was submitted to the journal at 1 September of 1967 [6]. This measurement was performed at the VEPP-2 electron-positron machine in Novosibirsk. In following years the number of different colliders was growing fast by productivity and dimensions, reaching LEP and LHC sizes.

[a] e-mail: ignatov@inp.nsk.su

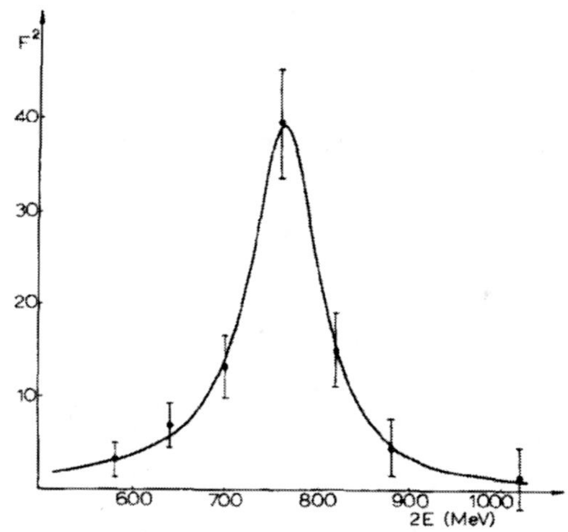

Figure 1. First measurement of hadrons production on colliders. Experimental values of $e^+e^- \to \rho \to \pi^+\pi^-$ cross section by VEPP-2 [6].

In case of the simplest $e^+e^- \to \pi^+\pi^-$ channel, many measurements were done over the years coming to current the state of art as can be seen in Fig. 2. The few milestones were along this path: the CMD and OLYA experiments at the VEPP-2M collider performed detail scan of ρ energy region with 2% systematic precision, to compare to level of >10% as was before 1985. The next breakthrough was improvement of the systematic precision to 0.6% which was achieved by the CMD-2 experiment [7–10]. The uncertainty in a_μ^{had} was improved by factor 3 as the result of VEPP-2M measurements, as can be seen in Fig. 3 for $\pi^+\pi^-$ contribution from central $0.6 < \sqrt{s} < 0.9$ GeV energy subrange.

The direct energy scanning approach was somehow limited by statistical precision at that time. After VEPP-2M results the new developed ISR method [11] at detec-

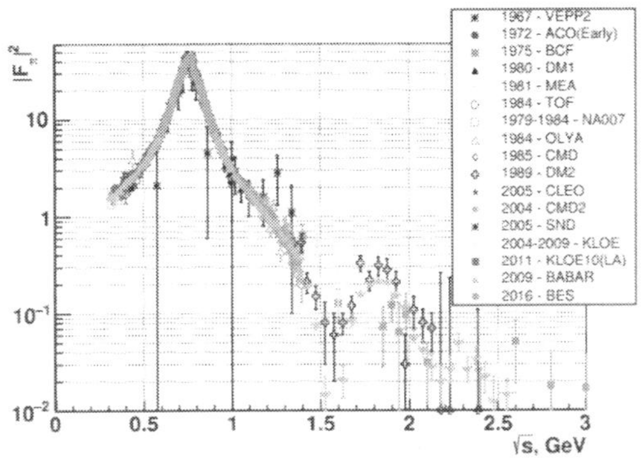

Figure 2. Status of $e^+e^- \to \pi^+\pi^-$ cross-section measurement with different experiments contributing over years.

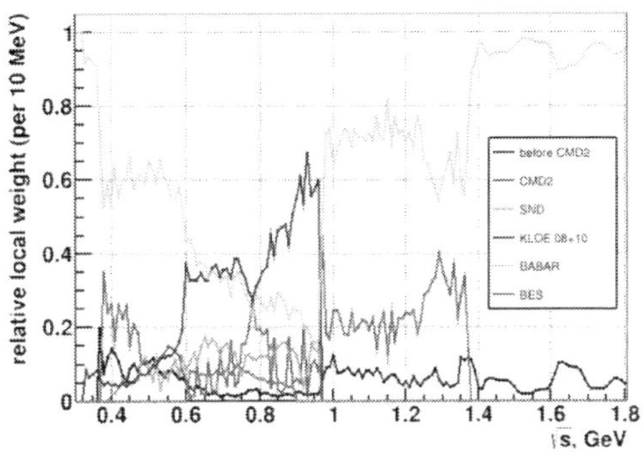

Figure 4. Relative local weight of different experiments to the $\pi^+\pi^-$ cross-section. Nowadays this data is statistically dominated by ISR (KLOE, BaBar).

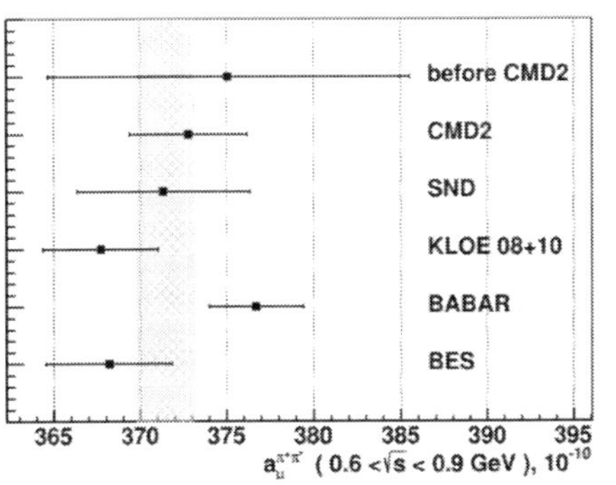

Figure 3. The $\pi^+\pi^-$ contribution to a_μ^{had} from energy range $0.6 < \sqrt{s} < 0.9$ GeV by most recent experiments.

tors KLOE (0.8% systematic precision) [12, 13], BaBar (0.5%) [14] and BES (0.9%) [15] provided successive progress in measurement of this cross-section with less limitations on statistics. Nowadays the $\pi^+\pi^-$ data is statistically dominated by the ISR method with main contribution from KLOE near the ρ-meson peak and from BaBar away from ρ (Fig. 4). The most comprehensive analysis with measurements of many various hadronic channels below 2 GeV was performed by the BABAR experiment.

New experiments on the measurement of the muon anomalous magnetic moment in FermiLab [16] and J-PARC [17] plan to improve accuracy of the experimental a_μ value by a factor of at least 4. Precise tests of electroweak theory in experiments at future colliders such as ILC, CLIC, FCC-ee will require knowledge of the hadronic contribution to $\alpha_{QED}(M_Z^2)$ at the level of $\sim 0.5 \div 0.3 \times 10^{-4}$ [18, 19] while the current value is $(276.26 \pm 1.38) \times 10^{-4}$ [5]. To provide comparable accuracies in the theoretical predictions, the accuracy of integral R measurement must be $\sim 0.2\%$, and particular it is very desirable to improve systematic precision of the $\pi^+\pi^-$ cross section by at least a factor of two of current level. These days there are three electron-positron colliders doing direct exclusive measurement of $\sigma(e^+e^- \to hadrons)$: BESIII and KEDR are doing scan of energy from 2 GeV to 5 GeV, and the VEPP-2000 is the only one doing scan below 2 GeV.

2 VEPP-2000, CMD-3 and SND

The electron-positron collider VEPP-2000 [20, 21] has been operating at Budker Institute of Nuclear Physics since 2010. The collider is designed to provide luminosity up to 10^{32}cm^{-2}s^{-1} at the maximum center-of-mass energy $\sqrt{s} = 2$ GeV. The VEPP-2000 is exploiting unique optics with round beams, which allow to reach higher luminosity. During the first 2011-2013 seasons it was collected integrated luminosity comparable to all BaBar effective luminosity from ISR measurements (Fig. 5). The luminosity at energies above 1 GeV was limited by a deficit of positrons and limited energy of the booster. In 2013-2016 the VEPP-2000 collider and the detectors have been upgraded. A new injection complex was commissioned at the BINP in 2016. Also the upgrade of the booster ring was done to increase the maximum energy, so it become possible to inject beams to the main ring in the whole available energy range without energy ramping. The VEPP-2000 collider has resumed data taking at the beginning of 2017. Significant improvement in luminosity by a factor of 2 to 8 compared to 2011-2013 was reached as seen in Fig. 6, but there are many things which can gives additional improvements on achievable collider luminosity.

Two detectors CMD-3 [22, 23] and SND [24] are installed at the interaction regions of the collider. In 2010 both experiments started data taking. The physics program [25] includes high precision measurements of the $e^+e^- \to hadrons$ cross sections in the wide energy range up to 2 GeV, where rich intermediate dynamics involved, studies of known and searches for new vector mesons, studies of $n\bar{n}$ and $p\bar{p}$ production cross sections near threshold and searches for exotic hadrons. It requires a detector with high efficiency for multiparticle events and good en-

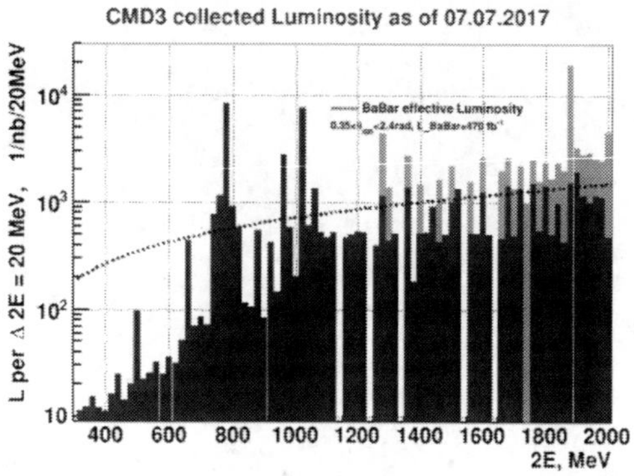

Figure 5. Integrated luminosity collected by the CMD-3 at different energies with 20 MeV energy bins. Blue: 2011-2013 runs, green: 2017 run. The line shows the effective ISR integrated luminosity at BaBar.

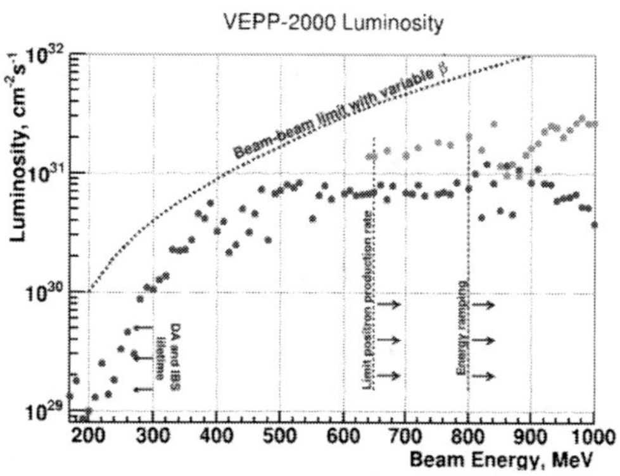

Figure 6. Average luminosity of the VEPP-2000 at different energies. Red points: 2011-2013 runs, green points: 2017 runs.

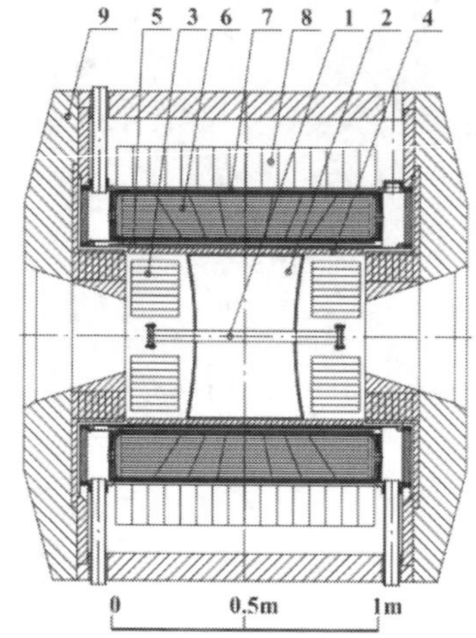

Figure 7. CMD-3 detector: 1 – beam pipe, 2 – drift chamber, 3 – BGO calorimeter, 4 – Z-chamber, 5 – SC solenoid ($0.13X_0$, 13 kGs), 6 – LXe calorimeter, 7 – TOF system, 8 – CsI electromagnetic calorimeter, 9 – yoke, not shown muon range system.

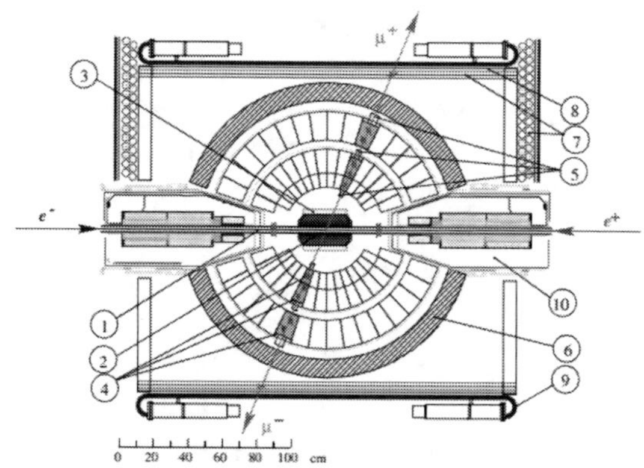

Figure 8. SND detector, section along the beams: 1 – beam pipe, 2 – tracking system, 3 – aerogel Cherenkov counter, 4 – NaI(Tl) crystals, 5 – phototriodes, 6 – iron absorber, 7 – streamer tubes, 8 – iron plates, 9 – scintillation counters, 10 – solenoid of VEPP-2000.

ergy and angular resolutions for charged particles as well as for photons.

The CMD-3 (Cryogenic Magnetic Detector) is a general-purpose detector, see Fig. 7. Coordinates, angles and momenta of charged particles are measured by the cylindrical drift chamber with a hexagonal cell for uniform reconstruction of tracks. The calorimetry is performed with the endcap BGO calorimeter and the barrel calorimeter. The barrel calorimeter, placed outside of the superconducting solenoid with 1.3 T magnetic field, consists of two systems: ionization Liquid Xenon calorimeter surrounded by the CsI scintillation calorimeter. The total thickness of the barrel calorimeter is about $13.5X_0$. The LXe calorimeter has seven layers with strip readout which give information about a shower profile and are also able to measure coordinates of photons with about a millimeter precision.

The spherical neutral detector SND is a general purpose nonmagnetic detector, see Fig. 8. The SND consists of the tracking system based on cylindrical drift and proportional chambers placed in a common gas volume, aerogel threshold counters, calorimeter and muon system based on proportional tubes and plastic scintillator. The main part of the SND is the three-layer spherical electromagnetic calorimeter based on NaI(Tl) crystals. The total calorimeter thickness for particles originating from the interaction region is $13.4X_0$ of NaI(Tl).

All major channels are under analysis including such as with up to 6 pions or 2 kaons and 2 pions in the final state. The CMD-3 and SND collaborations are already published many results such as: $e^+e^- \to \pi^0\gamma$, $n\bar{n}$, $\eta\gamma$, $\omega\pi^0$, $\eta\pi^+\pi^-$, $\pi^+\pi^-\pi^0$, $\omega\eta$, $\omega\eta\pi^0$, K^+K^- for SND; $e^+e^- \to p\bar{p}$, $3(\pi^+\pi^-)$, $2(\pi^+\pi^-)$, $\omega\eta$, $\omega\pi^+\pi^-\pi^0$, $K^+K^-\pi^+\pi^-$, K^+K^-, K_SK_L for CMD-3. Many others analyses are in progress, some of them will be discussed below.

3 $e^+e^- \to \pi^+\pi^-$ channel

The dominant contribution to production of hadrons in the energy range $\sqrt{s} < 1$ GeV comes from the $e^+e^- \to \pi^+\pi^-$ mode. This channel gives the main contribution to the hadronic term and overall theoretical precision of a_μ. It is the most challenging channel due to high precision requirement, which ideally should be known with systematic precision 0.2% to fulfill requirements of new g-2 experiments and physics on future electron-positron colliders. The CMD-3 has plans to reduce further systematic error which was achieved by the CMD-2. The first energy scan below 1 GeV for the $\pi^+\pi^-$ measurement was performed at VEPP-2000 in 2013. The collected data sample corresponds to about 18 pb^{-1} of integrated luminosity, and it is already higher than in the previous CMD-2 experiment and is similar or better than in the BaBar [14] and the KLOE [12, 13] experiments (Fig. 9).

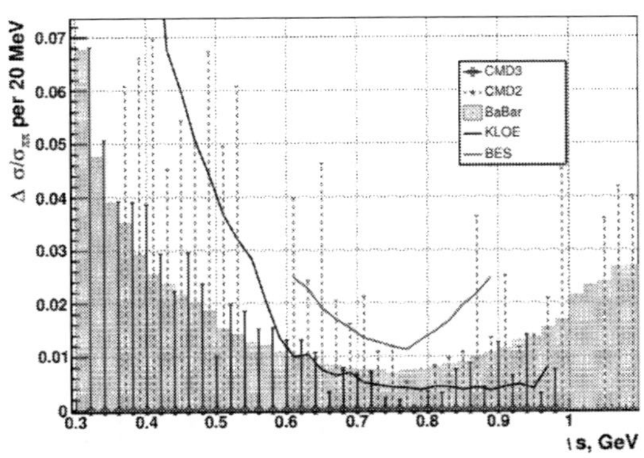

Figure 9. Statistical precision of $|F_\pi|^2$ from the CMD-3 data in comparison with CMD-2, BaBar, KLOE and BESIII results.

The crucial pieces of analysis to reach claimed goal includes stable $e/\mu/\pi$ separation, precise fiducial volume determination, theoretical precision of radiative corrections, etc. And important point is that many systematic studies rely on high collected statistics.

The $\pi^+\pi^-$ process has a simple event signature with 2 back-to-back charged particles. They can be selected by using the following criteria: two collinear well reconstructed charged tracks are detected, these tracks are close to the interaction point, both tracks are inside a good region of the drift chamber. The selected data sample includes events with $e^+e^-, \mu^+\mu^-, \pi^+\pi^-$ pairs and cosmic muons, and it practically doesn't contain any other physical background at energies $\sqrt{s} < 1$ GeV. These final states can be separated using either the information about energy deposition in the calorimeter or that about particle momenta in the drift chamber. At low energies momentum resolution of the drift chamber is sufficient to separate different types of particles. The pion momentum is well aside from the electron one up to energies $\sqrt{s} \lesssim 0.9\,\text{GeV}$, while the $\mu^+\mu^-$ events are separated from others up to $\sqrt{s} \lesssim 0.66\,\text{GeV}$. At higher energies the peak of electron

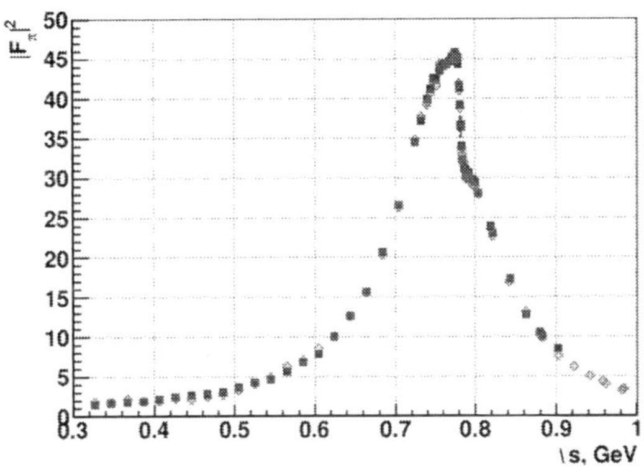

Figure 10. Preliminary results on F_π^2 from CMD-3. Open crosses – separation done on the calorimeter information, filled squares – on particle momentum. Some additional corrections, common to two methods (e.g., the trigger efficiency), are not applied.

shower in the calorimeter is far away from the peak of minimal ionization particles. The separation using energy deposition works better at higher energies and becomes less robust at lower energies.

Full energy deposition in combined LXe and CsI calorimeters is used at the moment. A further methods are under development to exploit full power of the layered barrel calorimeter. Additional information from independent measurements of energy deposition in 7 strip layers of the LXe and in the CsI calorimeter gives better discrimination power between different particles due to different interaction process involved (electromagnetic shower, ionization process, nuclear interaction).

The preliminary result on the pion formfactor measured by the CMD-3 is shown in Fig 10 with the comparison of two approaches using either momentum information or energy deposition. The additional corrections, common to two methods (e.g., the trigger efficiency), are not applied. These two methods overlap in the wide energy range and provide a cross-check of each other, allowing to reach in future a systematic error of event separation at the level of 0.2%.

The contribution to systematic precision from $e/\mu/\pi$ separation is estimated now as 0.1-0.5% when using momentum information and up to 1.5% in case of by energy deposition, and it should be greatly improved with exploiting of full power of the combined barrel calorimeter.

Determination of the fiducial volume can be made independently with help of the LXe and Z-chamber subsystems. It allows an efficient monitoring of detector operation stability during data taking. The relative difference in scale between this two subsystem was stable at level which corresponds to 0.1% systematic error of luminosity determination at $\theta_{track} = 1$ rad.

Measurement of beam energy by Compton backscattering of the laser photons with precision $\sigma_E < 50$ keV [26] helps to keep a systematic uncertainty from this source below 0.1%.

The systematic uncertainty coming from tracks reconstruction inefficiency mostly cancels out in normalization ratio of number of pions to number of electrons. A difference is still present because of the pion specific losses: nuclear interactions and decays in flight. This can be studied in comparisons of data to simulation prediction, and at the moment the agreement is at level ~10% of the effect, which contributes to total systematic error as 0.6-0.3%. This part will be improved with better drift chamber understanding, which include accounting of detail per drift cell inefficiencies, noise level, etc.

Another important source of systematics is a theoretical precision of radiative corrections [27]. Additional studies like crosschecks of different calculation approaches and further proof from comparison with experimental data are necessary in this field. The high statistics collected with the CMD-3 allowed to observe a discrepancy in momentum distribution of experimental data vs theoretical spectra from the MCGPJ generator[28], which is used in this analysis. The source of the discrepancy was understood: the collinear approximation for photon jets was not good enough to describe differential cross-section in the $P^+ \times P^-$ momentum distributions when both final leptons has energy much below initial value (when 2 final photons are involved in kinematics).

Comparison between the MCGPJ[28] and the BabaYaga@NLO[29] generators was performed. The integrated cross-section for the cuts applied is well consistent at the level better than 0.1% between both tools, and BabaYaga@NLO better describes the data in momentum distribution. While this discrepancy mostly doesn't affect analysis by energy deposition, it becomes crucial if momentum distribution information is used.

Several steps to improve the MCGPJ were done, main one was including of photon jets angular distribution in the generator. The consistency between both generators was improved, but there is still discrepancy at level ~ 10% in momentum tails, which gives additional $0.0 \div 0.4\%$ contribution to systematic uncertainty on number of $\pi^+\pi^-$ events extracted in analysis based on momentum information. As seen on effects from two photon contributions to momentum spectra, it becomes highly desirable to have exact NNLO $e^+e^- \to e^+e^-(\gamma\gamma)$ generator to reach precision $\lesssim 0.1\%$.

4 $e^+e^- \to K^+K^-, K_SK_L$ channels

The largest contribution to $R(s)$ at ϕ-resonance comes from 2 kaon production channels. The recent CMD-3 result for $e^+e^- \to K_SK_L$ cross section [30] is shown in Fig. 11. This is the most precise measurement of this cross section with reached 1.8% systematic uncertainty. The data is very well consistent with previous experiments, and the obtained parameters of ϕ resonance are in good agreement with PDG data.

The $e^+e^- \to K^+K^-$ channel was recently published by the SND experiment [31] for energies above 1.05 GeV, and by the CMD-3 [32] experiment below 1.06 GeV as shown in Fig. 12. The SND measurement agrees with the previous most accurate data obtained by the BaBaR experiment and has comparable or better systematic precision.

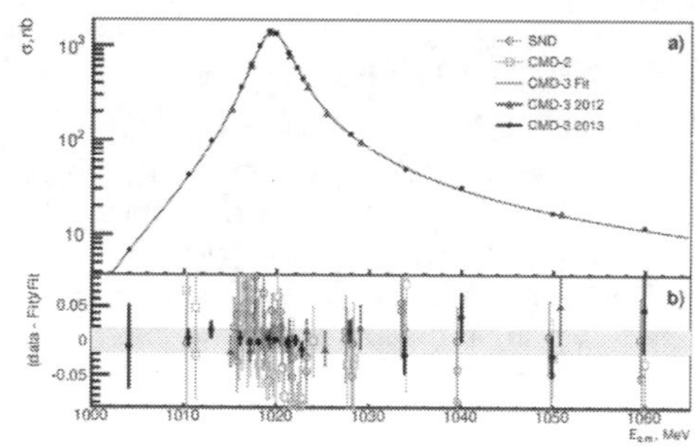

Figure 11. (a) Measured $e^+e^- \to K_S^0K_L^0$ cross section by CMD-3 experiment, the curve is the fit of data. (b) Relative difference between the data and fit. Comparison between different experimental data is shown. The width of the band shows the systematic uncertainties in CMD-3 experiment.

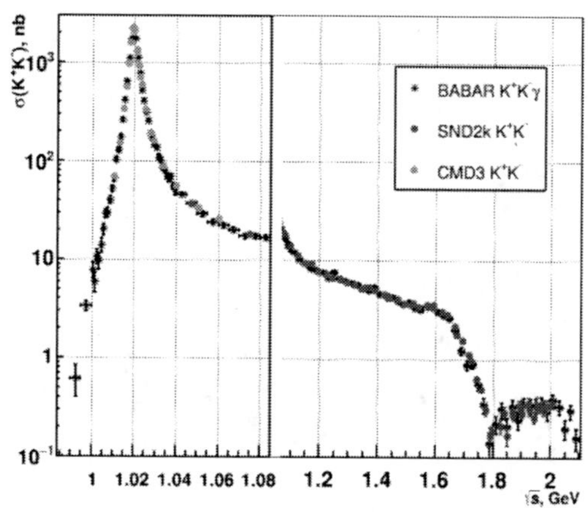

Figure 12. The $e^+e^- \to K^+K^-$ cross section measured by CMD-3 [32], SND [31] and BABAR [34].

The K^+K^- cross section was measured by CMD-3 at the phi-meson energy range with 2% systematic accuracy. As shown in Figs. 13 and 14, the obtained results have comparable accuracy, but they are not consistent, in general, with the previous data. In particular, an inconsistency between the old CMD-2 [33] result and more recent BaBar data [34] was confirmed. The new CMD-3 measurements of $e^+e^- \to K^+K^-$ and $e^+e^- \to K_SK_L$ mention above demonstrates perfect agreement with isospin symmetry: the ratio of coupling constants with the Coulomb factor taken into account is $g_{\phi K^+K^-}/g_{\phi K_SK_L}/\sqrt{Z(m_\phi^2)} = 0.990 \pm 0.017$.

The lower CMD-2 $e^+e^- \to K^+K^-$ cross-section is explained by overestimation of the value of the trigger efficiency for slow kaons in the previous experiment. The positive trigger decision of the CMD-2 required the pres-

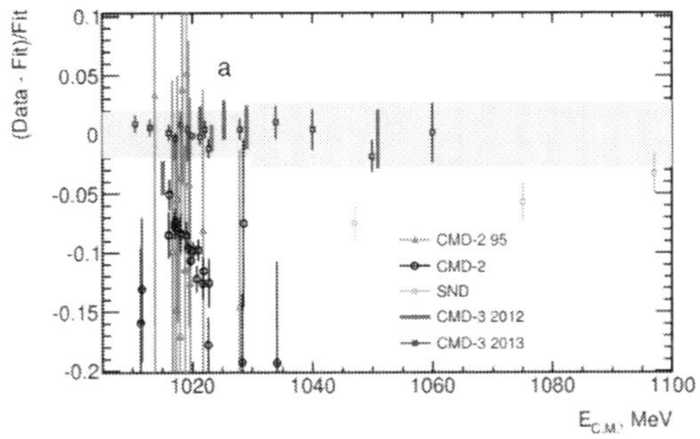

Figure 13. Comparison of the CMD-3 measurement of $e^+e^- \to K^+K^-$ to previous CMD-2 [33] and SND [31] measurements. The width of the band shows the systematic uncertainties in CMD-3 study.

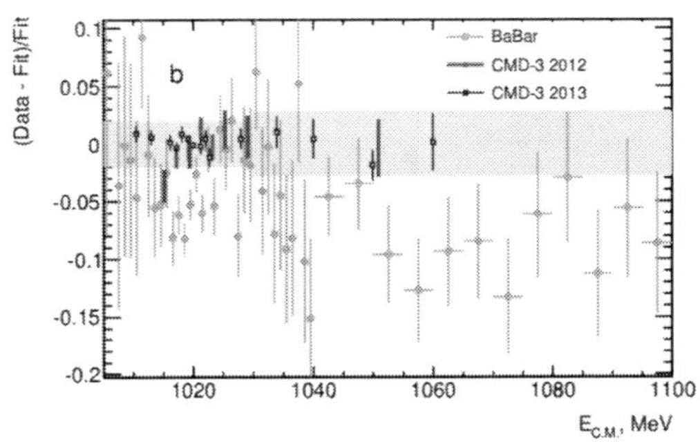

Figure 14. Comparison of the CMD-3 measurement of $e^+e^- \to K^+K^-$ to the BaBar measurement [34] using ISR approach. The width of the band shows the systematic uncertainties in the CMD-3 study.

ence of one charged track in the DCh in coincidence with the corresponding hits in the Z-chamber, where the last are not always reached by slow kaons. In contrast to the CMD-2 experiment, the new CMD-3 detector has two independent trigger systems, the Z-chamber is excluded from the decision, and a charged (total) trigger efficiency is close to 100%. The more accurate look on this type of events based on an CMD-3 experience gives a significantly larger trigger efficiency correction than the value obtained in the original CMD-2 analysis [33]. A reanalysis of CMD-2 data is expected.

5 $e^+e^- \to \pi^0\gamma$ channel

The cross section of $e^+e^- \to \pi^0\gamma$ process is the third largest cross section after $e^+e^- \to 2\pi$ and $e^+e^- \to 3\pi$ below 1 GeV. The SND group recently published analysis of this channel based on a full data sample collected on the previous VEPP-2M collider [35]. The previous most precise measurements were done by CMD-2 [36] and SND [37] at VEPP-2M, where the SND result was obtained using only 25% of collected data. The measured

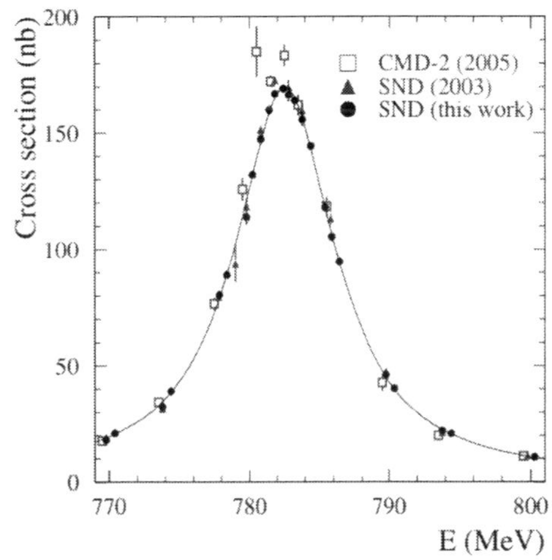

Figure 15. The $e^+e^- \to \pi^0\gamma$ cross section measured by the SND [35](it is shown only near ω-resonance energy range) together with the previous most accurate measurements by CMD-2(2005) [36] and SND(2003) [37]. The curve is VDM fit.

cross section (Fig. 15) is more precise than previous results and agrees with them within the systematic errors. The measurements of the $\rho \to \pi^0\gamma$ and $\omega \to \pi^0\gamma$ branching fractions have better accuracy than PDG values. Using PDG value for $B(\omega \to 3\pi)B(\omega \to e^+e^-)$ they obtained the ratio $\Gamma(\omega \to \pi^0\gamma)/\Gamma(\omega \to 3\pi) = 0.0992 \pm 0.0023$, which is higher than the value reported by KLOE [38] by 3.4σ.

6 Unaccounted modes

The total $e^+e^- \to hadron$ cross section below 2 GeV is calculated as a sum of exclusive cross sections for all possible hadronic modes. Some unmeasured channels are calculated using rough isospin relations, which can be very approximative. To fulfill precise $R(s)$ measurement program, it is very important to take into account all possible channels and to do measurements of unaccounted modes or accounted by indirect isospin relations. Some of them were recently studied by the CMD-3 and SND groups.

The process $e^+e^- \to \pi^+\pi^-\pi^0\eta$ was studied using $\pi^+\pi^-4\gamma$ final state [39]. This channel can go through the several intermediate states. At least four mechanisms of this reaction: $\omega\eta$, $\phi\eta$, $a_0(980)\rho$ and structureless $\pi^+\pi^-\pi^0\eta$ were observed. About 50% of the total cross section at the region below 1.8 GeV is due to the $\omega\eta$, $\phi\eta$ contributions. Above 1.8 GeV the dominant mechanism of the reaction is $a_0(980)\rho$. Only the $\omega\eta$ and $\phi\eta$ contributions were taken into account in total hadron cross-section, while not accounted parts gives about 3-5% of total $R(s)$ in this energy range. The published measurement by CMD-3 [39] and preliminary SND result on the total $e^+e^- \to \pi^+\pi^-\pi^0\eta$ cross section are shown in Fig. 16.

The process $e^+e^- \to \omega\pi^0\eta$ was studied by the SND in the energy range $\sqrt{s} = 1.45 \div 2.00$ GeV using fully

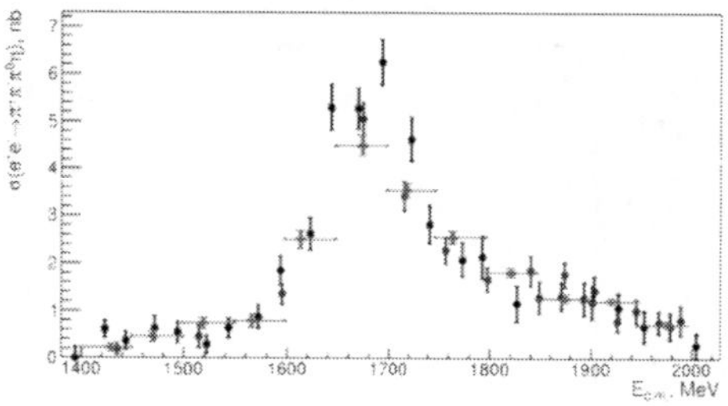

Figure 16. The $e^+e^- \to \pi^+\pi^-\pi^0\eta$ cross section measured by CMD3 and SND. Black and blue colors are for the 2011 and 2012 data from CMD-3. Red points - preliminary SND result.

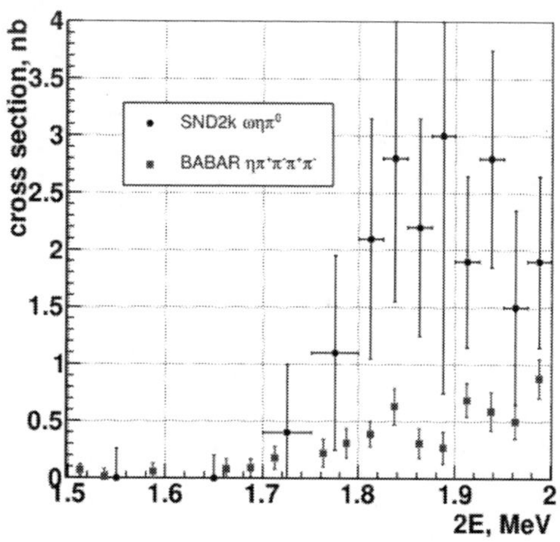

Figure 17. The $e^+e^- \to \omega\pi^0\eta$ cross section measured by SND [40]. The part of this cross-section was included before in $R(s)$ via "isospin relation" from the $e^+e^- \to \eta 2\pi^+ 2\pi^-$ cross-section measured by BaBar [41], as shown by red points.

neutral $e^+e^- \to \pi^0\pi^0\eta\gamma \to 7\gamma$ final state [40]. It is the first measurement of this process. Analysis of the $\pi^0\gamma$ invariant mass spectra showed that the $\omega\pi^0\eta$ mechanism is dominating in this reaction. From the analysis of the $\eta\pi^0$ invariant mass distribution it has been found that the $\omega a_0(980)$ intermediate state is dominating in this process. The cross section is about 2.5 nb above 1.8GeV, which takes 5% of the total hadronic cross section in this energy region. Only part of this cross-section was taken in to account before via "isospin relation" from $e^+e^- \to \eta 2\pi^+ 2\pi^-$ as shown in Fig. 17.

7 R(s) at $\sqrt{s} > 2$ GeV

The c.m. energy range from 2 to 5 GeV is almost asymptotic for *u*-, *d*-, and *s*-quarks. There are no resonances made of light quarks at these energies. The R function has slow dependence with energy from 2 GeV right up to threshold of D-mesons (3.73 GeV), except narrow regions around J/ψ and $\psi(2S)$. The values of $R \approx 2.2$ with current experimental precision are consistent with the pQCD prediction [42]. The 3.73-5.0 GeV energy region are covered by the c-quark resonances decayed to D-mesons. Numerous R measurements exist in the c.m. energy range between 2 and 5 GeV, where many experimental groups studied various states just above the charmonium threshold: Crystal Ball [43], PLUTO [44], DASP [45], Mark-I [46], BES [47], KEDR [48, 49]. In general, these measurements are consistent. The most detailed measurement was made by the BESII experiment in total at 165 energy points from 2 to 5 GeV with averaged systematic precision from 7% to 3.3%. The new BESIII result is expected soon, they did scan of the 2.00–4.59 GeV energy region in 2012–2015 years (125 energy points with the integrated luminosity 1.3 fb^{-1}). The best systematic precision 2% was reached by the KEDR experiment, with measurement of R in 20 energy points from 1.84 to 3.72 GeV (Fig. 18). The KEDR experiment has plans to do scan of 2E=4.5-8(10) GeV energy range, where there is no precise $R(s)$ data consistent with pQCD.

8 Conclusions

The precise low-energy $e^+e^- \to$ *hadrons* cross section are used in SM prediction of various fundamental quantities. New precise data are desirable to fulfill requirements of new g-2 experiments and physics on future electron-positron colliders. Direct scan experiments were always providing this $e^+e^- \to$ *hadrons* measurements, while alternative ISR method appeared during last decade. The two methods have owns independent systematic sources, and they are complementary to each other for cross-checking of possible inconsistencies.

The VEPP-2000 is the only collider able to scan energies below <2 GeV for measurement of exclusive $e^+e^- \to$ *hadrons* channels. A new positron injection complex was commissioned during 2013-2016 years, and the data taking was resumed in 2017 with increased luminosity by few times at $\sqrt{s} > 1.2$ GeV. The first scan below 1 GeV for $\pi^+\pi^-$ measurement was done in 2013. The already collected data sample for the cross section measurement has the same or better statistical precision than was achieved by other experiments. High statistics will allow to study and to control better different systematic contributions, and the improvement here is also expected.

Several previously unmeasured processes like $e^+e^- \to \pi^+\pi^-\pi^0\eta$ and $\omega\pi^0\eta$ contributed to the total hadronic cross section below 2 GeV has been studied.

New results are expected from CMD-3, SND, KEDR, BESIII with progress forward by luminosity and precision.

This work is supported in part by the Russian Science Foundation (project 14-50-00580) and Russian Foundation of Basic Research (17-02-00847, 17-02-00897).

References

[1] J. Kühn and M. Steinhauser, Nucl. Phys. B **619** (2001) 588; Erratum-ibid, B640 (2002) 415.

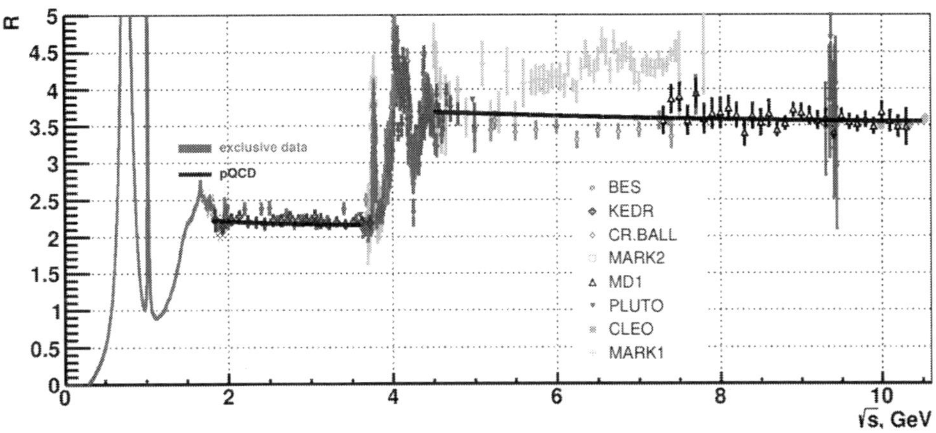

Figure 18. $R = \sigma(e^+e^- \to \text{hadrons})/\sigma(e^+e^- \to \mu^+\mu^-)$. Points — the BESII, KEDR and other inclusive data. Shaded area — sum of all exclusive cross-sections for $\sqrt{s} < 2$ GeV. Black line — pCQD prediction.

[2] M.A. Shifman, A.I. Vainshtein and V.I. Zakharov, Nucl. Phys. B **147** (1979) 385.

[3] F. Jegerlehner, Nucl.Phys.Proc.Suppl. **162** (2006): 22

[4] A. Czarnecki, S. I. Eidelman and S. G. Karshenboim, Phys. Rev. D **65** (2002) 053004

[5] K. Hagiwara *et al.*, J. Phys. G **38** (2011): 085003

[6] V. L. Auslander, G. I. Budker, J. N. Pestov, V. A. Sidorov, A. N. Skrinsky and A. G. Khabakhpashev, Phys. Lett. **25B** (1967) no.6, 433.

[7] R. R. Akhmetshin *et al.*, Phys. Lett. B **578** (2004):285

[8] V. M. Aulchenko *et al.*,JETP Lett. **82** (2005) 743

[9] R. R. Akhmetshin *et al.*, JETP Lett. **84** (2006): 413

[10] R. R. Akhmetshin *et al.*, Phys.Lett. B **648** (2007): 28

[11] A. Denig, ISR measurements, this proceedings

[12] F. Ambrosino *et al.*, Phys. Lett. B **670** (2009): 285

[13] F. Ambrosino *et al.*, Phys. Lett. B **700** (2011): 102

[14] J. P. Lees *et al.*, Phys. Rev. D **86** (2012): 032013

[15] M. Ablikim *et al.*, Phys. Lett. B **753** (2016) 629

[16] J. Grange *et al.*, FERMILAB-DESIGN-2014-02, arXiv:1501.06858 [physics.ins-det].

[17] M. Aoki, et al., KEK-J-PARC-PAC2009-12.

[18] M. Baak *et al.*,Eur. Phys. J. C **74** (2014) 3046

[19] P. Azzi *et al.*, arXiv:1703.01626 [hep-ph].

[20] P. Y. Shatunov *et al.*, Phys. Part. Nucl. Lett. **13** (2016) no.7, 995.

[21] D. Shwartz *et al.*, PoS ICHEP **2016** (2016) 054.

[22] V. M. Aulchenko *et al.*, BUDKER-INP-2001-45

[23] B. Khazin,Nucl. Phys. Proc. Suppl. **181-182** (2008): 376

[24] M. N. Achasov *et al.*, Nucl. Instrum. Meth. A **598** (2009) 31.

[25] S. Eidelman,Nucl.Phys.Proc.Suppl. **162** (2006): 323

[26] E. V. Abakumova *et al.*,Phys. Rev. Lett. **110** (2013): 140402

[27] S. Actis *et al.*, Eur. Phys. J. C **66** (2010): 585

[28] A. B. Arbuzov *et al.*, Eur. Phys. J. C **46** (2006): 689

[29] G. Balossini *et al.*, Nucl. Phys. B **758** (2006): 227

[30] E. A. Kozyrev *et al.*,Phys. Lett. B **760** (2016) 314

[31] M. N. Achasov *et al.*, Phys. Rev. D **94** (2016) no.11, 112006

[32] E. A. Kozyrev *et al.*, arXiv:1710.02989 [hep-ex].

[33] R. R. Akhmetshin *et al.*,Phys. Lett. B **669** (2008) 217

[34] J. P. Lees *et al.*,Phys. Rev. D **88** (2013) no.3, 032013

[35] M. N. Achasov *et al.*,Phys. Rev. D **93** (2016) no.9, 092001

[36] R. R. Akhmetshin *et al.*,Phys. Lett. B **605** (2005) 26

[37] M. N. Achasov *et al.*, Phys. Lett. B **559** (2003) 171

[38] F. Ambrosino *et al.*,Phys. Lett. B **669** (2008) 223

[39] R. R. Akhmetshin *et al.*,Phys. Lett. B **773** (2017) 150

[40] M. N. Achasov *et al.*, Phys. Rev. D **94** (2016) no.3, 032010

[41] B. Aubert *et al.*,Phys. Rev. D **76** (2007) 092005, Erratum: [Phys. Rev. D **77** (2008) 119902]

[42] P. A. Baikov *et al.*, Phys. Lett. B **714**, 62 (2012).

[43] A. Osterheld *et al.*, SLAC-PUB-4160, 1986.

[44] J. Burmeister *et al.*, Phys. Lett. **66B** (1977) 395.

[45] R. Brandelik *et al.*, Phys. Lett. **76B** (1978) 361.

[46] J.L. Siegrist *et al.*, Phys. Rev. D **26** (1982) 969.

[47] M. Ablikim *et al.*,Phys. Lett. B **677** (2009) 239

[48] V. V. Anashin *et al.*, Phys. Lett. B **753** (2016) 533

[49] V. V. Anashin *et al.*, Phys. Lett. B **770** (2017) 174

Flavour physics and extra-dimensions

Abhishek M. Iyer[1,*]

[1]INFN-Sezione di Napoli, Via Cintia, 80126 Napoli, Italia

Abstract. Randall-Sundrum (RS) model of warped extra-dimensions were originally proposed to explain the Planck-weak scale hierarchy. It was soon realised that modifications of the original setup, by introducing the fields in the bulk, has several interesting features. In particular it imbues a rich flavour structure to the fermionic sector thereby offering an understanding of the Yukawa hierarchy problem. This construction is also useful in explaining the recently observed deviations in the decay of the B mesons. We consider two scenarios to this effect : A) Right handed muon fields coupled more to NP that the corresponding muon doublets (unorthodox case). Non-universality exists in the right handed sector. B) Standard scenario with anomalies explained primarily by non-universal couplings to the lepton doublets. Further, we establish correlation with the parameter space consistent with the flavour anomalies in the neutral current sector and obtain predictions for rare K- decay which are likely to be another candle for NP with increased precision. The prediction for rare K-decays are different according to the scenario, thereby serving as a useful discriminatory tool. We also discuss the large flavour violation in the lepton sector and present an example with the implementation of bulk leptonic MFV which is essential to realize the model with low KK scales. Further we consider a radical solution, called GUT RS models, where the RS geometry can work as theory of flavour in the absence of flavour symmetries. In this case the low energy brane corresponds to the GUT scale as a result of which RS is no longer solution to the gauge hierarchy problem. The Kaluza Klein (KK) modes in this setup are naturally heavy due to which the low energy constraints can be easily avoided. We use this framework to discuss the supersymmetric version of the RS model and provide means to test this scenario by considering rare lepton decays like $\tau \to \mu\gamma$.

1 Introduction

Flavour physics offers an indirect probe towards the possible existence of new physics (NP) effects which may be characterized by a flavour structure different from that of the SM. Process like $\mu \to e\gamma$, $\tau \to \mu\gamma$ in leptonic sector and rare K decays in the hadronic sector are characterised by small contributions in the SM but relatively weaker experimental bounds. More recently, the anomalous decays of the B mesons has generated a lot of interest. Specifically the measurement of $\mathcal{B}(B^+ \to K^+\mu^+\mu^-)$ and $\mathcal{B}(B^+ \to K^+e^+e^-)$ was quoted in form of the following ratio [1]

$$R_K = \left.\frac{\mathcal{B}(B^+ \to K^+\mu^+\mu^-)}{\mathcal{B}(B^+ \to K^+e^+e^-)}\right|_{q^2=1-6\ GeV^2} = 0.745^{+0.090}_{-0.074}\,(stat) \pm 0.036\,(syst) \tag{1}$$

while the SM expectation is $R_K^{SM} = 1.003$ [2]. This implies a $\sim 2.6\ \sigma$ deviation as a possible evidence of lepton non-universality. This observation was further validated by the following measurement

$$\begin{aligned} R_{K^*} &= \frac{\mathcal{B}(B^0 \to K^{*0}\mu^+\mu^-)}{\mathcal{B}(B^0 \to K^{*0}e^+e^-)} \\ &= 0.660^{+0.110}_{-0.070}(stat) \pm 0.024(syst), \quad \text{low } q^2 \\ &= 0.685^{+0.113}_{-0.069}(stat) \pm 0.047(syst), \quad \text{mid } q^2 \end{aligned} \tag{2, 3}$$

where low q^2 corresponds to $0.045 \le q^2 \le 1.1$ GeV^2 while mid q^2 corresponds to $1.1 \le q^2 \le 6.0$ GeV^2 The SM values for the corresponding q^2 bins are: $R_{K^*}^{SM} \simeq 0.93$ for low q^2 while $R_{K^*}^{SM} = 1$ elsewhere. This corresponds to a 2.4σ deviation for low q^2 and $\sim 2.5\ \sigma$ for medium q^2. The deviations from the SM values can parametrized by additional contributions to the Wilson coefficients C_i of the following effective operators: [3]:

$$\mathcal{L} \supset \frac{G_F\alpha}{\sqrt{2}\pi}\sum_i C_i O_i \tag{4}$$

where $C_i = C_i^{SM} + \Delta C_i$.

$$\begin{aligned} O_9 &= (\bar{s}_L\gamma^\mu b_L)(\bar{l}\gamma_\mu l) \qquad O_{9'} = (\bar{s}_R\gamma^\mu b_R)(\bar{l}\gamma_\mu l) \\ O_{10} &= (\bar{s}_L\gamma^\mu b_L)(\bar{l}\gamma_\mu\gamma^5 l) \quad O_{10'} = (\bar{s}_R\gamma^\mu b_R)(\bar{l}\gamma_\mu\gamma^5 l) \end{aligned} \tag{5}$$

Here ΔC_i determines the NP contributions to the Wilson coefficients and l denotes a lepton. In the following we will consider a scenario where NP effects exits in both the muon and the electron sector.

*e-mail: iyera@na.infn.it

In this talk we discuss the explanation of these anomalies in a model with a single warped extra-dimension with the following line element [4]:

$$ds^2 = e^{-2A(y)}\eta_{\mu\nu}dx^\mu dx^\nu - dy^2 \quad (6)$$

where $A(y) = k|y|$, $k \sim \frac{M_{Pl}}{4\pi}$ and $0 \leq y \leq \pi R$. The coordinates $y = 0, \pi R$ correspond to the location of the (UV, IR) brane respectively. A generalization of this framework setup with a bulk custodial symmetry [5] in considered to evade constraints from $Z \to b\bar{b}$. In addition to the KK states of the SM W, Z, the model is characterized by additional 'custodial' heavy gauge bosons thereby leading to a distinct phenomenology, in the flavour sector in particular. A comprehensive analysis of different flavour transitions in this setup was considered in [6–9]. We explore the parameter space admitted by the current anomalies and offer predictions for the $K \to \pi\nu\nu$ decays, in the $s \to d$ sector [10]. Fits with two scenarios are demonstrated:
Scenario **A**. Right handed leptons are more composite (closer to the IR brane) than the left handed leptons. This is particularly true for the muon and tau. The left handed lepton couple universally to the NP. Non-universality in this case exists in the right-sector. ,
Scenario **B**. Left handed leptons are more composite than the right handed leptons.
For Scenario **A**, the NP contribution to $\Delta C^\mu_{9,10}$ is dominant with a smaller contribution to $\Delta C^e_{9,10}$, thereby resulting a 4-D fit along the lines of [11]. The primed operators do not contribute as we assume universality in the bulk wavefunctions of the right handed quarks. Both these scenarios are characterized by different predictions for the the K decays thereby making it a useful discriminant.

The note is organized as follows: In Section 2 we compute the fits for the anomalies in $b \to sll$ processes for two different 1-D hypotheses. In Section 3 we consider the rare kaon decays and demonstrate how it can be utilised to possibly distinguish between the two scenarios used to fit the $b \to sll$ anomalies. In Section 4 we give an explicit example with MFV implemented in the lepton sector, in particular for the first scenario. in Section 5 we discuss a radical solution with the implementation of RS model at the GUT as a means to avoid the introduction of additional flavour symmetries and we finally conclude.

2 $b \to sll$ processes : B anomalies

In bulk custodial models in RS , neutral currents at tree level receive contribution from $X \in Z_{SM, Z_X, Z_H, \gamma^{(1)}}$. The expression for the coupling of the SM fermions to these NP states can be written as:

$$\mathcal{L}_{NP} \subset X_\mu \Big[\alpha^{bs}_L(X)(\bar{s}_L\gamma^\mu b_L) + \alpha^{bs}_R(X)(\bar{s}_R\gamma^\mu b_R) + \bar{\mu}\left(\alpha^l_V(X)\gamma^\mu - \alpha^l_A(X)\gamma^\mu\gamma^5\right)\mu\Big] \quad (7)$$

where $\alpha^l_{V,A}(X) = \frac{\alpha^l_L(X) \pm \alpha^l_R(X)}{2}$. α gives the coupling in the zero mode fermions to the gauge bosons in the mass basis of the latter. Using these expressions, the Wilson coefficients for each gauge field X are written as:

$$\Delta C_9 = -\frac{\sqrt{2}\pi}{M_X^2 G_F \alpha}\alpha^{bs}_L(X)\alpha^l_V(X),$$
$$\Delta C_{10} = \frac{\sqrt{2}\pi}{M_X^2 G_F \alpha}\alpha^{bs}_L(X)\alpha^l_A(X), \quad (8)$$

For the primed operators the expressions are similar with $L \leftrightarrow R$. In the above we assumed that the up-sector quark are in the mass-diagonal basis and down quark rotation matrices are $D_{L,R} \sim V_{CKM}$. We now discuss the following two possibilities for the fits to the data:

1) Scenario **A**: This scenario is characterized by the relatively larger contribution of the lepton singlets to the NP than the doublets. The doublets are chosen to have to have universal bulk wavefunction with $c > 0.5$. This choice is particularly helpful in obtaining an anarchic neutrino mixing matrix which is roughly given as $U^{ij}_{PMNS} \sim \frac{f(c^i_L)}{f(c^j_L)}$
The contributions to $\Delta C'_{9,10}$ are made consistent with zero by assuming that the right handed down quarks couple similarly to the NP. Numerically this implies $c_{d_R, s_R, b_R} > 0.55$. The ranges chosen for c parameter scan are chosen to be: $c_{Q_3} \in [0, 0.5]$, $c_{\mu_L} = c_L \in [0.51, 0.6]$ and $c_{\mu_R} \in [0.45, 0.55]$. Further by choosing the c parameters for the first two quark generations $c_{Q_{1,2}} > 0.55$ ensures a universality of their coupling to the NP states. Fig. 1 gives the results of the scan: The top plot gives the correlation between $\Delta C^\mu_9 - \Delta C^e_9$ while the bottom gives the correlation between $\Delta C^\mu_{10} - \Delta C^e_{10}$. The 2-$\sigma$ regions for a 4D fit to the data is [11]

$$C^\mu_9 \in [-0.33, 0.06] \quad C^e_9 \in [-2.23, 0.74]$$
$$C^\mu_{10} \in [-0.29, 0.14] \quad C^e_{10} \in [-2.60, 0.60] \quad (9)$$

The non-negligible values of the ΔC^e is due to left doublets having $c \sim 0.5$ thereby resulting in a mildly larger coupling to the NP states than would be expected of states having $c \geq 0.55$. For these choices of c parameters corresponding to the values in Fig.1, fitting the muon mass requires choosing the $O(1)$ Yukawa ~ 0.03. Though slightly fine tuned with regards to the fit to the muon mass, this scenario is more favorable with regards to an anarchic neutrino mixing matrix but also in suppressing FCNC in the lepton sector through the implementation of 5D MFV.

2) Scenario **B**: In this case the non-universality is now transferred to the lepton doublets while the singlets are closer to the UV brane and their coupling to the NP is universal. Further, without loss of generality we assume that the τ doublets are closer to the IR brane than the μ doublets ($c_{\tau_L} < c_{\mu_L}$). All the lepton singlets and the electron doublet satisfy $c > 0.55$. This results in $\Delta C^e_{9,10}$ much smaller than $\Delta C^\mu_{9,10}$, with its magnitude being at most ~ 0.2. For most of the region, it effectively reduces this to a 2-D fit where the value of $C^e_{9,10}$ is an order of magnitude less,

Top left plot of Fig.2 gives the correlation in the ΔC^μ_9-ΔC^μ_{10} plane which correspond to points which sat-

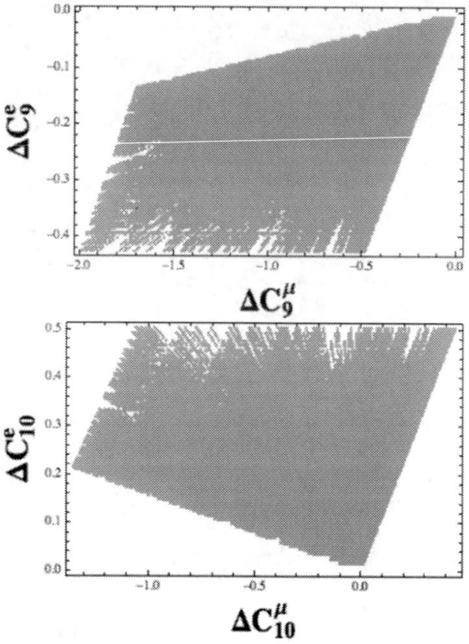

Figure 1. Scenario **A**: Plot show the correlation in the C_9 and C_{10} parameter plane for both the electron and the muon. We use $M_{KK} = 3\,TeV$

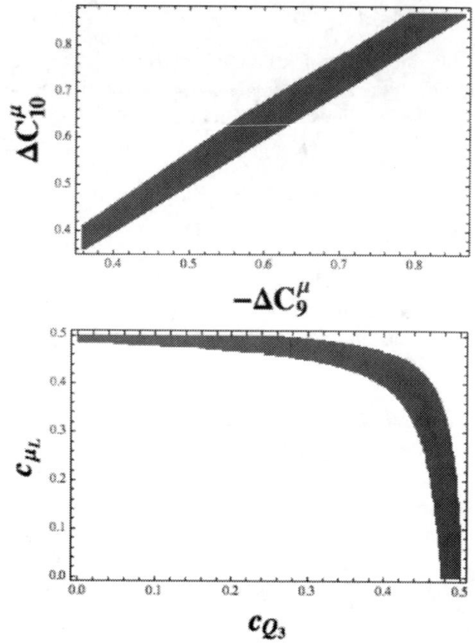

Figure 2. Scenario **B**: Top plot gives the distribution for ΔC_9 and ΔC_{10}. The corresponding c parameters ranges are given in the bottom plot.

isfy $0.36 < |\Delta C^{\mu}_{9,10}| < 0.87$. There also exist solutions for which $\Delta C^{\mu}_9 = -\Delta C^{\mu}_{10}$ further reducing it to a 1-D fit as discussed in [18, 19]. The only two relevant parameters for the fits to the B-anomalies are $c_{Q_3} - c_{\mu_L}$ and the correlation is shown in right plot of Fig. 2.

3 Kaon decays

In this section we use rare K decays to differentiate between the two scenarios considered earlier. $K^+ \to \pi^+\nu\nu$ and $K_L \to \pi^0\nu\nu$ are likely to constitute the next probe towards the possible existence of NP. They correspond to $s \to d\nu\nu$ form of transitions and are likely to be correlated to $b \to sll$ transitions in most NP scenarios. The SM value for the $K^+ \to \pi^+\nu\bar{\nu}$ and $K_L \to \pi^0\nu\bar{\nu}$ is [12, 13, 21]:

$$\mathcal{B}(K^+ \to \pi^+\nu\bar{\nu}) = 8.3 \pm 0.3 \pm 0.3 \times 10^{-11}$$
$$\mathcal{B}(K_L \to \pi^0\nu\bar{\nu}) = 2.9 \pm 0.2 \pm 0.0 \times 10^{-11} \quad (10)$$

where the first error is due to the uncertainty in the parameters of V_{CKM} mixing matrix and the second one is due to the remaining theoretical uncertainties. The current experimental bound is [22]

$$\mathcal{B}(K^+ \to \pi^+\nu\bar{\nu}) = 17.3^{+11.5}_{-10.5} \times 10^{-11}$$
$$\mathcal{B}(K_L \to \pi^0\nu\bar{\nu}) \leq 2.6 \times 10^{-8} \quad (90\%\ \text{C.L.}) \quad (11)$$

These measurements are likely to be significantly improved in the future. The NA62 experiment at CERN [14, 15] is pursuing to reach a precision of 15% compared to the SM in 2018. while 5% accuracy will be achieved with more time. With regards to the $K_L \to \pi^0\nu\bar{\nu}$, the KOTO experiment at J-PARC aims at measuring $\mathcal{B}(K_L \to \pi^0\nu\bar{\nu})$ around the SM sensitivity in the first instance [16, 17]. Moreover, the KOTO-step2 experiment will aim at 100 events for the SM branching ratio. This implies a precision of 10% of this measurement.

These processes can be described by the following effective Lagrangian parameterizing for $s \to d\nu\nu$ transitions:

$$\mathcal{L} = \frac{4G_F\alpha}{2\sqrt{2}\pi} V^*_{ts}V_{td}C_{ds,l}\left(\bar{s}_L\gamma_\mu d_L\right)\left(\bar{\nu}_l\gamma^\mu\nu_l\right) \quad (12)$$

The Wilson co-efficient $C_{ds,l}$ in the SM is given as:

$$C^{SM}_{ds,l} = -\frac{1}{s^2_{\theta_w}}\left(X_t + \frac{V^*_{cs}V_{cd}}{V^*_{ts}V_{td}}X^l_c\right) \quad (13)$$

where X_t and X^l_c are the loop functions for the top and charm contribution respectively and given as: $X_t = 1.481 \pm 0.009$ and $\frac{1}{3}\sum_l \frac{X^l_c}{\lambda^4} = 0.365 \pm 0.012$ [20] the branching ratio for $K^+ \to \pi^+\nu\nu$ is given as:

$$\mathcal{B}(K^+ \to \pi^+\nu\nu) = \frac{\kappa_+(1+\Delta_{em})}{3}\sum_{l=e,\mu,\tau}\left|\frac{V^*_{ts}V_{td}}{\lambda^5}X_t + \frac{V^*_{cs}V_{cd}}{\lambda}\left(\frac{X^l_c}{\lambda^4} + \delta P^l_c\right)\right|^2$$
$$\mathcal{B}(K_L \to \pi^0\nu\nu) = \frac{\kappa_L}{3}\sum_{l=e,\mu,\tau}\left(\frac{\text{Im}(V^*_{ts}V_{td})}{\lambda^5}X_t\right)^2 \quad (14)$$

where $\kappa_L = 2.231 \pm 0.013 \times 10^{-10}(\lambda/0.225)$, $\kappa_+ = 5.173 \pm 0.025 \times 10^{-11}(\lambda/0.225)$, $\Delta_{em} = -0.003$ [23] and $\delta P^l_{c,u} = 0.04 \pm 0.02$ [24]. The individual values of X^l were obtained from Table 1 of [25]: $X^{e,\mu} = 11.18 \times 10^{-4}$, $X^\tau = 7.63 \times 10^{-4}$.

We now consider the NP contributions to the process $s \to dll$ given in Eq. 12. In the bulk custodial model under consideration, the effective lagrangian for the process is given as

$$\mathcal{L}_{s\to d\nu\nu} \equiv \left[\alpha^{sd}_L(\bar{s}_L\gamma^\mu d_L) + \alpha^{sd}_R(\bar{s}_R\gamma^\mu d_R)\right](\nu_L\bar{\gamma}_\mu \nu)\alpha^l_L \quad (15)$$

In general this includes both the left handed and the right handed current in the quark sector signaling a possible deviation from the $(V-A)(V-A)$ structure given in Eq. 12. This aspect was explored in great detail in [7]. We discuss this process in the context of the two scenarios discussed in Section 2. It is worth stressing at this point that the $s \to d\nu\nu$ transitions only depend on the left handed c_L parameters for the leptons, while both c_{Q_3} and c_{b_R} play a role. However, since we assumed only the third generation doublets to have $c_{Q_3} < 0.5$, there are no tree-level FCNC in the right handed sector. The contribution can be quantified by making the following change to the X_t in Eq. 13:

$$X_t \to X_t + \sum_{X=Z_X,H,Y^{(1)}} \frac{\sqrt{2}2\pi}{4G_f\alpha}\frac{\alpha^{sd}(X)\alpha^l_L(X)}{M^2_{KK}} \quad (16)$$

where as earlier the contribution due to Z_X is suppressed. The SM limit is computed in the limit $M_{KK} \to \infty$.

We consider the following ratio for both the decays $\mathcal{B}^i_{total}/\mathcal{B}^i_{SM}$ for $i = K_L, K^+$ and evaluate it for the two scenarios discussed earlier:

1) Scenario **A**: This case is characterized by the universality in the left handed lepton sector. Since neutrinos in the final state are left handed, only c_L (parameter for the lepton doublets) will play a role its computation. To stress the fact that B anomalies are explained purely due to non-universality in the right handed sector for leptons we choose : $c_{\mu_R} \sim 0.48$ for the muon singlet while $c_L \sim 0.51$ for all three generations. Fig. 3 gives plot of $\mathcal{B}^i_{total}/\mathcal{B}^i_{SM}$ computed as a function of c_{Q_3} and evaluated for $c_L = 0.51$. This corresponds to the parameter space of the hypothesis under consideration. It can be seen that for both the decays, the ratio is very close to the SM prediction thereby predicting no net enhancement. In principle one can choose to reduce c_{Q_3} lesser than 0 at the cost of increasing its compositeness and possible tension with $Z \to b\bar{b}$ constraints.

2) Scenario **B**: This case is characterized by non-universality in the left handed lepton sector while the NP coupling to the right handed singlets are universal. Fig. 4 gives the ratio $\mathcal{B}^i_{total}/\mathcal{B}^i_{SM}$ for both the kaon decays as a function of c_{μ_L} and c_{Q_3}. We note that for this scenario where $c_{\mu_L} < 0.5$, the region consistent with the $b \to sll$ leads to enhancement of $\sim 1.2 - 1.6$, depending on the value of c_{μ_L} and c_{Q_3}. This is an useful example where a more accurate measurement of certain process may help in narrowing down the NP parameter space.

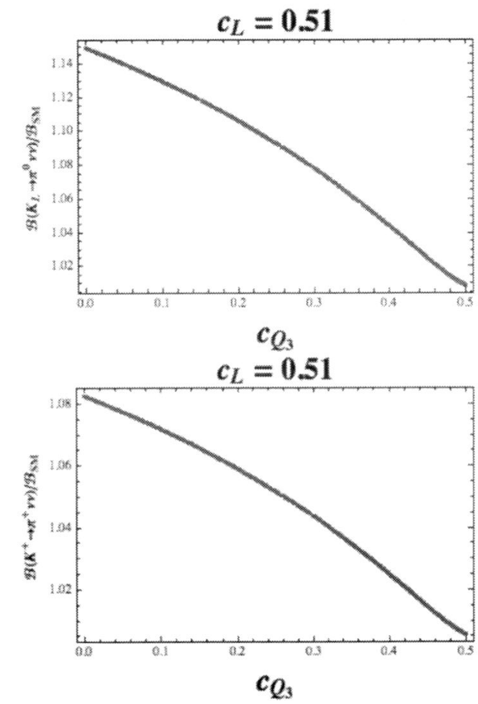

Figure 3. Scenario **A**: Plots depicting the excess over the SM expectation for the K decays modes. The c parameters for the doublets is universal and chosen to be $c_L = 0.51$.

Process	Experimental(Upper Bound)
$\mathcal{B}(\mu \to e\gamma)$	4.2×10^{-13} [27]
$\mathcal{B}(\mu \to eee)$	1.1×10^{-12}[28]
$\mathcal{B}(\mu - e)$ Conv(Ti)	6.1×10^{-13} [29]

Table 1. Experimental upper bound for the branching fraction of leptonic flavour observables in the 1-2 sector.

4 Leptonic MFV

The localization of the fermions at different points in the bulk leads to their non-universal coupling to the gauge KK states, These typically give rise to additional contributions to different to FCNC processes at tree level. Thus this is an example of a scenario where lepton non-universality leads to flavour violation. In minimal setup and for KK scales within the reach of LHC, these contributions can be particularly large in the lepton sector. This is mainly due to the strong upper bounds on processes in the 1-2 sector $\mu \to e\gamma$, $\mu \to eee$, $\mu - e$ conversion [26]. The current experimental upper bounds on these processes are given in Table 1. The large contributions can be attributed to the misalignment between the Yukawa coupling matrix and the bulk mass parameters which determine the nature of the fermionic profiles in the bulk. A complete model explaining anomalies in B sector with a relatively low NP scale must also satisfy constraints given in Table 1.

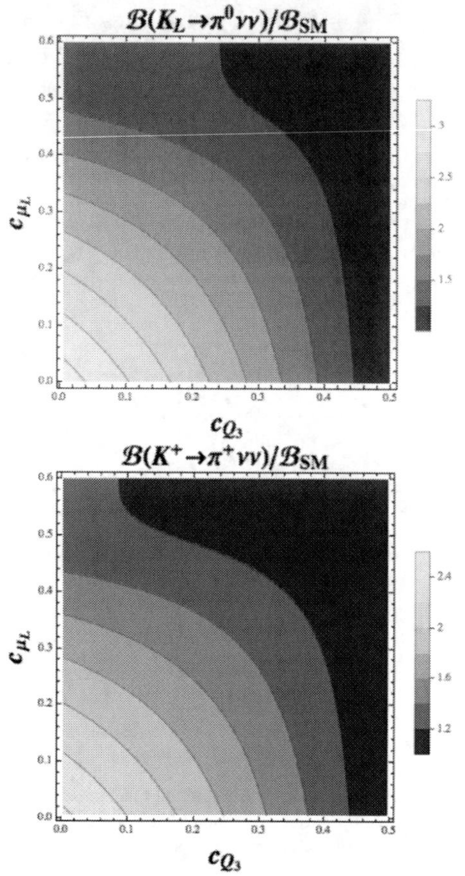

Figure 4. Scenario **B**: Plots depicting the excess over the SM expectation for the K decays modes. $c_{\tau_L} = 0.4$ and $c_{e_L} = 0.6$ are fixed for the computation while c_{μ_L} is varied.

One possibility to alleviate these constraints is to consider the implementation of MFV in 5D [30]. The large contributions to FCNC are primarily due to the misalignment between the flavour violating combination $YY^\dagger$ and the mass-squared matrix $m^2 \sim (F(c)YY^\dagger F(c))$. Here $F(c)$ is the diagonal matrix of bulk profiles given as $F(c) = Diag(f(c_e), f(c_\mu), f(c_\tau))$. Since Y and $F(c)$ do not commute in general, diagonalization of m^2 does not necessarily imply diagonalization of $YY^\dagger$. This misalignment can be reduced by making the specific choice where the 5D Yukawa parameters are written in terms of the bulk mass parameters as follows:

$$\begin{aligned} c_L &= a_L I + b_L Y_E Y_E^\dagger + d_L Y_N Y_N^\dagger \\ c_E &= a_E I + b_E Y_E^\dagger Y_E \\ c_N &= a_N + b_N Y_N^\dagger Y_N \end{aligned} \tag{17}$$

where $a_i, b_i \in \Re$. We also assume the presence of a bulk flavour symmetry group in the leptonic sector as $SU(3)_L \times SU(3)_E \times SU(3)_N$. Using the flavour symmetry we can work in a basis in which Y_E is diagonal. Thus flavour violations are embedded in Y_N which transforms under the flavour group as $Y_N \to V_5 Y_N$ where V_5 is the anarchic mixing matrix. Without loss of generality we can choose $V_5 = V_{PMNS}$. To enable the explanation of the B anomalies, we assume that the corresponding bulk flavour symmetry in the hadronic sector is broken with the corresponding c parameters not satisfying to similar relations like Eq. 17 in the hadronic sector. We now discuss its implementation for the two scenarios discussed in Section 2.
a) Scenario **A**: In this scenario the lepton doublets are localized closer to the UV brane and are assumed to have universal coupling to NP. The μ_R is relatively closer to the IR brane We assume $c_L \simeq 0.51$ for all three generations. The c values for the charged lepton singlets are chosen as $c_E = \{0.4, 0.48, 0.764\}$ and corresponding $O(1)$ Yukawa couplings are $Y_E = \{0.77, 0.03, 0.32\}$. The c_E are written in terms of Y_E by choosing $a_E = 0.49, b_E = -0.84$ in Eq. 17. For c_L we choose $a_L \gg b_L, d_L$ to preserve the universality for the left handed doublets.

To fit the neutrino mass date we make choices for c_N and Y_N as: $c_N = \{1.17, 1.17, 1.21\}$ and $Diag(Y_N) \equiv \{0.1, 0.1, 0.134\}$ leading to the following fit for the neutrino oscillation data:

$$\Delta m^2_{sol} = 7.7 \times 10^{-5}\, eV^2 \quad \Delta m^2_{atm} = 1.98 \times 10^{-3}\, eV^2 \tag{18}$$

which corresponds to an inverted hierarchy spectrum.

Corresponding to these choices, the $\mathcal{B}(\mu \to e\gamma)$ is given as [31]:

$$\mathcal{B}(\mu \to e\gamma) = 4 \times 10^{-8} \times (Y_N Y_N^\dagger)^2_{12} \frac{3TeV}{M_{KK}} \tag{19}$$

where

$$Y_N = \begin{bmatrix} 0.0823679 & 0.0548227 & 0.0194184 \\ -0.0477093 & 0.0531867 & 0.0937521 \\ 0.0306488 & -0.0645418 & 0.0937521 \end{bmatrix} \tag{20}$$

leading to $\mathcal{B}(\mu \to e\gamma) \sim 2.5 \times 10^{-14}$. Similarly the other $1-2$ transitions are also within the experimental bounds quoted in Table 1 for $M_{KK} = 3$ TeV.
b) Scenario **B**: In the earlier case we saw that the universality of the lepton doublet parameters c_L was essential to implement the MFV ansatz and obtain satisfactory fits to the neutrino mases. Universality ensures that the values of the rotation matrix $U_{ij} \sim \frac{f_{L_i}}{f_{L_j}} \sim O(1)$ which roughly correspond to the elements of the PMNS rotation matrix. Since the c_L is not universal in this scenario, fits to the neutrino data also require hierarchical choices in c_N. This makes the implementation of c_N proportional to $Y_N^\dagger Y_N$ in 17 extremely challenging and difficult to achieve. This possibly requires a more complicated parameter scan and will not be discussed here.

5 Randall Sundrum model at the GUT scale

Given these strong constraints on the RS set up at the weak scale, one can ask the question whether RS is suitable to be a theory of flavour as well as a solution to the hierarchy problem simultaneously. It might be that RS as a theory of a flavour might be better suited at the GUT scale rather than at the weak scale. The Froggatt-Nielsen models are typically defined at scales closer to the Planck scale, so perhaps flavour physics might have its origins at the Planck scale. With this point of view, we will now consider RS to span between the Planck and the GUT scales. As a result the lowest lying KK scales are $O(M_{GUT})$ and are decoupled from the low energy theory. Introducing bulk supersymmetry serves as a solution to the hierarchy problem at the weak scale. The supersymmetric case has the added advantage that it could lead to observable signatures at the weak scale and will be the focus of attention in this section [35].

In addition to the matter fields, we consider the the case with bulk Higgs doublets ($H_{u,d}$) and a SUSY breaking spurion X. The Kähler terms, leading to non-negative contributions to the scalar masses are given as [32, 33]

$$\mathcal{K} = \sum_{M=Q,U,D,L,E} e^{-2\sigma(y)}\delta(y - y_{0,\pi R})\frac{\hat{C}_{ij}}{M_{pl}^4}X^\dagger X M^\dagger M \quad (21)$$

Replacing $X \equiv \langle F \rangle$, leads to the non-tachyonic soft masses as

$$\begin{aligned} m_{ij}^2 &= m_{3/2}^2 r^2 \hat{m}_{ij}\Big(\xi_{UV}(c_i)\xi_{UV}(c_j)\xi_{UV}^2(c_S) \\ &+ \frac{1}{\epsilon^2}\xi_{IR}(c_i)\xi_{IR}(c_j)\xi_{IR}^2(c_S)\Big) \end{aligned} \quad (22)$$

and where

$$\xi_{UV}(c) = \sqrt{\frac{1-2c}{\epsilon^{2c-1}-1}} \qquad \xi_{IR}(c) = \sqrt{\frac{1-2c}{1-\epsilon^{2c-1}}} \quad (23)$$

and where the dimensionless parameter r is defined as $r = \frac{k}{M_5}$ with M_5 being the fundamental 5D Planck scale. Typically $r < 1$ so that the assumption of a classical gravity is valid. The gravitino mass $m_{3/2} = \frac{\langle F \rangle}{M_5}$ is the ratio of the vacuum expectation value of the F-term of the SUSY breaking field X to the Planck scale. In addition to this, the theory also leads to tachyonic masses at M_{GUT} and are given as [34]

$$m_{tachyonic}^2(c_m, c_s) = -2m_{3/2}^2(1 + 2\alpha_{ms}) \quad (24)$$

where $1 + 2\alpha_{ms} = \frac{(1-2c_m)(2-2c_s)}{2(4-2c_m-2c_s)}\left(\frac{(1-\epsilon^{3-2c_m})(1-\epsilon^{3-2c_s})}{\epsilon^2(1-\epsilon^{1-2c_m})(1-\epsilon^{1-2c_s})} - 1\right)$. Owing to the presence of these tachyonic terms, which are more prominent for the third generation quarks and less mildly for the thrid generation lepton, large values of M_3 are required so as to drive the masses to positive values by RGE effects. Further, owing to the tachynic masses of $\tilde{\tau}$, M_2 must be greater that M_2 to avoid the presence of tachyonic staus in the model. This can be seen from Fig. 5 Which predominantly prefers $M_2 > M_1$. This figure was obtained using analytic expressions for one-loop RGE in the limit $Y_\tau \to 0$.

The flavoured masses in Eq. 22 has interesting phenomenological implications. It leads to mixing between different fermionic generations possibly leading to new contributions to FCNC. Defining the flavour violating parameter as $\delta_{ij} = \frac{\tilde{m}_{ij}^2}{\sqrt{\tilde{m}_i^2 \tilde{m}_j^2}}$ $\quad i \neq j$, which controls the extent

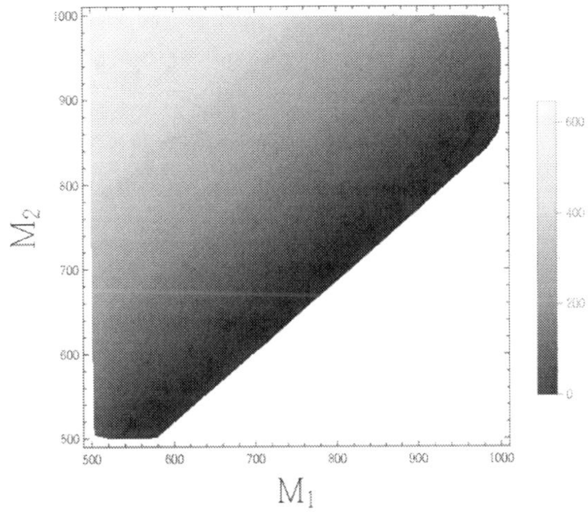

Figure 5. Lightest stau eigenstate a a function of $M_{1,2}$

of flavour violating in the theory. Fig. 6 gives the running of δ in the leptonic setor.

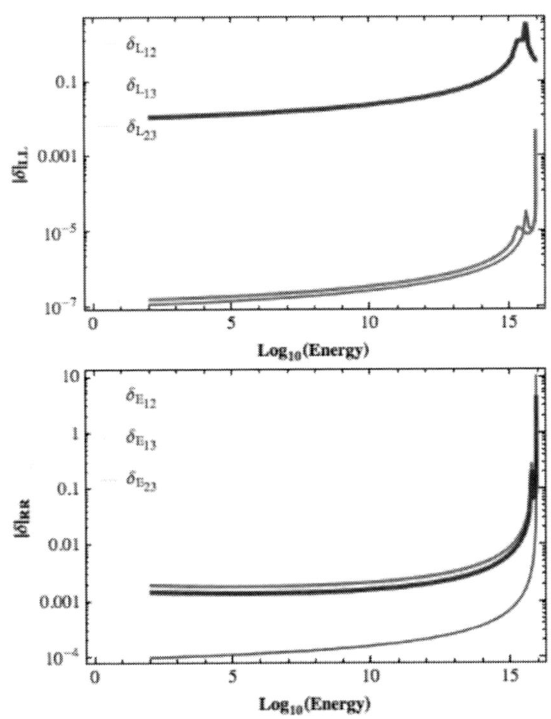

Figure 6. Running of δ for the leptonic sector.

Though $O(1)$ at the high scale, it runs down to acceptably small values leading to possible signatures of these scenarios through FCNC processes. For instance Fig. 7 gives the correlation between $\tau \to \mu\gamma$, $\mu \to e\gamma$ and

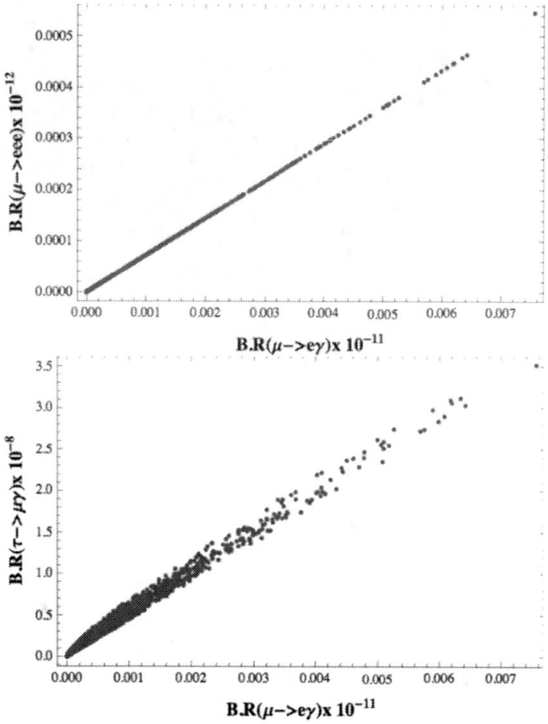

Figure 7. Correlation between $\mu \to e\gamma$ and $\mu \to eee$ (left) and $\mu \to e\gamma$ and $\tau \to \mu\gamma$

$\mu \to eee$. Since the squarks are decoupled from the theory, there are no observable FCNC effects in the squark sector.

6 Conclusions

Custodial RS models offers an interesting perspective to explore the recently observed anomlies in the decays of the B mesons. We considered two scenarios which differ in the relative localization (compositeness) of the lepton doublets and singlets in the bulk. For Scenario **A**, characterized by universal lepton doublet wavefunction, the implementation of anarchic neutrino mixing and bulk leptonic MFV is simpler. Thereby the flavour violation in the lepton sector are under control thereby paving the way for a more complete setup. The rare K decays are however consistent with the SM expectation. Finally we discuss a scenario with GUT RS where the model serves a theory of flavour while the effective low energy theory is MSSM. This has interesting implications for flavour in the lepton sector.

Acknowledgements

I would like to thank the organizers of FCCP 2017 for the opportunity to present my work in the workshop. I am also grateful to my collaborators: Giancarlo D'Ambrosio, Emilian Dudas, Sudhir Vempati for the fruitful collaborations which have lead to the interesting results presented in the workshop.

References

[1] R. Aaij *et al.* [LHCb Collaboration], Phys. Rev. Lett. **113**, 151601 (2014) doi:10.1103/PhysRevLett.113.151601 [arXiv:1406.6482 [hep-ex]].

[2] M. Bordone, G. Isidori and A. Pattori, Eur. Phys. J. C **76**, no. 8, 440 (2016) doi:10.1140/epjc/s10052-016-4274-7 [arXiv:1605.07633 [hep-ph]].

[3] G. Buchalla, A. J. Buras and M. E. Lautenbacher, Rev. Mod. Phys. **68**, 1125 (1996) doi:10.1103/RevModPhys.68.1125 [hep-ph/9512380].

[4] L. Randall and R. Sundrum, Phys. Rev. Lett. **83**, 3370 (1999) doi:10.1103/PhysRevLett.83.3370 [hep-ph/9905221].

[5] K. Agashe, A. Delgado, M. J. May and R. Sundrum, JHEP **0308**, 050 (2003) doi:10.1088/1126-6708/2003/08/050 [hep-ph/0308036].

[6] M. Blanke, A. J. Buras, B. Duling, S. Gori and A. Weiler, JHEP **0903**, 001 (2009) doi:10.1088/1126-6708/2009/03/001 [arXiv:0809.1073 [hep-ph]].

[7] M. Blanke, A. J. Buras, B. Duling, K. Gemmler and S. Gori, JHEP **0903**, 108 (2009) doi:10.1088/1126-6708/2009/03/108 [arXiv:0812.3803 [hep-ph]].

[8] M. Bauer, S. Casagrande, U. Haisch and M. Neubert, JHEP **1009**, 017 (2010) doi:10.1007/JHEP09(2010)017 [arXiv:0912.1625 [hep-ph]].

[9] S. Casagrande, F. Goertz, U. Haisch, M. Neubert and T. Pfoh, JHEP **1009**, 014 (2010) doi:10.1007/JHEP09(2010)014 [arXiv:1005.4315 [hep-ph]].

[10] G. D'Ambrosio and A. M. Iyer, arXiv:1712.08122 [hep-ph].

[11] T. Hurth, F. Mahmoudi and S. Neshatpour, Nucl. Phys. B **909**, 737 (2016) doi:10.1016/j.nuclphysb.2016.05.022 [arXiv:1603.00865 [hep-ph]].

[12] A. J. Buras, M. Gorbahn, U. Haisch and U. Nierste, JHEP **0611**, 002 (2006) Erratum: [JHEP **1211**, 167 (2012)] doi:10.1007/JHEP11(2012)167, 10.1088/1126-6708/2006/11/002 [hep-ph/0603079].

[13] A. J. Buras, D. Buttazzo, J. Girrbach-Noe and R. Knegjens, JHEP **1511**, 033 (2015) doi:10.1007/JHEP11(2015)033 [arXiv:1503.02693 [hep-ph]].

[14] A. Romano, arXiv:1411.6546 [hep-ex].

[15] F. Newson *et al.*, arXiv:1411.0109 [hep-ex].

[16] T. K. Komatsubara, Prog. Part. Nucl. Phys. **67**, 995 (2012) doi:10.1016/j.ppnp.2012.04.001 [arXiv:1203.6437 [hep-ex]].

[17] K. Shiomi [KOTO Collaboration], arXiv:1411.4250 [hep-ex].

[18] B. Capdevila, A. Crivellin, S. Descotes-Genon, J. Matias and J. Virto, arXiv:1704.05340 [hep-ph].

[19] W. Altmannshofer, P. Stangl and D. M. Straub, Phys. Rev. D **96**, no. 5, 055008 (2017) doi:10.1103/PhysRevD.96.055008 [arXiv:1704.05435 [hep-ph]].

[20] G. Buchalla and A. J. Buras, Nucl. Phys. B **548**, 309 (1999) doi:10.1016/S0550-3213(99)00149-2 [hep-ph/9901288].

[21] J. Brod, M. Gorbahn and E. Stamou, Phys. Rev. D **83**, 034030 (2011) doi:10.1103/PhysRevD.83.034030 [arXiv:1009.0947 [hep-ph]].

[22] C. Patrignani *et al.* [Particle Data Group], Chin. Phys. C **40**, no. 10, 100001 (2016). doi:10.1088/1674-1137/40/10/100001

[23] F. Mescia and C. Smith, Phys. Rev. D **76**, 034017 (2007) doi:10.1103/PhysRevD.76.034017 [arXiv:0705.2025 [hep-ph]].

[24] G. Isidori, F. Mescia and C. Smith, Nucl. Phys. B **718**, 319 (2005) doi:10.1016/j.nuclphysb.2005.04.008 [hep-ph/0503107].

[25] G. Buchalla and A. J. Buras, Nucl. Phys. B **412**, 106 (1994) doi:10.1016/0550-3213(94)90496-0 [hep-ph/9308272].

[26] K. Agashe, A. E. Blechman and F. Petriello, Phys. Rev. D **74**, 053011 (2006) doi:10.1103/PhysRevD.74.053011 [hep-ph/0606021].

[27] A. M. Baldini *et al.* [MEG Collaboration], Eur. Phys. J. C **76**, no. 8, 434 (2016) doi:10.1140/epjc/s10052-016-4271-x [arXiv:1605.05081 [hep-ex]].

[28] U. Bellgardt *et al.* [SINDRUM Collaboration], Nucl. Phys. B **299**, 1 (1988). doi:10.1016/0550-3213(88)90462-2

[29] J. Kaulard *et al.* [SINDRUM II Collaboration], Phys. Lett. B **422**, 334 (1998). doi:10.1016/S0370-2693(97)01423-8

[30] A. L. Fitzpatrick, G. Perez and L. Randall, Phys. Rev. Lett. **100**, 171604 (2008) doi:10.1103/PhysRevLett.100.171604 [arXiv:0710.1869 [hep-ph]].

[31] G. Perez and L. Randall, JHEP **0901**, 077 (2009) doi:10.1088/1126-6708/2009/01/077 [arXiv:0805.4652 [hep-ph]].

[32] K. w. Choi, D. Y. Kim, I. W. Kim and T. Kobayashi, Eur. Phys. J. C **35**, 267 (2004) doi:10.1140/epjc/s2004-01749-9 [hep-ph/0305024].

[33] D. Marti and A. Pomarol, Phys. Rev. D **64**, 105025 (2001) doi:10.1103/PhysRevD.64.105025 [hep-th/0106256].

[34] E. Dudas and G. von Gersdorff, JHEP **1210**, 100 (2012) doi:10.1007/JHEP10(2012)100 [arXiv:1207.0815 [hep-th]].

[35] E. Dudas and A. M. Iyer and S. K. Vempati, To appear
[arXiv:1801.xxxx].

Standard Model and New physics for $\varepsilon'_K/\varepsilon_K$

Teppei Kitahara[1,2,a]

[1] *Institute for Theoretical Particle Physics (TTP), Karlsruhe Institute of Technology, Engesserstraße 7, D-76128 Karlsruhe, Germany*

[2] *Institute for Nuclear Physics (IKP), Karlsruhe Institute of Technology, Hermann-von-Helmholtz-Platz 1, D-76344 Eggenstein-Leopoldshafen, Germany*

Abstract. The first result of the lattice simulation and improved perturbative calculations have pointed to a discrepancy between data on $\varepsilon'_K/\varepsilon_K$ and the standard-model (SM) prediction. Several new physics (NP) models can explain this discrepancy, and such NP models are likely to predict deviations of $\mathcal{B}(K \to \pi\nu\bar{\nu})$ from the SM predictions, which can be probed precisely in the near future by NA62 and KOTO experiments. We present correlations between $\varepsilon'_K/\varepsilon_K$ and $\mathcal{B}(K \to \pi\nu\bar{\nu})$ in two types of NP scenarios: a box dominated scenario and a Z-penguin dominated one. It is shown that different correlations are predicted and the future precision measurements of $K \to \pi\nu\bar{\nu}$ can distinguish both scenarios.

1 Introduction and Standard-Model predictions of ε'_K

Charge-parity (CP) violating flavour-changing neutral current decays of K mesons are extremely sensitive to new physics (NP) and can probe radiative corrections of particles with masses far above the reach of the Large Hadron Collider. Prime examples of such observables are a direct CP violation in $K \to \pi\pi$ decays and $\mathcal{B}(K_L \to \pi^0\nu\bar{\nu})$.

In $K \to \pi\pi$ decays, one distinguishes between two types of CP violations: direct and indirect CP violations which are parametrized by ε'_K and ε_K, respectively. Both types of CP violations have been quantified by many kaon experiments precisely. While ε_K is a per mille effect in the Standard Model (SM), ε'_K is smaller by another three orders of magnitude: $\varepsilon'_K \sim \mathcal{O}(10^{-6})$. This strong suppression comes from the smallness of the isospin-3/2 amplitude compared to the isospin-1/2 amplitude ($\Delta I = 1/2$ rule) and an accidental cancellation of leading contributions in the SM. Their suppressions lead to high sensitivity to the physics beyond the SM.

Until recently, large theoretical uncertainties precluded reliable predictions for ε'_K. Although SM predictions of ε'_K using chiral perturbation theory are consistent with the experimental value, their theoretical uncertainties are large. In contrast, calculation by the dual QCD approach [1] finds the SM value much below the experimental one. A major breakthrough has been the recent lattice-QCD calculation of the hadronic matrix elements by RBC-UKQCD collaboration [2, 3], which gives support to the latter result.

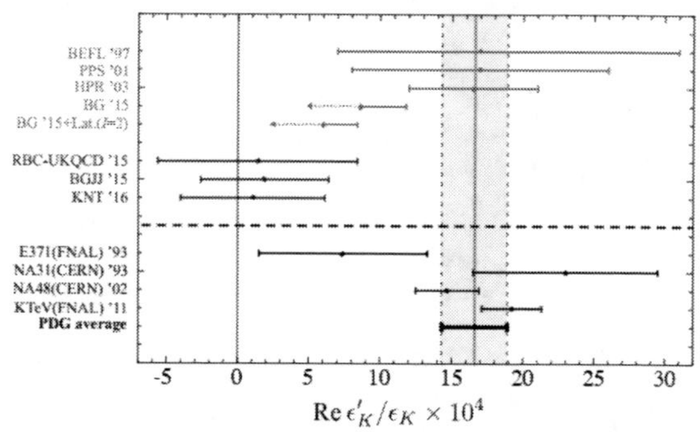

Figure 1. Compilation of representative SM predictions and the experimental values for Re($\varepsilon'_K/\varepsilon_K$). All error bars represent 1 σ range. The SM predictions are taken from Bertolini *et al.* (BEFL '97) [4], Pallante *et al.* (PPS '01) [5], Hambye *et al.* (HPR '03) [6], Buras and Gérard (BG '15) [7, 8], RBC-UKQCD lattice result [2, 3], Buras *et al.* (BGJJ '15) [9], and Kitahara *et al.* (KNT '16) [10], where magenta bars are based on analytic approaches to hadronic matrix elements, while blue bars are based on lattice results. The black thick one is the world average of the experimental values [15].

The compilation of representative SM predictions and the experimental values for Re($\varepsilon'_K/\varepsilon_K$) is given in Fig. 1. The SM predictions (colored bars) are taken from: Bertolini *et al.* (BEFL '97) [4], Pallante *et al.* (PPS '01) [5], Hambye *et al.* (HPR '03) [6], Buras and Gérard (BG '15) [7, 8] with lattice result for $I = 2$ (BG '15+Lat.), RBC-UKQCD lattice result [2, 3], Buras *et al.* (BGJJ '15) [9], and Kitahara *et al.* (KNT '16) [10]. The ex-

[a] e-mail: teppei.kitahara@kit.edu

perimental values (black bars) are taken from: E371 [11], NA31 [12], NA48 [13] and KTeV [14] collaborations, and the black thick one is the world average of the experimental values [15]

$$\mathrm{Re}\,(\varepsilon'_K/\varepsilon_K)_{\mathrm{exp}} = (16.6 \pm 2.3) \times 10^{-4}. \tag{1}$$

In order to predict ε'_K in the SM, one has to calculate the hadronic matrix elements of four-quark operators with nonperturbative methods. The magenta bars in Fig. 1 have utilized analytic approaches to calculating them: chiral quark model (BEFL '97), chiral perturbation theory (PPS '01) with minimal hadronic approximation (HPR '03), and the dual QCD approach (BG '15). Note that the dual QCD approach predicts an upper bound on $\varepsilon'_K/\varepsilon_K$. Recently, the result of the chiral perturbation theory has been updated [16]:

$$(\varepsilon'_K/\varepsilon_K)_{\mathrm{ChPT}} = (15 \pm 7) \times 10^{-4}. \tag{2}$$

On the other hand, a determination of all hadronic matrix elements from lattice QCD has been obtained only recently by the RBC-UKQCD collaboration [2, 3], and the blue bars are based on the lattice result:

$$\varepsilon'_K/\varepsilon_K = \begin{cases} (1.9 \pm 4.5) \times 10^{-4} & \text{(BGJJ '15)}, \\ (1.06 \pm 5.07) \times 10^{-4} & \text{(KNT '16)}. \end{cases} \tag{3}$$

These results are obtained by next-to-leading order (NLO) calculations exploiting *CP*-conserving data to reduce hadronic uncertainties and include isospin-violating contributions [17] which are not included in the lattice result. Furthermore, the latter result includes an additional $O(\alpha_{EM}^2/\alpha_s^2)$ correction, which appears only in this order, and also utilizes a new analytic solution of the renormalization group (RG) equation which avoids the problem of singularities in the NLO terms. The two numbers in Eq. (3) disagree with the experimental value in Eq. (1) by $2.9\,\sigma$ [9] and $2.8\,\sigma$ [10], respectively. The uncertainties are dominated by the lattice statistical and systematic uncertainties for the $I = 0$ amplitude.

In Fig. 2, the contributions of individual operators to $\varepsilon'_K/\varepsilon_K$ are shown, where y_i for $i = 3, 4, \ldots, 10$ are the corresponding Wilson coefficients. These values are based on Ref. [10]. Q_3–Q_6 are called QCD penguin operators,

$$Q_3 = (\bar{s}d)_{V-A} \sum_q (\bar{q}q)_{V-A}, \tag{4}$$

$$Q_4 = \left(\bar{s}_\alpha d_\beta\right)_{V-A} \sum_q \left(\bar{q}_\beta q_\alpha\right)_{V-A}, \tag{5}$$

$$Q_5 = (\bar{s}d)_{V-A} \sum_q (\bar{q}q)_{V+A}, \tag{6}$$

$$Q_6 = \left(\bar{s}_\alpha d_\beta\right)_{V-A} \sum_q \left(\bar{q}_\beta q_\alpha\right)_{V+A}, \tag{7}$$

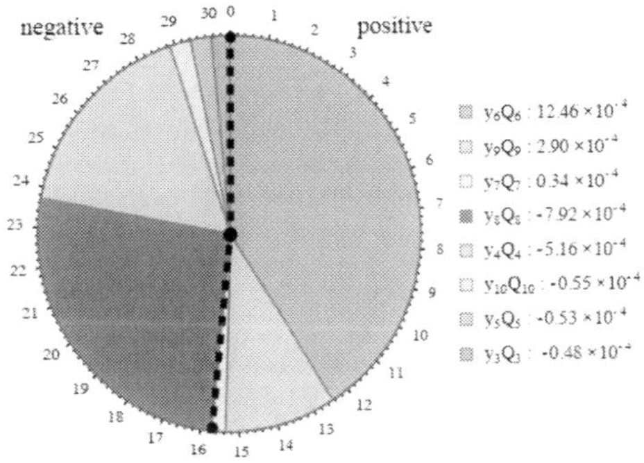

Figure 2. The composition of $\varepsilon'_K/\varepsilon_K$ with respect to the operator basis. The right and left side of the dashed lines represent positive and negative contributions to $\varepsilon'_K/\varepsilon_K$, respectively. This figure is based on the result of Ref. [10].

while Q_7–Q_{10} are called electroweak penguin operators,

$$Q_7 = \frac{3}{2} (\bar{s}d)_{V-A} \sum_q e_q (\bar{q}q)_{V+A}, \tag{8}$$

$$Q_8 = \frac{3}{2} \left(\bar{s}_\alpha d_\beta\right)_{V-A} \sum_q e_q \left(\bar{q}_\beta q_\alpha\right)_{V+A}, \tag{9}$$

$$Q_9 = \frac{3}{2} (\bar{s}d)_{V-A} \sum_q e_q (\bar{q}q)_{V-A}, \tag{10}$$

$$Q_{10} = \frac{3}{2} \left(\bar{s}_\alpha d_\beta\right)_{V-A} \sum_q e_q \left(\bar{q}_\beta q_\alpha\right)_{V-A}, \tag{11}$$

where $V \mp A$ represents $\gamma_\mu(1 \mp \gamma_5)$, α and β denote color indices, and e_q is the electric charge of the quark q. The leading contributions come from Q_6 and Q_8, having opposite sign, and thus a cancellation emerges. Remarkably, this figure also shows that even if one includes sub-leading contributions, the cancellation still exists with high precision.

The main difference between each result of analytic approaches and the lattice result comes from the hadronic parameter $B_6^{(1/2)} \propto \langle(\pi\pi)_{I=0}|Q_6|K^0\rangle$, which controls the largest positive contribution to $\varepsilon'_K/\varepsilon_K$ (the $y_6 Q_6$ component in the Fig. 2). In chiral perturbation theory, typically large values are obtained: $B_6^{(1/2)} \sim 1.6$ (BEFL '97), ~ 1.6 (PPS '01), and ~ 3 (HPR '03, see Ref. [7]). On the other hand, the dual QCD approach predicts a smaller number, $B_6^{(1/2)} \leq B_8^{(3/2)} \sim 0.8$ (BG '15). The lattice result is consistent with the latter result: $B_6^{(1/2)} = 0.56 \pm 0.20$ [3, 10]. Although the lattice simulation [3] includes final-state interactions partially along the line of Ref. [18], final-state interactions have to be still fully included in the calculations in light of a discrepancy of a strong phase shift δ_0 [19, 20]. In the near future, the increasing precision of lattice calculations with improved methods will further sharpen the SM predictions in Eq. (3) and answer the question about NP in $\varepsilon'_K/\varepsilon_K$ [21].

We also should comment on the $\Delta I = 1/2$ rule $(\mathrm{Re}A_0/\mathrm{Re}A_2)_{\mathrm{exp.}} = 22.45 \pm 0.05$. Although none of the analytic approaches can explain such a large value, the first lattice calculation has found a consistent value within $1\,\sigma$, $(\mathrm{Re}A_0/\mathrm{Re}A_2)_{\mathrm{Lat.}} = 31.0 \pm 11.1$ [2, 3, 8].

Several NP models including supersymmetry (SUSY) can explain the discrepancy of $\varepsilon'_K/\varepsilon_K$. It is known that such NP models are likely to predict deviations of branching ratios of the kaon rare decay from the SM predictions, especially $\mathcal{B}(K \to \pi\nu\bar{\nu})$ which includes CP-violating flavour-changing neutral current decay and can be probed precisely in the near future by NA62 and KOTO experiments.[1] In this contribution, based on the lattice result of $\varepsilon'_K/\varepsilon_K$ and Eq. (3), we present correlations between $\varepsilon'_K/\varepsilon_K$ and $\mathcal{B}(K \to \pi\nu\bar{\nu})$ in two types of NP scenarios: a box dominated scenario and a Z-penguin dominated one, and discuss how to distinguish between them.

2 Box dominated (Trojan penguin) scenario

We first focus on the box dominated scenario, where all NP contributions to $|\Delta S| = 1$ and $|\Delta S| = 2$ processes are dominated by the four-fermion box diagrams. Such a situation is realized in the minimal supersymmetric standard model (MSSM) [23]. The desired effect in ε'_K is generated via gluino-squark box diagrams which are shown in Fig. 3, when the mass difference between the right-handed up and down squarks exists. Such a contribution is called *Trojan penguin* because its effect is parameterized by the electroweak penguin operator at low energy scale [24].

While the sizable effects in ε'_K are obtained by the Trojan penguin contributions, a simultaneous efficient suppression of the SUSY QCD contributions to ε_K can also be achieved, which is shown in the following subsection.

2.1 Suppression of contributions to ε_K

An explanation of the puzzle between Eq. (1) and Eq. (3) by physics beyond the SM requires a NP contribution which is seemingly even larger than the SM contribution. However, it is known that once constraints from the corresponding $|\Delta S| = 2$ transition are taken into account, one expects that NP effects in a $|\Delta S| = 1$ four-quark process are highly suppressed. To explain the NP hierarchy in $|\Delta S| = 1$ vs $|\Delta S| = 2$ transitions, we specify to ε'_K and ε_K: The SM contributions are governed by the combination $\tau = -V_{td}V^*_{ts}/(V_{ud}V^*_{us}) \sim (1.5 - i0.6) \times 10^{-3}$ with $\varepsilon'^{\mathrm{SM}}_K \propto \mathrm{Im}\,\tau/M_W^2$ and $\varepsilon^{\mathrm{SM}}_K \propto \mathrm{Im}\,\tau^2/M_W^2$. If the NP contribution enters through the $|\Delta S = 1|$ parameter δ and is mediated by heavy particles of mass M, one obtains $\varepsilon'^{\mathrm{NP}}_K \propto \mathrm{Im}\,\delta/M^2$, $\varepsilon^{\mathrm{NP}}_K \propto \mathrm{Im}\,\delta^2/M^2$, and therefore the experimental constraint $|\varepsilon^{\mathrm{NP}}_K| \leq |\varepsilon^{\mathrm{SM}}_K|$ leads to

$$\left|\frac{\varepsilon'^{\mathrm{NP}}_K}{\varepsilon'^{\mathrm{SM}}_K}\right| \leq \frac{\left|\varepsilon'^{\mathrm{NP}}_K/\varepsilon'^{\mathrm{SM}}_K\right|}{\left|\varepsilon^{\mathrm{NP}}_K/\varepsilon^{\mathrm{SM}}_K\right|} = O\left(\frac{\mathrm{Re}\,\tau}{\mathrm{Re}\,\delta}\right). \quad (12)$$

[1]The correlations between $\varepsilon'_K/\varepsilon_K$, $\mathcal{B}(K \to \pi\nu\bar{\nu})$, and ε_K through the CKM components in the SM are discussed in Ref. [22].

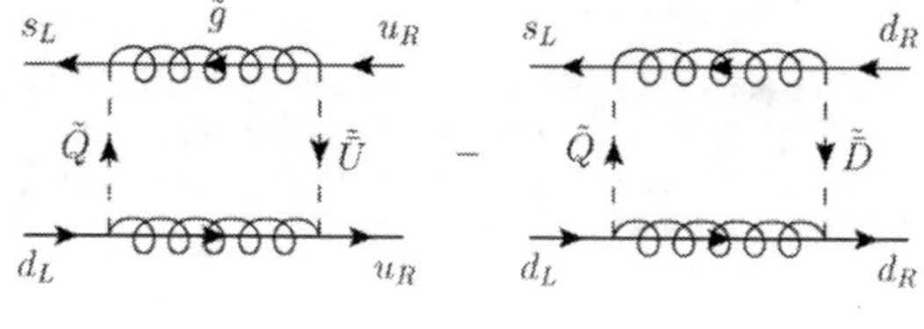

Figure 3. Trojan penguin contributions to $\mathrm{Im}\,A_2$ for $m_{\bar{U}} \neq m_{\bar{D}}$.

If NP enters through a loop with particles of mass $M \gtrsim 1$ TeV, the NP effects can be relevant only for $|\delta| \gg |\tau|$, and thus Eq. (12) seemingly forbids detectable NP contributions to ε'_K.

In the MSSM, there is a bypass to Eq. (12). The Majorana nature of the gluino leads to a suppression of gluino-squark box contributions to ε_K. This is so, because there are two such diagrams (crossed and uncrossed boxes) with opposite signs. If the gluino mass $m_{\tilde{g}}$ equals roughly 1.5 times the average down squark mass M_S, both contributions to $\varepsilon^{\mathrm{SUSY}}_K$ cancel [25]. For $m_{\tilde{g}} > 1.5M_S$, the gluino-box contribution approximately behaves as $[m_{\tilde{g}}^2 - (1.5M_S)^2]/m_{\tilde{g}}^4$, and the $1/m_{\tilde{g}}^2$ decoupling sets in. Note that this suppression appears only when a hierarchy $\Delta_{Q,12} \gg \Delta_{\bar{D},12}$ or $\Delta_{Q,12} \ll \Delta_{\bar{D},12}$ is satisfied, where the following notation is used for the squark mass matrices: $M^2_{X,ij} = m^2_X\left(\delta_{ij} + \Delta_{X,ij}\right)$, with $X = Q,\ \bar{U}$, or $\bar{D}$.

2.2 Contributions to ε'_K

The master equation for $\varepsilon'_K/\varepsilon_K$ (see e.g. Ref. [9]) reads:

$$\frac{\varepsilon'_K}{\varepsilon_K} = \frac{\omega_+}{\sqrt{2}|\varepsilon^{\mathrm{exp}}_K|\mathrm{Re}A_0^{\mathrm{exp}}}\left[\frac{\mathrm{Im}A_2}{\omega_+} - \left(1 - \hat{\Omega}_{\mathrm{eff}}\right)\mathrm{Im}A_0\right], \quad (13)$$

with $\hat{\Omega}_{\mathrm{eff}} = (14.8 \pm 8.0) \times 10^{-2}$, the measured $|\varepsilon^{\mathrm{exp}}_K|$, $\omega_+ = (4.53 \pm 0.02) \times 10^{-2}$, and the amplitudes $A_I = \langle(\pi\pi)_I|\mathcal{H}^{|\Delta S|=1}|K^0\rangle$ involving the effective $|\Delta S| = 1$ Hamiltonian $\mathcal{H}^{|\Delta S|}$. $I = 0, 2$ represents the strong isospin of the final two-pion state.

The MSSM contributions to $\varepsilon'_K/\varepsilon_K$ have been widely studied in the past. However, the SUSY-breaking scale M_S was considered in the ballpark of the electroweak scale, so that the suppression mechanism inferred from Eq. (12) is avoided. The low-energy Hamiltonian in the case of small left-right squark mixing reads

$$\mathcal{H}^{|\Delta S|=1}_{\mathrm{eff,\,SUSY}} = \frac{G_F}{\sqrt{2}}\sum_q\left[\sum_{i=1}^{2} c^q_i(\mu)Q^q_i(\mu) + \sum_{i=1}^{4}[c'^q_i(\mu)Q'^q_i(\mu) + \tilde{c}'^q_i(\mu)\tilde{Q}'^q_i(\mu)]\right] + \mathrm{H.c.}, \quad (14)$$

where G_F is the Fermi constant and

$$Q_1^{q=u,c,t} = \left(\bar{s}_\alpha q_\beta\right)_{V-A}\left(\bar{q}_\beta d_\alpha\right)_{V-A}, \quad (15)$$

$$Q_2^{q=u,c,t} = (\bar{s}q)_{V-A}(\bar{q}d)_{V-A}, \quad (16)$$

$$Q_1^{\prime q=\text{all}} = (\bar{s}d)_{V-A}(\bar{q}q)_{V+A}, \quad (17)$$

$$Q_2^{\prime q=\text{all}} = \left(\bar{s}_\alpha d_\beta\right)_{V-A}\left(\bar{q}_\beta q_\alpha\right)_{V+A}, \quad (18)$$

$$Q_3^{\prime q=\text{all}} = (\bar{s}d)_{V-A}(\bar{q}q)_{V-A}, \quad (19)$$

$$Q_4^{\prime q=\text{all}} = \left(\bar{s}_\alpha d_\beta\right)_{V-A}\left(\bar{q}_\beta q_\alpha\right)_{V-A}. \quad (20)$$

Here, opposite-chirality operators $\tilde{Q}_i^{\prime q}$ are given by interchanging $V-A \leftrightarrow V+A$.

In this section, we focus on a case that the dominant SUSY contribution comes from Trojan penguin diagrams in Fig. 3. The other SUSY solution focusing the Z-penguin contributions is investigated in the next section. The box diagrams contribute to $\mathrm{Im}A_2$ when $m_{\bar{U}} \neq m_{\bar{D}}$. Because these contributions are governed by the strong interaction and there is an enhancement factor $1/\omega_+ = 22.1$ for the $\mathrm{Im}A_2$ term in (13), they easily become the largest contribution to $\varepsilon'_K/\varepsilon_K$ [24]. In order to obtain the desired large effect in ε'_K, one needs a contribution to the operators $Q'_{1,2}$ with $(V-A)\times(V+A)$ Dirac structure, whose matrix elements are chirally enhanced by a factor $(m_K/m_s)^2$. Hence, the flavour mixing has to be in the left-handed squark mass matrix. The opposite situation with right-handed flavour mixing and $\tilde{u}_L$-$\tilde{d}_L$ mass splitting is not possible because of the $SU(2)_L$ invariance.

For the calculation of SUSY contributions to $\varepsilon'_K/\varepsilon_K$, one has to use the RG equations to evolve the Wilson coefficients calculated at the high scale M_S down to the hadronic scale $\mu_h = 1.3$ GeV at which the hadronic matrix elements are calculated [2, 3, 10]. To use the well-known NLO 10×10 anomalous dimensions for the SM four-fermion operator basis [26], we switch from Eq. (14) to

$$\mathcal{H}_{\text{eff, SUSY}}^{|\Delta S|=1} = \frac{G_F}{\sqrt{2}}\sum_{i=1}^{10}[C_i(\mu)Q_i(\mu) + \tilde{C}_i(\mu)\tilde{Q}_i(\mu)] + \text{H.c.}, \quad (21)$$

where $Q_{1,...,10}$ are given in Eqs. (4)–(11), (15), and (16), and

$$C_{1,2}(\mu) = c^u_{1,2}(\mu), \quad (22)$$

$$\tilde{C}_{1,2}(\mu) = 0, \quad (23)$$

$$C_{3,4,5,6}(\mu) = \frac{1}{3}[c'^u_{3,4,1,2}(\mu) + 2c'^d_{3,4,1,2}(\mu)], \quad (24)$$

$$C_{7,8,9,10}(\mu) = \frac{2}{3}[c'^u_{1,2,3,4}(\mu) - c'^d_{1,2,3,4}(\mu)], \quad (25)$$

and the coefficients $\tilde{C}_{3,...10}$ for the opposite-chirality operators can be obtained from $C_{3,...10}$ by replacing $c'^q_i \to \tilde{c}'^q_i$. Note that the contribution of Fig. 3 is collected into the coefficients $C_{7,8}$. For the RG evolution of the coefficients, we use the new analytic solution of the RG equations discussed in Ref. [10].

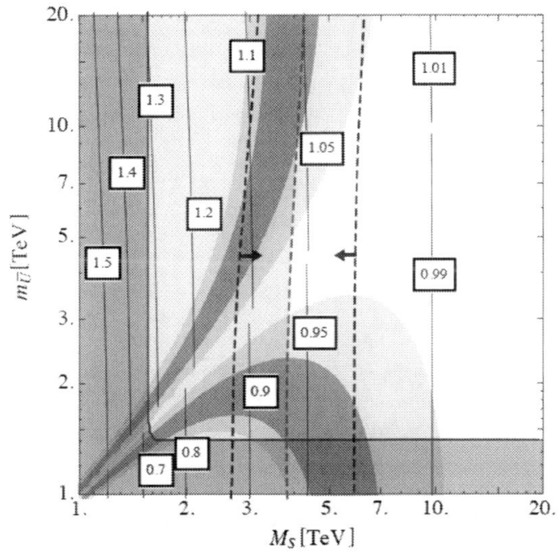

Figure 4. The $\varepsilon'_K/\varepsilon_K$ discrepancy between Eq. (1) and Eq. (3) is resolved at the $1\,\sigma$ ($2\,\sigma$) level within the dark (light) green region. The red shaded region is excluded by ε_K at 95 % C.L. using the inclusive value $|V_{cb}|$, while the region between the blue-dashed lines can explain the ε_K discrepancy which is present if the exclusive determination of $|V_{cb}|$ is used [27]. The blue shaded region is excluded by the current LHC results from CMS and ATLAS [28–30]. The only nonzero off-diagonal element of the squark mass matrices is $\Delta_{Q,12} = 0.1\exp(-i\pi/4)$ for $m_{\bar{U}} > m_{\bar{D}} = M_S$ (upper branch) and $\Delta_{Q,12} = 0.1\exp(i3\pi/4)$ for $m_{\bar{U}} < m_{\bar{D}} = M_S$ (lower branch). Black contour represents $\mathcal{B}(K_L\to\pi^0\nu\bar{\nu})/\mathcal{B}^{\text{SM}}(K_L\to\pi^0\nu\bar{\nu})$.

In Fig. 4, the portion of the squark mass plane which simultaneously explains $\varepsilon'_K/\varepsilon_K$ discrepancy and ε_K constraint is shown. As input, we take the grand-unified theory (GUT) relation for gaugino masses, $\alpha_s(M_Z) = 0.1185$, $m_{\tilde{g}}/M_S = 1.5$ for the suppressed ε_K, and $m_{\bar{Q}} = m_{\bar{D}} = \mu_{\text{SUSY}} = M_S$ with varying $m_{\bar{U}}$. The universal slepton mass is set to be $m_L = 300$ GeV. Furthermore, the trilinear SUSY-breaking matrices A_q are set to zero, $\tan\beta = 10$, and the only nonzero off-diagonal element of the squark mass matrices is $\Delta_{Q,12} = 0.1\exp(-i\pi/4)$ for the left-handed squark sectors for $m_{\bar{U}} > m_{\bar{D}} = M_S$ (upper branch) and $\Delta_{Q,12} = 0.1\exp(i3\pi/4)$ for $m_{\bar{U}} < m_{\bar{D}} = M_S$ (lower branch). We have calculated all relevant one-loop contributions to the coefficients in Eq. (14) in the squark mass eigenbasis. The $\varepsilon'_K/\varepsilon_K$ discrepancy between Eq. (1) and Eq. (3) can be resolved at $1\,\sigma$ ($2\,\sigma$) in the dark (light) green region. The red region is excluded by the measurement of ε_K at 95 % C.L. in combination with the inclusive $|V_{cb}|$, while the region between the blue dashed lines can explain the ε_K discrepancy at 95 % C.L. for the exclusive value of $|V_{cb}|$ [27]. Note that $\theta = \pm\pi/4$ maximizes the effect in $\varepsilon_K^{\text{SUSY}}$, while the SUSY contributions to $\varepsilon'_K/\varepsilon_K$ is maximized at $\theta = \pm\pi/2$ resulting instead in a vanishing effect in $\varepsilon_K^{\rm SUSY}$. Therefore, Fig. 4 is a conservative result in light of the CP-violating phase.

The blue shaded region is excluded by the current LHC results [28–30]. Here, in order to be conservative, we use the most stringent one, i.e. we maximize the bound which is a function of the neutralino mass.

The black contour represents $\mathcal{B}(K_L \to \pi^0\nu\bar{\nu})$ which is normalized by $\mathcal{B}^{\rm SM}(K_L \to \pi^0\nu\bar{\nu}) = (2.9\pm0.2)\times10^{-11}$ [31]. In this setup, we find that $\mathcal{B}(K_L \to \pi^0\nu\bar{\nu})/\mathcal{B}^{\rm SM}(K_L \to \pi^0\nu\bar{\nu}) \simeq 1.05$–$1.1$ is predicted in light of the $\varepsilon'_K/\varepsilon_K$ discrepancy (and the potential ε_K discrepancy) if $m_{\bar{U}} > m_{\bar{D}}$. More detailed investigation of $\mathcal{B}(K_L \to \pi^0\nu\bar{\nu})$ is shown in the next subsection.

2.3 $\mathcal{B}(K_L \to \pi^0\nu\bar{\nu})$ and $\mathcal{B}(K^+ \to \pi^+\nu\bar{\nu})$

The SUSY contributions to ε_K can be suppressed by the crossed and uncrossed box diagrams when the gluino mass is heavier than the squark mass, while there is no such cancellation in a chargino box contribution to $K_L \to \pi^0\nu\bar{\nu}$ and $K^+ \to \pi^+\nu\bar{\nu}$ which permits potentially large effects. We investigate the correlation between ε'_K and $\mathcal{B}(K \to \pi\nu\bar{\nu})$ varying the following parameters:

$$|\Delta_{Q,12}|,\ \theta,\ M_3,\ m_{\bar{U}}/m_{\bar{D}}, \qquad (26)$$

with $0 < |\Delta_{Q,12}| < 1$ and $0 < \theta < 2\pi$. Here, defining the bilinear terms for the squarks as $\theta \equiv \arg(\Delta_{Q,12})$. We fix the slepton mass and the lightest squark mass close to the experimental limit ($m_L = 300\,\mathrm{GeV}$ and $m_{\tilde{q}_1} = 1.5\,\mathrm{TeV}$) and use GUT relations among all three gaugino masses.

Figure 5 shows the correlations between ε'_K and $\mathcal{B}(K \to \pi\nu\bar{\nu})$ in the $\mathcal{B}(K_L \to \pi^0\nu\bar{\nu})$–$\mathcal{B}(K^+ \to \pi^+\nu\bar{\nu})$ plane which are normalized by their SM predictions [31]. We find that the necessary amount of the tuning in the gluino mass and the CP violating phase in order to suppress contributions to ε_K determines deviations of $\mathcal{B}(K \to \pi\nu\bar{\nu})$ from the SM values. A quantity which parameterizes the fine-tuning parameter is defined in Ref. [31]. The current $\varepsilon'_K/\varepsilon_K$ discrepancy between Eq. (1) and Eq. (3) is resolved at $1\,\sigma\,(2\,\sigma)$ within the dark (light) green region. In the top (bottom) panel we used $m_{\bar{D}}/m_{\bar{U}} = 1.1$ (2) with $m_{\bar{U}} = m_Q$ for $0 < \theta < \pi$, and $m_{\bar{U}}/m_{\bar{D}} = 1.1$ (2) with $m_{\bar{D}} = m_Q$ for $\pi < \theta < 2\pi$. Numerically, we observe $\mathcal{B}(K_L \to \pi^0\nu\bar{\nu})/\mathcal{B}^{\rm SM}(K_L \to \pi^0\nu\bar{\nu}) \lesssim 2\,(1.2)$ and $\mathcal{B}(K^+ \to \pi^+\nu\bar{\nu})/\mathcal{B}^{\rm SM}(K^+ \to \pi^+\nu\bar{\nu}) \lesssim 1.4\,(1.1)$ in light of $\varepsilon'_K/\varepsilon_K$ discrepancy, if all squarks are heavier than 1.5 TeV and if a 1 (10) % fine-tuning is permitted.

We also observe a strict correlation between $\mathcal{B}(K_L \to \pi^0\nu\bar{\nu})$ and $m_{\bar{U}}/m_{\bar{D}}$: $\mathrm{sgn}\,[\mathcal{B}(K_L \to \pi^0\nu\bar{\nu}) - \mathcal{B}^{\rm SM}(K_L \to \pi^0\nu\bar{\nu})] = \mathrm{sgn}\,[m_{\bar{U}} - m_{\bar{D}}]$. Thus, $\mathcal{B}(K_L \to \pi^0\nu\bar{\nu})$ can indirectly determine whether the right-handed up or down squark is the heavier one.

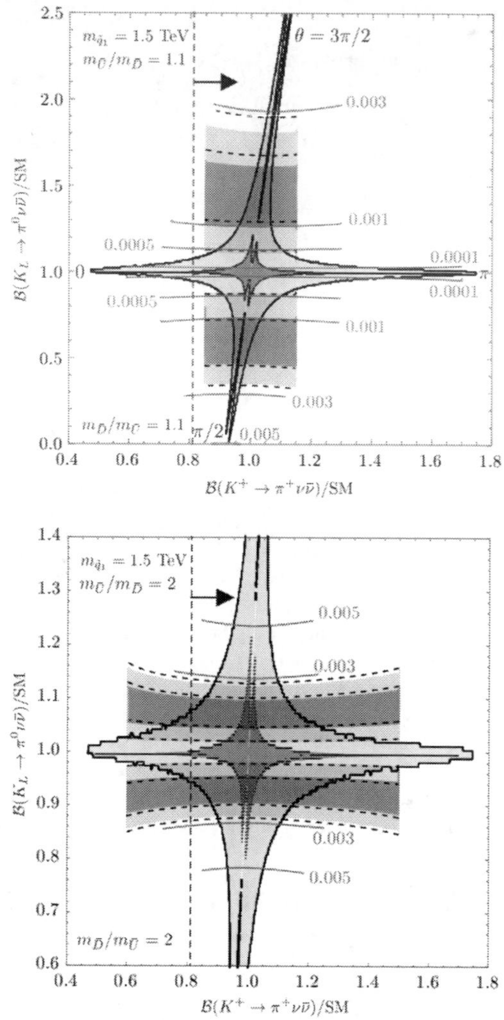

Figure 5. The correlation is shown in the Trojan penguin scenario. The light (dark) blue region requires a milder parameter tuning than 1 % (10 %) of the gluino mass and the CP violating phase in order to suppress contributions to ε_K. The red contour represents the SUSY contributions to $\varepsilon'_K/\varepsilon_K$, and the $\varepsilon'_K/\varepsilon_K$ discrepancy is resolved at $1\,\sigma\,(2\,\sigma)$ within the dark (light) green region. The lightest squark mass is fixed to 1.5 TeV. In the *top* panel, $m_{\bar{D}}/m_{\bar{U}} = 1.1$ ($m_{\bar{U}}/m_{\bar{D}} = 1.1$) is used for $0 < \theta < \pi$ ($\pi < \theta < 2\pi$) to obtain a positive SUSY contribution to $\varepsilon'_K/\varepsilon_K$. While, $m_{\bar{D}}/m_{\bar{U}} = 2$ ($m_{\bar{U}}/m_{\bar{D}} = 2$) is used for $0 < \theta < \pi$ ($\pi < \theta < 2\pi$) in the *bottom* panel. The region on the right side of the blue dashed lines are allowed by the current experimental measurements.

3 Z-penguin dominated (modified Z-coupling) scenario

Next, we focus on the Z-penguin dominated scenario. The negative dominant contribution to $\varepsilon'_K/\varepsilon_K$ comes from Z-penguin diagrams in the SM as y_8Q_8 (see Fig. 2). Since in the SM there is a large numerical cancelation between QCD-penguin and the Z-penguin contributions to $\varepsilon'_K/\varepsilon_K$, a modified Z flavour-changing (s–d) interaction from NP can explain the current $\varepsilon'_K/\varepsilon_K$ easily [32]. Then, the decay, $s \to d\nu\bar{\nu}$, proceeding through an intermediate Z boson, is modified by NP. Therefore, the branching ratios of $K \to \pi\nu\bar{\nu}$ are likely to deviate from the SM predictions once the $\varepsilon'_K/\varepsilon_K$ discrepancy between Eq. (1) and Eq. (3) is explained by the modified Z-coupling. They could be a signal to test the scenario. In the MSSM, such a scenario is also realized when the off-diagonal components of the trilinear SUSY-breaking couplings are large [33–35].

Such a signal is constrained by ε_K. The modified Z couplings affect ε_K via the so-called double penguin diagrams; the Z boson mediates the transition with two flavour-changing Z couplings. Such a contribution is enhanced when there are both left-handed and right-handed couplings because of the chiral enhancement of the hadronic matrix elements. An important point is that since the left-handed coupling is already present in the SM, the right-handed coupling must be constrained even without NP contributions to the left-handed one. Such interference contributions between the NP and the SM have been overlooked in the literature. References [35–37] have revisited the modified Z-coupling scenario including the interference contributions, and found the parameter regions allowed by the indirect CP violation change significantly.

We find that similar to the previous section, the deviations of $\mathcal{B}(K \to \pi\nu\bar{\nu})$ from the SM values are determined by the necessary amount of the tuning in NP contributions to ε_K. We parametrize it by ξ: A degree of the NP parameter tuning is represented by $1/\xi$, e.g., $\xi = 10$ means that the model parameters are tuned at the 10% level. The definition of ξ is given in Ref. [36].

In Fig. 6, contours of the tuning parameter ξ are shown for the simplified scenarios: LHS (all NP effects appear as left-handed), RHS (all NP effects appear as right-handed), ImZS (NP effects are purely imaginary), and LRS (left-right symmetric scenario) on the plane of the branching ratios of $K \to \pi\nu\bar{\nu}$ which are normalized by their SM predictions. We scanned the whole parameter space of the modified Z-coupling in each scenario, and selected the parameters where $\varepsilon'_K/\varepsilon_K$ is explained at the 1σ level. The experimental bounds from ε_K, ΔM_K, and $\mathcal{B}(K_L \to \mu^+\mu^-)$ are satisfied. In most of the allowed parameter regions, $\xi = O(1)$ is obtained. Thus, one does not require tight tunings in these simplified scenarios. In the figures, $\mathcal{B}(K_L \to \pi^0\nu\bar{\nu})$ is smaller than the SM value by more than 30%. On the other hand, $\mathcal{B}(K^+ \to \pi^+\nu\bar{\nu})$ depends on the scenarios. In LHS, we obtain $0 < \mathcal{B}(K^+ \to \pi^+\nu\bar{\nu})/\mathcal{B}(K^+ \to \pi^+\nu\bar{\nu})_{\rm SM} < 1.8$. In RHS, $\mathcal{B}(K^+ \to \pi^+\nu\bar{\nu})$ is comparable to or larger than the SM value, but cannot be twice as large. In ImZS, the branching ratios are perfectly correlated and $\mathcal{B}(K^+ \to \pi^+\nu\bar{\nu})$ does not deviate from the SM one. In LRS, $\mathcal{B}(K_L \to \pi^0\nu\bar{\nu})$ does not exceed about a half of the SM value. The more general situation is discussed in Ref. [36].

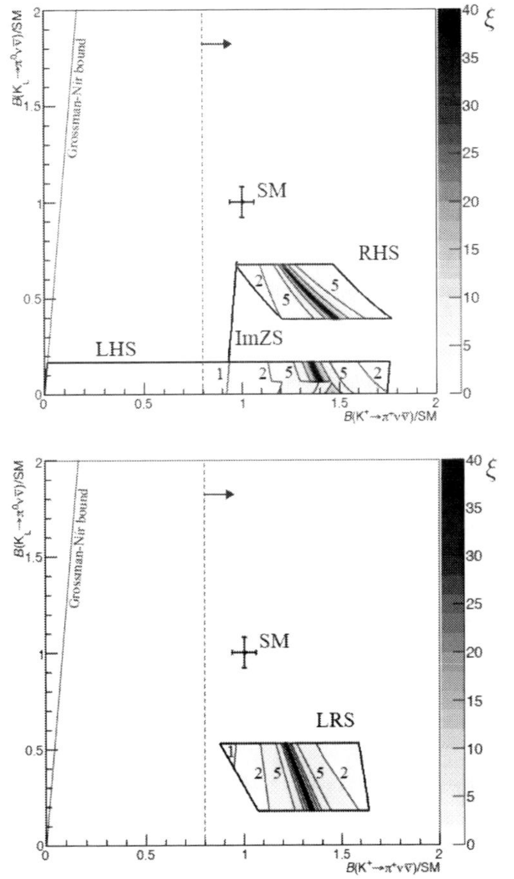

Figure 6. Contours of the tuning parameter ξ are shown in the simplified modified Z-coupling scenarios: LHS, RHS, and ImZS (*top* panel) and LRS (*bottom*). In the colored regions, $\varepsilon'_K/\varepsilon_K$ is explained at 1σ, and the experimental bounds of ε_K, ΔM_K, and $\mathcal{B}(K_L \to \mu^+\mu^-)$ are satisfied. The right region of the blue dashed line is allowed by the measurement of $\mathcal{B}(K^+ \to \pi^+\nu\bar{\nu})$ at 1σ. The NP scale is set to be $\mu = 1$ TeV.

4 Discussion and conclusions

In this contribution, we introduced the current situation for the $\varepsilon'_K/\varepsilon_K$ within the SM. The first lattice result and the improved perturbative calculations have shown the discrepancy between the predicted value and the data. Several NP models can explain the discrepancy of $\varepsilon'_K/\varepsilon_K$, and then $\mathcal{B}(K \to \pi\nu\bar{\nu})$ are predicted to deviate from the SM predictions. We have presented the correlations between $\varepsilon'_K/\varepsilon_K$, $\mathcal{B}(K_L \to \pi^0\nu\bar{\nu})$, and $\mathcal{B}(K^+ \to \pi^+\nu\bar{\nu})$ in the box dominated scenario and the Z-penguin dominated one. It is shown that the constraint from ε_K produces different correlations between two NP scenarios. In the future, measurements of $\mathcal{B}(K \to \pi\nu\bar{\nu})$ will be significantly improved. The NA62 experiment at CERN measuring $\mathcal{B}(K^+ \to \pi^+\nu\bar{\nu})$ is aiming to reach a precision of 10 % compared to the SM value already in 2018 [38]. In order to achieve 5% accuracy more time is needed. Concerning $K_L \to \pi^0\nu\bar{\nu}$, the KOTO experiment at J-PARC aims in a first step at measuring $\mathcal{B}(K_L \to \pi^0\nu\bar{\nu})$ around the SM sensitivity. Furthermore, the KOTO-step2 experiment will aim at 100 events for the SM branching ratio, implying a precision of 10 % of this measurement [39]. Therefore, we conclude that when the $\varepsilon'_K/\varepsilon_K$ discrepancy is explained by the NP contribution, NA62 experiment could probe whether a modified Z-coupling scenario is realized or not, and KOTO-step2 experiment can distinguish the box dominated scenario and the simplified modified Z-coupling scenario.

We should also comment on $K_S \to \mu^+\mu^-$, which proceeds via long-distance CP-conserving P-wave and short-distance CP-violating S-wave processes. Since the decay rate is dominated by the former, whose uncertainty is large, the sensitivity to the short-distance contributions is diminished. However, it is pointed out that the short-distance contribution is enhanced through an interference between the K_L and K_S states in the neutral kaon beam [40]. Therefore, one can also distinguish the NP scenarios using the correlation with $K_S \to \mu^+\mu^-$. Such a correlation has been investigated in the box dominated scenario (with large $\tan\beta$) [41] and the modified Z-coupling scenario [35, 40].

Acknowledgment

I would like to thank Veronika Chobanova, Andreas Crivellin, Giancarlo D'Ambrosio, Motoi Endo, Toru Goto, Miriam Lucio Martinez, Diego Martinez Santos, Satoshi Mishima, Ulrich Nierste, Isabel Suarez Fernandez, Paul Tremper, Daiki Ueda, and Kei Yamamoto for fruitful collaborations on the presented work. I also want to warmly thank the organizers of FCCP 2017 for inviting and giving me the opportunity to present these results in a great conference.

References

[1] A. J. Buras and J. M. Gérard, Nucl. Phys. B **264**, 371 (1986); A. J. Buras, J. M. Gérard and W. A. Bardeen, Eur. Phys. J. C **74**, 2871 (2014) [arXiv:1401.1385 [hep-ph]].

[2] T. Blum *et al.*, Phys. Rev. Lett. **108**, 141601 (2012) [arXiv:1111.1699 [hep-lat]]; T. Blum *et al.*, Phys. Rev. D **86**, 074513 (2012) [arXiv:1206.5142 [hep-lat]]; T. Blum *et al.*, Phys. Rev. D **91**, no. 7, 074502 (2015) [arXiv:1502.00263 [hep-lat]];

[3] Z. Bai *et al.* [RBC and UKQCD Collaborations], Phys. Rev. Lett. **115**, no. 21, 212001 (2015) [arXiv:1505.07863 [hep-lat]].

[4] S. Bertolini, J. O. Eeg, M. Fabbrichesi and E. I. Lashin, Nucl. Phys. B **514**, 93 (1998) [hep-ph/9706260].

[5] E. Pallante and A. Pich, Nucl. Phys. B **592**, 294 (2001) [hep-ph/0007208]; E. Pallante, A. Pich and I. Scimemi, Nucl. Phys. B **617**, 441 (2001) [hep-ph/0105011].

[6] T. Hambye, S. Peris and E. de Rafael, JHEP **0305**, 027 (2003) [hep-ph/0305104].

[7] A. J. Buras and J. M. Gérard, JHEP **1512**, 008 (2015) [arXiv:1507.06326 [hep-ph]].

[8] A. J. Buras and J. M. Gerard, Eur. Phys. J. C **77**, no. 1, 10 (2017) [arXiv:1603.05686 [hep-ph]].

[9] A. J. Buras, M. Gorbahn, S. Jäger and M. Jamin, JHEP **1511**, 202 (2015) [arXiv:1507.06345 [hep-ph]].

[10] T. Kitahara, U. Nierste and P. Tremper, JHEP **1612**, 078 (2016) [arXiv:1607.06727 [hep-ph]].

[11] L. K. Gibbons *et al.*, Phys. Rev. Lett. **70**, 1203 (1993).

[12] G. D. Barr *et al.* [NA31 Collaboration], Phys. Lett. B **317**, 233 (1993).

[13] V. Fanti *et al.* [NA48 Collaboration], Phys. Lett. B **465**, 335 (1999) [hep-ex/9909022]; J. R. Batley *et al.* [NA48 Collaboration], Phys. Lett. B **544**, 97 (2002) [hep-ex/0208009].

[14] A. Alavi-Harati *et al.* [KTeV Collaboration], Phys. Rev. Lett. **83**, 22 (1999) [hep-ex/9905060]; E. Abouzaid *et al.* [KTeV Collaboration], Phys. Rev. D **83**, 092001 (2011) [arXiv:1011.0127 [hep-ex]].

[15] C. Patrignani *et al.* [Particle Data Group], Chin. Phys. C **40**, no. 10, 100001 (2016).

[16] H. Gisbert and A. Pich, arXiv:1712.06147 [hep-ph].

[17] V. Cirigliano, A. Pich, G. Ecker and H. Neufeld, Phys. Rev. Lett. **91**, 162001 (2003) [hep-ph/0307030].

[18] L. Lellouch and M. Luscher, Commun. Math. Phys. **219**, 31 (2001) [hep-lat/0003023].

[19] G. Colangelo, J. Gasser and H. Leutwyler, Nucl. Phys. B **603**, 125 (2001) [hep-ph/0103088].

[20] R. Garcia-Martin, R. Kaminski, J. R. Pelaez, J. Ruiz de Elvira and F. J. Yndurain, Phys. Rev. D **83**, 074004 (2011) [arXiv:1102.2183 [hep-ph]].

[21] X. Feng, arXiv:1711.05648 [hep-lat].

[22] C. Lehner, E. Lunghi and A. Soni, Phys. Lett. B **759**, 82 (2016) [arXiv:1508.01801 [hep-ph]].

[23] T. Kitahara, U. Nierste and P. Tremper, Phys. Rev. Lett. **117**, no. 9, 091802 (2016) [arXiv:1604.07400 [hep-ph]].

[24] A. L. Kagan and M. Neubert, Phys. Rev. Lett. **83**, 4929 (1999) [hep-ph/9908404]; Y. Grossman, M. Neubert and A. L. Kagan, JHEP **9910**, 029 (1999) [hep-ph/9909297].

[25] A. Crivellin and M. Davidkov, Phys. Rev. D **81**, 095004 (2010) [arXiv:1002.2653 [hep-ph]].

[26] A. J. Buras, M. Jamin, M. E. Lautenbacher and P. H. Weisz, Nucl. Phys. B **400**, 37 (1993) [hep-ph/9211304]; A. J. Buras, M. Jamin and M. E. Lautenbacher, Nucl. Phys. B **400**, 75 (1993) [hep-ph/9211321]; M. Ciuchini, E. Franco, G. Martinelli and L. Reina, Nucl. Phys. B **415**, 403 (1994) [hep-ph/9304257].

[27] J. A. Bailey *et al.* [SWME Collaboration], Phys. Rev. D **92**, no. 3, 034510 (2015) [arXiv:1503.05388 [hep-lat]].

[28] G. Aad *et al.* [ATLAS Collaboration], JHEP **1510** (2015) 054 [arXiv:1507.05525 [hep-ex]].

[29] The ATLAS collaboration [ATLAS Collaboration], ATLAS-CONF-2016-078.

[30] CMS Collaboration [CMS Collaboration], CMS-PAS-SUS-16-015.

[31] A. Crivellin, G. D'Ambrosio, T. Kitahara and U. Nierste, Phys. Rev. D **96**, no. 1, 015023 (2017) [arXiv:1703.05786 [hep-ph]].

[32] A. J. Buras, F. De Fazio and J. Girrbach, Eur. Phys. J. C **74**, no. 7, 2950 (2014) [arXiv:1404.3824 [hep-ph]];

A. J. Buras, D. Buttazzo and R. Knegjens, JHEP **1511**, 166 (2015) [arXiv:1507.08672 [hep-ph]]; A. J. Buras, JHEP **1604**, 071 (2016) [arXiv:1601.00005 [hep-ph]].

[33] M. Tanimoto and K. Yamamoto, PTEP **2016**, no. 12, 123B02 (2016) [arXiv:1603.07960 [hep-ph]].

[34] M. Endo, S. Mishima, D. Ueda and K. Yamamoto, Phys. Lett. B **762**, 493 (2016) [arXiv:1608.01444 [hep-ph]].

[35] M. Endo, T. Goto, T. Kitahara, S. Mishima, D. Ueda and K. Yamamoto, arXiv:1712.04959 [hep-ph].

[36] M. Endo, T. Kitahara, S. Mishima and K. Yamamoto, Phys. Lett. B **771**, 37 (2017) [arXiv:1612.08839 [hep-ph]].

[37] C. Bobeth, A. J. Buras, A. Celis and M. Jung, JHEP **1707**, 124 (2017) [arXiv:1703.04753 [hep-ph]].

[38] E. Cortina Gil *et al.* [NA62 Collaboration], JINST **12**, no. 05, P05025 (2017) [arXiv:1703.08501 [physics.ins-det]].

[39] Talk by H. Nanjo in *International workshop on physics at the extended hadron experimental facility of J-PARC*, KEK Tokai, Japan, 5-6 March 2016; Talk by G. Ruggiero in *EPS Conference on High Energy Physics*, Venice, Italy, 5-12 July 2017.

[40] G. D'Ambrosio and T. Kitahara, Phys. Rev. Lett. **119**, no. 20, 201802 (2017) [arXiv:1707.06999 [hep-ph]].

[41] V. Chobanova, G. D'Ambrosio, T. Kitahara, M. Lucio Martinez, D. Martinez Santos, I. S. Fernandez and K. Yamamoto, arXiv:1711.11030 [hep-ph].

General discussion on g-2

Marc Knecht[1,a]

[1] Centre de Physique Théorique, CNRS/Aix-Marseille Univ./Univ. du Sud Toulon-Var (UMR 7332)
CNRS-Luminy Case 907, 13288 Marseille Cedex 9, France

Abstract. Progress made on the theoretical aspects of the standard model contributions to the anomalous magnetic moment of the charged leptons since the first FCCP Workshop on Capri in 2015 is reviewed. Emphasis is in particular given to the various cross-checks that have already become available, or might become available in the future, for several important contributions.

1 Introduction

The anomalous magnetic moments of the electron (a_e) and of the muon (a_μ) are among the most precisely measured quantities in particle physics. The latest available experimental results [1–3] read

$$\begin{aligned} a_e^{\rm exp} &= 1159652180.91(0.26)\cdot 10^{-12}\ [0.28\text{ppb}], \\ a_\mu^{\rm exp} &= 11\,659\,208.9(6.3)\cdot 10^{-10}\ [0.54\text{ppm}]. \end{aligned} \quad (1)$$

These measurements therefore represent stringent tests of the standard model, and constitute indirect windows towards possible degrees of freedom beyond it. Actually, due to the muon's larger mass, its anomalous magnetic moment is expected to be more sensitive to new physics than the electron's by a factor of $(m_\mu/m_e)^2 \sim 4\cdot 10^4$ [in a strict sense, this argument holds in this simple version provided that i) the new physics is decoupling and that ii) it couples universally to the various lepton flavours]. According to the same argument, the anomalous magnetic moment of the τ lepton (a_τ) would even be more sensitive to new physics scales. Unfortunately, the very short lifetime of the τ, $\tau_\tau = 290.3(5)\cdot 10^{-15}\ s$, has so far precluded any precision measurement of a_τ. The case of the τ will therefore not be mentioned any further in this review, and for more details on the present status of both experimental and theoretical aspects, the reader is referred to the well-documented existing literature, see for instance Refs. [3–9] and the works quoted therein.

On the theoretical side, the high precision achieved by the experiments has triggered a continuous effort in order to match the level of accuracy of the measurements of a_e and a_μ. The contributions from quantum electrodynamics (QED) belong to the realm of perturbation theory, and the difficulty here consists in computing all high orders which are relevant at the level of precision displayed in (1). The present status achieved in these computations is quite remarkable, since they have been pushed up to the five-loop level. For the weak interactions, the present level of precision is at two loops, with some leading three-loop effects included, which is quite sufficient in view of the present (and future) experimental precision. Finally, contributions from the strong interactions represent the hard core of the theoretical calculations. Here perturbation theory is of little help, since the main contribution comes from the low-energy regime. Understandably, this is therefore the sector where most recent efforts have gone, and where more progress is needed.

The present review will start from the situation as it was left at, say, the previous Capri Workshop [10] from two years ago, summarizing where progress has been achieved, and trying to point out issues that still need to be improved. For a more general overview, see [11] and references therein.

2 The present experimental situation

The value presently predicted by the standard model for a_μ is somewhat lower than the experimental one. This discrepancy now reaches the level of 3.5 standard deviation, and could easily exceed the 5σ level, should the new experiments [12–16] confirm the central value in Eq. (1) while reducing its uncertainty by a factor of four. As to the latter point, the first progress made since the workshop held in Capri in 2015 is that there are new data. The commissioning run of the E989 experiment at FNAL collected successfully about 700 000 positrons in June 2017. Provided the experiment further runs on schedule, as it did so far, a statistical sample comparable in size to the one collected by the E821 experiment at BNL should become available and be analysed during 2018 [14]. This will already provide a quite valuable experimental cross-check of the value given in Eq. (1). The final target of collecting a statistical sample more than 20 times larger than the

[a] e-mail: marc.knecht@cpt.univ-mrs.fr

one collected by the BNL E821 experiment, could then be reached in the year 2020 [14].

Experiment E34 at J-PARC aims at measuring the anomalous magnetic moment of the muon with a comparable improvement in precision, but with a different method, and hence with quite different systematics [16]. Depending on availability of budget and resources, the first results would become available only several years after the planed release of the final results of the FNAL experiment. Nevertheless, since by the year 2020 the anomalous magnetic moment of the muon could well be the only observable showing a deviation from its standard model prediction at a level exceeding 5 standard deviations, this experimental cross-check, although somewhat delayed in time, is highly important and most welcome.

In the case of the electron, the standard model prediction of the anomalous magnetic moment matches the experimental measurement, both in value and in precision:

$$a_e^{\text{exp}} - a_e^{\text{SM}} = (-1.30 \pm 0.77) \cdot 10^{-12}. \tag{2}$$

Notice that an explanation of the 3.5σ discrepancy in $a_\mu^{\text{exp}} - a_\mu^{\text{SM}}$ by new physics scenarios which obey the naive scaling argument described after Eq. (1) would not upset the agreement between a_e^{exp} and a_e^{SM} at the present level of precision.

3 Progress in the evaluation of the QED contributions

Three-flavour QED provides more than 99.99% of the standard-model value of a_μ, and it is thus important that the theoretical evaluations are under good control. Here, perturbation theory can be applied [$\ell = e, \mu$],

$$a_\ell^{\text{QED}} = \sum_n C_\ell^{(2n)} \left(\frac{\alpha}{\pi}\right)^n, \tag{3}$$

but the challenge lies in the high orders in the powers of α, the fine-structure constant, which need to be considered. Up to and including three loops, all contributions are known analytically (references can be found in [17] or [11]), so that the only limitation in precision for second and third orders lies in the precision with which the ratios of the masses of the charged leptons are known. The contributions at fourth and fifth orders are known numerically, and the limitation in precision in these cases comes from the numerical uncertainties in the evaluation of multi-dimensional integrals over Feynman parameters. The present situation reads [18–20]:

$$\begin{aligned} &C_\mu^{(2)} = 1/2 \quad C_\mu^{(4)} = 0.765\,857\,425(17) \\ &C_\mu^{(6)} = 24.050\,509\,96(32) \\ &C_\mu^{(8)} = 130.878\,0(61) \quad C_\mu^{(10)} = 750.72(93). \end{aligned} \tag{4}$$

Notice that the uncertainty in a_μ^{QED} induced by the uncertainties in the values of these coefficients are smaller than those of the present and future experiments. On the other hand, $C_\mu^{(8)}(\alpha/\pi)^4 \sim 3.8 \cdot 10^{-9}$ and $C_\mu^{(10)}(\alpha/\pi)^5 \sim 0.5 \cdot 10^{-10}$, so that at least the value of $C_\mu^{(8)}$ should be cross-checked by an independent calculation. This has recently been done [21, 22], thanks to the important fact that it is largely dominated by the diagrams containing electron loops, which are enhanced by factors like $\pi^2 \ln(m_\mu/m_e)$. These diagrams can then be evaluated using asymptotic expansion techniques [23, 24] in powers (modulo logarithms!) of the small ratio m_e/m_μ. In this way, all four-loop contributions containing at least one electron and/or tau loop [25] have now been checked, at a level of precision below the one of the value given in Eq. (4), but sufficient to match the level of precision reached by the present experiment value or expected for the future one. The QED prediction for a_μ thus rests on a very safe basis. Adding also the contribution from the weak interactions, $a_\mu^{\text{weak}} = 15.4(1) \cdot 10^{-10}$, see [26] and references therein, and using the value of α from Ref. [27] [corrected for the recent shift in the value of the Rydberg constant], one obtains $a_\mu^{\text{exp}} - a_\mu^{\text{QED}} - a_\mu^{\text{weak}} = 721.65(6.38) \cdot 10^{-10}$.

In the case of the electron, the expansion coefficients read [17, 19, 20, 28]:

$$\begin{aligned} &C_e^{(2)} = 1/2 \quad C_e^{(4)} = -0.328\,478\,444\,00\ldots \\ &C_e^{(6)} = 1.181\,234\,016\ldots \\ &C_e^{(8)} = -1.911\,321\,390\ldots \quad C_e^{(10)} = 6.595(223). \end{aligned} \tag{5}$$

and in several respects the situation looks very different from the muon. First, the contribution to a_e from QCD is even much more important than in the case of the muon:

$$a_e^{\text{QED}} = 1159652180.277(15)_{\alpha^5}(720)_{\alpha(Rb11)} \cdot 10^{-12}, \tag{6}$$

i.e. $a_e^{\text{exp}} - a_e^{\text{QED}} = 0.434(772) \cdot 10^{-12}$, where the error is dominated by the uncertainty in the determination of α [27]. Second, the contributions at order α^5 is larger than the experimental uncertainty on a_e, $C_e^{(10)}(\alpha/\pi)^5 \sim 4.4 \cdot 10^{-13}$, so here its value really matters. Third, the diagrams with loops involving only photons and electrons [the so-called mass-independent contributions] dominate the result, and all diagrams involving muon and/or tau loops are heavily suppressed, by powers of m_e/m_μ and m_e/m_τ. Their values have been computed numerically [19, 20] and cross-checked [25] with the asymptotic expansion method. This however leaves out the mass-independent part of the $O(\alpha^4)$ and $O(\alpha^5)$ coefficients, which require an independent cross-check. For the former, such a calculation in semi-analytical form has recently been achieved by S. Laporta [17, 28], a quite remarkable *tour de force*! Only $C_e^{(10)}$ remains therefore unchecked so far.

4 Progress in the evaluation of the hadronic vacuum polarization (HVP)

The contribution known as hadronic vacuum polarization (HVP) occurs first at order $O(\alpha^2)$, cf. Fig. 1. This lowest-order HVP contribution ($a_\mu^{\text{HVP-LO}}$) provides the largest hadronic correction to a_μ, but also, to date, one of the largest contribution to the theoretical uncertainty (see Table 1). $a_\mu^{\text{HVP-LO}}$ can be expressed in the following way [29–31]

$$a_\ell^{\text{HVP-LO}} = \frac{1}{3}\left(\frac{\alpha}{\pi}\right)^2 \int_{4M_\pi^2}^\infty \frac{ds}{s}\frac{m_\ell^2}{s} K(m_\ell^2/s) R^{\text{had}}(s), \tag{7}$$

Table 1. The contributions to a_μ (in units of 10^{-10}) using the latest available values

QED	+11 658 471.9	[18]
HVP-LO	+693.1(3.4)	[37]
	+692.23(2.54)	[38]
	+688.07(4.14)	[39]
HVP-NLO	−9.83(4)	[38]
	−9.93(7)	[39]
HVP-NNLO	+1.24(1)	[40]
	+1.22(1)	[39]
HLxL	+10.3(2.9)	[39]
EW 1 loop	+19.48(1)	[41]
EW 2 loops	−4.12(10)	[26]

with

$$K(m_\ell^2/s) = \int_0^1 dx \frac{x^2(1-x)}{(1-x) + \frac{m_\ell^2}{s}x^2}, \qquad (8)$$

and $R^{\rm had}(s)$ represents the R-ratio of the cross section for $e^+e^- \to$ hadrons. This contribution can therefore be evaluated using available experimental input. Actually, as illustrated by the second diagram (*b*) in Fig. 1, what is usually refered to as HVP-LO contains some contributions at order $O(\alpha^3)$, for instance the radiative modes $e^+e^- \to \pi^0\gamma, \eta\gamma$. The remaining $O(\alpha)$ [$O(\alpha^2)$] QED corrections to Fig. 1 are descibed as next-to-leading HVP (HVP-NLO) [next-next-to-leading HVP (HVP-NNLO)].

Previous data-based evaluations of $a_\mu^{\rm LO-HVP}$ had reached a precision around 0.6% [32–34]. Recently, three new evaluations have been made. Their results, shown in Table 1, are in agreement within the uncertainties given. They include new data (for the experimental aspects, see

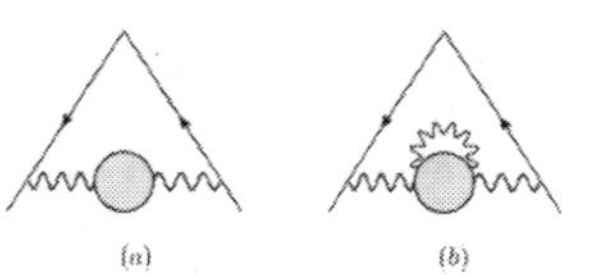

Figure 1. Diagrammatic representation of $a_\mu^{\rm HVP-LO}$, the leading hadronic vacuum polarization contribution to a_μ. In diagram (*a*), the shaded blob represents the $\langle VV \rangle$ QCD two-point function. Also shown is a next-to-leading contribution, diagram (*b*), actually included in $a_\mu^{\rm HVP-LO}$ (see text).

Table 2. The contribution to $a_\mu^{\rm HVP-LO}$ (in units of 10^{-10}) coming from the measurement of the $e^+e^- \to \pi^+\pi^-$ cross section in the region between 600 and 900 MeV. Values taken from Fig. 7 in Ref. [42].

Experiment	$a_\mu^{\rm HVP-LO\,2\pi}(600-900\ {\rm MeV})$
BaBar(09)	376.7(2.7)
KLOE(comb)	366.9(2.1)
BESIII(15)	368.2(4.2)
SND(04)	371.7(5.0)
CMD-2(comb)	372.4(3.0)

Table 3. Some recent lattice evaluations of $a_\mu^{\rm HVP-LO}$ (in units of 10^{-10}). The first error is the statistical one, the second the systematic one.

$654 \pm 32^{+21}_{-23}$	[43]
$667 \pm 6 \pm 12$	[44]
$711.0 \pm 7.5 \pm 17.3$	[45]
$715.4 \pm 16.3 \pm 9.2$	[46]

[35, 36]), and reach now a precision around 0.4%. However, some tension remains between the high-precision $e^+e^- \to \pi^+\pi^-$ data collected in the region of the ρ resonance, for instance, by BaBar on the one side, and by KLOE and BESIII on the other side, as shown in Table 2 (see also [35]). From a general point of view, given the very high precision achieved, and even more so in view of this tension in the data, some cross-checks would be welcome here also. Several proposal in this direction have been put forward, and are worth investigating.

For one thing, it has been proposed to evaluate the vacuum polarization integral using data not from the time-like region, but from the space-like one, i.e. by considering either Bhabha scattering [48], or muon-electron scattering [49]. While the required statistical accuracy could be reached upon operating for two years at, for instance, the muon beam of 150 GeV available at the CERN North Area, getting the systematic uncertainties below the required level of about 10 ppm constitutes a real challenge. The issue of controling higher-order radiative corrections will also be crucial. The latter aspect has been discussed in several presentations [50–53] at this Worshop, to which I refer the interested reader.

Staying in the Euclidian region, there has been growing interest in the lattice-QCD community to compute $a_\mu^{\rm HVP-LO}$ through numerical simulations. Several results have appeared since the first FCCP meeting in 2015, and are summarized in Table 3. The precision is still too low, around 2.5% or more, in order to be competitive with the determinations based on the $e^+e^- \to$ hadrons data. Moreover, the systematic uncertainties indicated cover different realities as far as the control over lattice artefacts and/or isospin-breaking effects (see e.g. [47]) is concerned. Nevertheless, the situation looks promising, and one might also consider, as done by the authors of Ref. [46], replacing experimental data by lattice data in the energy regions where the latter are more accurate. Supplementing lattice data for very short and very long distances with experimental measurements of $R^{\rm had}(s)$, the value shown in the last entry of Table 3 becomes $a_\mu^{\rm HVP-LO} = 692.5(1.4)(0.5)(0.7)(2.1) \cdot 10^{-10}$, where the two first errors come from the lattice, the two last ones from experiment. Combining all these uncertainties in quadrature leads to an overall relative error of less than 0.4%!

A third approach to the evaluation of $a_\mu^{\rm HVP-LO}$ has been proposed in Refs. [54–56]. In order to illustrate it, let us come back to Eq. (7), and denote as $f(x;\kappa)$, with $\kappa \equiv m_\ell^2/s$ and $0 < \kappa \le \kappa_0 \equiv m_\ell^2/4M_\pi^2$, the integrand of the function

$K(\kappa)$ in (8). From the simple inequalities

$$f(x;\kappa_0) < f(x;\kappa) < \kappa x^2[1-\kappa x^2(1-x)], \qquad (9)$$

valid within the ranges of variation of both x and κ, one obtains the bounds (keeping only the first term on the right-hand side of the second inequality reproduces the upper bound obtained long ago in Ref. [57])

$$\frac{\alpha}{\pi}K(\kappa_0)\mathcal{M}(0) < a_\ell^{\text{HVP-LO}} < \frac{\alpha}{\pi}\frac{1}{3}[\mathcal{M}(0) - \frac{11}{10}\mathcal{M}(-1)], \qquad (10)$$

in terms of the moments of the Mellin-Barnes transform of the imaginary part of the vacuum polarization function [54],

$$\mathcal{M}(y) = \int_{4M_\pi^2}^{\infty}\frac{ds}{s}\left(\frac{m_\ell^2}{s}\right)^{y-1}\frac{1}{\pi}\text{Im}\Pi(s),\ \frac{1}{\pi}\text{Im}\Pi = \frac{1}{3}\frac{\alpha}{\pi}R^{\text{had}}. \qquad (11)$$

Using, for illustrative purposes, the simple toy model for $R^{\text{had}}(s)$ given in Ref. [58], one finds, in the case of the electron, very stringent lower and upper bounds, which in both cases deviate from the actual value given by the toy model by less than 0.1%. The situation in less favourable in the case of the muon. The lower bound amounts to about only 70% of the actual value, and the upper bound exceeds it by 12%. In order to move beyond the simple bounds of the type shown above requires to use more elaborate mathematical techniques. This has been done by the authors of Refs. [54–56], who obtain a rapidly converging representation in terms of a series of the type

$$a_\mu^{\text{HVP-LO}} = \sum_{n\geq 0}[c_n\mathcal{M}(-n) + c_n'\mathcal{M}'(-n)], \qquad (12)$$

involving not only $\mathcal{M}(s)$, but also its derivative $\mathcal{M}'(s)$. Only a few terms of this series can already provide an evaluation that differs from the full result by less that 1%. The corresponding moments can be determined from phenomenology [59], or computed on the lattice [60].

5 Progress in the evaluation of the hadronic light-by-light (HLxL)

Hadronic light-by-light scattering is the next important hadronic contribution to a_μ. It involves the fourth-rank vacuum polarization tensor, as shown on Fig. 2. In contrast to the two-point hadronic vacuum polarization tensor involved in $a_\mu^{\text{HVP-LO}}$, there is, in the case of a_μ^{HLxL}, no simple and direct link to an experimental observable, similar to Eq. (7). Other strategies must then be devised in order to overcome the difficulties in handling strong-interaction effects in the non-perturbative regime, while keeping the amount of model dependence at the lowest possible level. Two approaches that can potentially fulfill this condition have been developped in order to evaluate a_μ^{HLxL}: dispersion relations and numerical simulations of QCD on the lattice.

The dispersive framework set up in Refs. [61, 62] uses a decomposition of the fourth-rank vacuum polarization

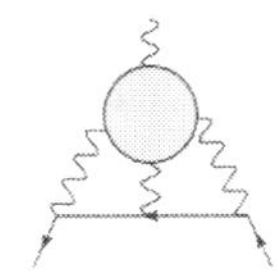

Figure 2. The hadronic light-by-light contribution to a_μ. The blob represents the hadronic fourth-rank polarization tensor, i.e. vaccun expectation value of the time-ordered product of four hadronic electromagnetic currents.

tensor into independent invariant functions free of kinematic zeroes and of singularities, for which dispersion relations can be written. These dispersion relations can then be saturated by one-, two- or more meson states and expressed in terms of corresponding form factors. The latter can furthermore be obtained from experiment [63, 64] or again... from lattice QCD [65, 66]. Applying this formalism to the $\pi\pi$ intermediate state [67] [68] provides not only a new evaluation of the so-called pion box contribution (see also Ref. [69] for a recent discussion of this contribution), but includes also S-wave rescattering effects. The result obtained this way is very accurate, at the 4% level. Results concerning other contributions are to be expected in the future. What is so far missing in this dispersive approach, is how QCD short-distance properties [70] (see also the discussion of this issue in [11]) of the fourth-rank vacuum polarization tensor will eventually be implemented in this formalism, that is how the various individual contributions will be put together in order to reproduce the correct high-energy behaviour.

Two groups are at present actively involved in the computation of a_μ^{HLxL} from lattice QCD [71–75]. The authors of Ref. [72] obtain $a_\mu^{\text{HLxL}} = 5.35(1.35)\cdot 10^{-10}$, where the error is statistical only (finite-volume effects are discussed in Ref. [73]). Although the central value lies somewhat on the low side as compared to current phenomenological estimates [34, 76], this new result opens promising prospects for the future.

6 Conclusion

The Fermilab g-2 experiment will very soon measure the anomalous magnetic moment of the muon with a precision comparable to the Brookhaven experiment, and, within a couple of years will reduce its uncertainty by a factor or four. Hopefully, the J-PARC experiment E34, which is based on a completely different set-up, will later on provide a cross-check of this important measurement.

QED provides, by far, the largest contribution to a_μ, and even more so to a_e. All numerical evaluations of the contributions at fourth order have now been satisfactorily cross-checked by analytical tools, including the mass-independent one, so crucial for a_e. The only potential problem could come, in the case of the electron, from the fifth-order contribution, which has only been determined numerically, and by one group only.

The data-based determinations of the contribution to a_μ from hadronic vacuum polarization have been improved thanks to new high-precision experimental input, and more data should become available in the near future, hopefully helping in removing some tensions between different experiments. Cross-checks coming from different approaches seem to be possible. Extracting the vacuum polarization function from μe scattering constitutes an interesting proposal in this direction, although the control of the systematic effects represents a quite challenging issue. Interesting developments have also come from numerical simulations of QCD on the lattice. Improvements in the statistical precision and increased control over systematic effects are still needed in order to compete with the data-based determinations. Of particular interest is the possibility to combine data and lattice results, taking advantage of the performance of each in various energy regions.

Progress in the determination of the HLxL contribution has been made both in the dispersive approach, and in lattice QCD. The dispersive approach offers the possibility to make use of experimental data or lattice results for various form factors. The results obtained so far are encouraging, and confirm that no important physical effect has been overlooked by the existing phenomenological evaluations. The issue here will rather consist in improving on the present precision and in obtaining reliable estimate of the uncertainties due to lattice artefacts.

Acknowledgements

I would like to thank the organizers of the Workshop "Flavour changing and conserving processes 2017" for the kind invitation to participate to this second FCCP Workshop, and for choosing the perfect place for such a meeting. This work has been partially supported by the OCEVU Labex (ANR-11-LABX-0060) and the A*MIDEX project (ANR-11-IDEX-0001-02) funded by the "Investissements d'Avenir" French government program managed by the ANR.

References

[1] D. Hanneke et al., Phys. Rev. Lett. **100**, 120801 (2008).

[2] G. W. Bennett et al., Phys. Rev D **73**, 072003 (2006).

[3] C. Patrignani et al. [Particle Data Group], Chin. Phys. C **40**, 100001 (2016).

[4] G.A. González-Sprinberg et al., Nucl. Phys. B **582**, 3 (2000).

[5] S. Narison, Phys. Lett. B **513**, 53 (2001); Err. B **526**, 414 (2002)

[6] J. Abdallah et al. [DELPHI Coll.], Eur. Phys. J. C **35**, 158 (2004).

[7] S. Eidelman, M. Passera, Mod. Phys. Lett. A **22**, 159 (2007).

[8] J. Bernabéu et al., Nucl. Phys. B **790**, 160 (2008).

[9] S. Eidelman et al., (2016).

[10] G. D'Ambrosio, M. Iacovacci, M. Passera, G. Venanzoni, P. Massarotti and S. Mastroianni, EPJ Web Conf. **118** (2016).

[11] M. Knecht, EPJ Web Conf. **118**, 01017 (2016).

[12] R. M. Carey et al.. The New g-2 experiment - 2009. Fermilab. Proposal 0989.

[13] B. Lee Roberts, Chin. Phys. C **34**, 741 (2010).

[14] See the contribution of T. Gorringe to this workshop.

[15] T. Mibe [J-PARC g-2 Collaboration], Chin. Phys. C **34**, 745 (2010).

[16] See the contribution of G. Marshall to this workshop.

[17] See the contribution of S. Laporta to this workshop.

[18] T. Aoyama et al, Phys. Rev. Lett. **109**, 111808 (2012) [arXiv:1205.5370 [hep-ph]].

[19] T. Aoyama et al, Prog. Theor. Exp. Phys. **2012**, 01A107 (2012).

[20] T. Aoyama et al, Phys. Rev. D **91**, 033006 (2015); Phys. Rev. D **96**, 019901(E) (2017).

[21] A. Kurz et al., Phys. Rev. D **92**, no. 7, 073019 (2015) [arXiv:1508.00901 [hep-ph]].

[22] A. Kurz et al., Phys. Rev. D **93**, no. 5, 053017 (2016) [arXiv:1602.02785 [hep-ph]].

[23] M. Beneke and V. A. Smirnov, Nucl. Phys. B **522**, 321 (1998) [hep-ph/9711391].

[24] V. A. Smirnov, Springer Tracts Mod. Phys. **177**, 1 (2002).

[25] A. Kurz et al., Nucl. Phys. B **879**, 1 (2014) [arXiv:1311.2471 [hep-ph]].

[26] C. Gnendiger et al., Phys. Rev. D **88**, 053005 (2013) [arXiv:1306.5546 [hep-ph]].

[27] R. Bouchendira et al., Phys. Rev. Lett. **106**, 080801 (2011).

[28] S. Laporta, Phys. Lett. B **772**, 232 (2017).

[29] C. Bouchiat and L. Michel, J. Phys. Radium 22, 121 (1961).

[30] L. Durand, Phys. Rev. **128**, 441 (1962); Err.-ibid. **129**, 2835 (1963).

[31] M. Gourdin and E. de Rafael, Nucl. Phys. B **10**, 667 (1969).

[32] M. Davier et al., Eur. Phys. J. C **71**, 1515 (2011) [Eur. Phys. J. C **72**, 1874 (2012)] [arXiv:1010.4180 [hep-ph]].

[33] K. Hagiwara et al., J. Phys. G **38**, 085003 (2011) [arXiv:1105.3149 [hep-ph]].

[34] F. Jegerlehner and R. Szafron, Eur. Phys. J. C **71**, 1632 (2011) [arXiv:1101.2872 [hep-ph]].

[35] See the contribution of F. Ignatov to this workshop.

[36] See the contribution of A. Denig to this workshop.

[37] M. Davier et al., Eur. Phys. J. C **77**, 827 (2017) [arXiv:1706.09436 [hep-ph]].

[38] See the contribution of D. Nomura to this workshop.

[39] F. Jegerlehner, EPJ Web Conf. **166**, 00022 (2018) [arXiv:1705.00263 [hep-ph]].

[40] A. Kurz et al., Phys. Lett. B **734**, 144 (2014) [arXiv:1403.6400 [hep-ph]].

[41] R. Jackiw and S. Weinberg, Phys. Rev. D **5**, 2396 (1972); I. Bars and M. Yoshimura, Phys. Rev. D **6**, 374 (1972); K. Fujikawa et al., Phys. Rev. D **6**, 2923 (1972); G. Altarelli et al., Phys. Lett. B **40**, 415 (1972); W. A. Bardeen et al., Nucl. Phys. B **46**, 315 (1972).

[42] A. Anastasi *et al.* [KLOE-2 Collaboration], arXiv:1711.03085 [hep-ex].
[43] M. Della Morte *et al.*, JHEP **1710**, 020 (2017) [arXiv:1705.01775 [hep-lat]].
[44] B. Chakraborty et al. [HPQCD coll.], Phys. Rev. D **96**, no. 3, 034516 (2017) [arXiv:1601.03071 [hep-lat]].
[45] S. Borsanyi *et al.* [Budapest-Marseille-Wuppertal coll.], arXiv:1711.04980 [hep-lat].
[46] T. Blum *et al.*, arXiv:1801.07224 [hep-lat].
[47] P. Boyle et al., arXiv:1710.07190 [hep-lat].
[48] C. M. Carloni Calame et al., Phys. Lett. B **746**, 325 (2015) [arXiv:1504.02228 [hep-ph]].
[49] G. Abbiendi et al., Eur. Phys. J. C **77**, 139 (2017) [arXiv:1609.08987 [hep-ex]].
[50] See the contribution of L. Trentadue to this workshop.
[51] See the contribution of U. Marconi to this workshop.
[52] See the contribution of P. Mastrolia to this workshop.
[53] See the contribution of F. Piccinini to this workshop.
[54] E. de Rafael, Phys. Lett. B **736**, 522 (2014) [arXiv:1406.4671 [hep-lat]].
[55] E. de Rafael, Phys. Rev. D **96**, no. 1, 014510 (2017) [arXiv:1702.06783 [hep-ph]].
[56] J. Charles et al., arXiv:1712.02202 [hep-ph].
[57] J. S. Bell and E. de Rafael, Nucl. Phys. B **11**, 611 (1969).
[58] D. Bernecker and H. B. Meyer, Eur. Phys. J. A **47**, 148 (2011) [arXiv:1107.4388 [hep-lat]].
[59] M. Benayoun et al, arXiv:1605.04474 [hep-ph].
[60] S. Borsanyi *et al.*, Phys. Rev. D **96**, 074507 (2017) [arXiv:1612.02364 [hep-lat]].
[61] G. Colangelo et al.,JHEP **1409**, 091 (2014) [arXiv:1402.7081 [hep-ph]].
[62] G. Colangelo et al., JHEP **1509**, 074 (2015) [arXiv:1506.01386 [hep-ph]].
[63] G. Colangelo et al., Phys. Lett. B **738**, 6 (2014) [arXiv:1408.2517 [hep-ph]].
[64] A. Nyffeler, Phys. Rev. D **94**, no. 5, 053006 (2016) doi:10.1103/PhysRevD.94.053006 [arXiv:1602.03398 [hep-ph]].
[65] X. Feng et al., Phys. Rev. D **91**, no. 5, 054504 (2015) [arXiv:1412.6319 [hep-lat]].
[66] A. Gérardin et al., Phys. Rev. D **94**, no. 7, 074507 (2016) [arXiv:1607.08174 [hep-lat]].
[67] G. Colangelo et al., Phys. Rev. Lett. **118**, no. 23, 232001 (2017) [arXiv:1701.06554 [hep-ph]].
[68] G. Colangelo et al., JHEP **1704**, 161 (2017) [arXiv:1702.07347 [hep-ph]].
[69] J. Bijnens and J. Relefors, JHEP **1609**, 113 (2016) [arXiv:1608.01454 [hep-ph]].
[70] K. Melnikov and A. Vainshtein, Phys. Rev. D **70**, 113006 (2004).
[71] J. Green et al., PoS LATTICE **2015**, 109 (2016) [arXiv:1510.08384 [hep-lat]].
[72] T. Blum et al, Phys. Rev. Lett. **118**, 022005 (2017) [arXiv:1610.04603 [hep-lat]].
[73] T. Blum et al., Phys. Rev. D **96**, 034515 (2017) [arXiv:1705.01067 [hep-lat]].
[74] See the contribution of C. Lehner to this workshop.
[75] See the contribution of A. Nyffeler to this workshop.
[76] See the contribution of J. Bijnens to this workshop.

 https://doi.org/10.1051/epjconf/201817901009

NA62 and NA48/2 results on search for Heavy Neutral Leptons

Gianluca Lamanna[1,a] for the NA62 collaboration[b]

[1] INFN & Physics Department, University of Pisa, Largo Pontecorvo,3 - 50131, Pisa

Abstract. In this paper we present new results on upper limits for the search of Heavy Neutral Leptons (HNL) with data collected by NA48/2 (2003-2004), NA62-RK (2007) and NA62 (2015) CERN experiments. The data collected with different trigger configuration allow to search for both long and short living heavy neutrinos in the mass range below the kaon mass. In addition the status of the search for $K^+ \to \pi^+ \nu\bar{\nu}$ with the NA62 detector will be briefly presented.

1 Introduction

Although neutrinos in Standard Model (SM) are mass-less, the experimental observation of neutrinos oscillations is a clear indication that neutrinos are massive. Several extensions of SM try to accommodate this experimental evidence by introducing new degrees of freedom, as, for instance, the possibility to have right-handed sterile neutrinos.

One of the simplest model to extend the SM in this sector is the Neutrino Minimal Standard Model (νMSM) [1]. In this model 18 new parameters are introduced in the Lagrangian to explain simultaneously neutrinos oscillation, dark matter and baryon asymmetry in the Universe. The new Lagrangian generates three new massive right-handed neutrinos. The lightest, with mass O(1 keV) is considered as a possible dark matter candidate, while the others, with mass ranging from hundreds of MeV to few GeV, give masses to the SM neutrinos through the see-saw mechanism and introduce next phases in SM, with respect to the CKM phases, becoming a new source of CP violation. The masses of these Heavy Neutral Leptons (HNL) are chosen to be below the electroweak scale and, in principle, can be produced in meson decays. The search in the kaon sector is one of the most promising for large part of the parameter phase space.

The NA48/2 and NA62 experiments, located at the Super Proton Synchrotron (SPS) at CERN, collected, starting from 2003, large samples of kaon decays. This report presents three different and complementary high statistics searches, performed in three different experimental setup with charged kaons decaying in flight.

2 The NA48/2 and NA62 experiments at CERN.

The primary goal of the NA48/2 experiment was to search for CP violation in charged kaon sector through the study of the Dalitz Plot in the three pions decay [2]. The detector based on the NA48 setup, a part some minor changes, is discussed in details in [3]. The experiment used 400 GeV/c protons extracted from the SPS accelerator, impinging on a beryllium target to produce a secondary hadron beam. A system of magnets and collimators selects charged beams with central momentum of 60 GeV (± 3 GeV). Both negative and positive charged particles are transported to a 114 m long fiducial decay region in

[a] e-mail: gianluca.lamanna@cern.ch

[b] R. Aliberti, F. Ambrosino, R. Ammendola, B. Angelucci, A. Antonelli, G. Anzivino, R. Arcidiacono, M. Barbanera, A. Biagioni, L. Bician, C. Biino, A. Bizzeti, T. Blazek, B. Bloch-Devaux, V. Bonaiuto, M. Boretto, M. Bragadireanu, D. Britton, F. Brizioli, M.B. Brunetti, D. Bryman, F. Bucci, T. Capussela, A. Ceccucci, P. Cenci, V. Cerny, C. Cerri, B. Checcucci, A. Conovaloff, P. Cooper, E. Cortina Gil, M. Corvino, F. Costantini, A. Cotta Ramusino, D. Coward, G. D'Agostini, J. Dainton, P. Dalpiaz, H. Danielsson, N. De Simone, D. Di Filippo, L. Di Lella, N. Doble, B. Dobrich, F. Duval, V. Duk, J. Engelfried, T. Enik, N. Estrada-Tristan, V. Falaleev, R. Fantechi, V. Fascianelli, L. Federici, S. Fedotov, A. Filippi, M. Fiorini, J. Fry, J. Fu, A. Fucci, L. Fulton, E. Gamberini, L. Gatignon, G. Georgiev, S. Ghinescu, A. Gianoli, M. Giorgi, S. Giudici, F. Gonnella, E. Goudzovski, C. Graham, R. Guida, E. Gushchin, F. Hahn, H. Heath, T. Husek, O. Hutanu, D. Hutchcroft, L. Iacobuzio, E. Iacopini, E. Imbergamo, B. Jenninger, K. Kampf, V. Kekelidze, S. Kholodenko, G. Khoriauli, A. Khotyantsev, A. Kleimenova, A. Korotkova, M. Koval, V. Kozhuharov, Z. Kucerova, Y. Kudenko, J. Kunze, V. Kurochka, V.Kurshetsov, G. Lanfranchi, G. Lamanna, G. Latino, P. Laycock, C. Lazzeroni, M. Lenti, G. Lehmann Miotto, E. Leonardi, P. Lichard, L. Litov, R. Lollini, D. Lomidze, A. Lonardo, P. Lubrano, M. Lupi, N. Lurkin, D. Madigozhin, I. Mannelli, G. Mannocchi, A. Mapelli, F. Marchetto, R. Marchevski, S. Martellotti, P. Massarotti, K. Massri, E. Maurice, M. Medvedeva, A. Mefodev, E. Menichetti, E. Migliore, E. Minucci, M. Mirra, M. Misheva, N. Molokanova, M. Moulson, S. Movchan, M. Napolitano, I. Neri, F. Newson, A. Norton, M. Noy, T. Numao, V. Obraztsov, A. Ostankov, S. Padolski, R. Page, V. Palladino, C. Parkinson, E. Pedreschi, M. Pepe, M. Perrin-Terrin, L. Peruzzo, P. Petrov, F. Petrucci, R. Piandani, M. Piccini, J. Pinzino, I. Polenkevich, L. Pontisso, Yu. Potrebenikov, D. Protopopescu, M. Raggi, A. Romano, P. Rubin, G. Ruggiero, V. Ryjov, A. Salamon, C. Santoni, G. Saracino, F. Sargeni, V. Semenov, A. Sergi, A. Shaikhiev, S. Shkarovskiy, D. Soldi, V. Sougonyaev, M. Sozzi, T. Spadaro, F. Spinella, A. Sturgess, J. Swallow, S. Trilov, P. Valente, B. Velghe, S. Venditti, P. Vicini, R. Volpe, M. Vormstein, H. Wahl, R. Wanke, B. Wrona, O. Yushchenko, M. Zamkovsky, A. Zinchenko.

vacuum, where K^+ and K^- decays are registered simultaneously with the same detectors.

A spectrometer, housed in a helium vessel located just after the vacuum tank, thanks to a thin Kevlar window, is composed by four drift chambers (DCH) and a dipole magnet. The Spectrometer allows to measure the charged kaon decay products momentum with a resolution of $\sigma_p/p = (1.02 \oplus 0.044\ p)\%$ where p is in GeV/c. The photon detection is performed by a quasi-homogeneous liquid Krypton calorimeter (LKr), with a depth of 27 X_0. The LKr has a resolution of $\sigma(E)/E = 0.032/\sqrt{E} \oplus 0.09/E \oplus 0.0042$ with E in GeV. Other detectors, for particle identification and trigger purpose, complete the setup. The data taking took place in 2003-2004 collecting a total of about 3×10^9 $K^\pm \to \pi^\pm\pi^\pm\pi^\mp$, about 9×10^7 $K^\pm \to \pi^\pm\pi^0\pi^0$ and other decays for rare processes studies.

The data taking in 2007 was devoted to the measurement of the R_K ratio [4]. The NA62-RK collaboration used a slightly modified beam line and detector with respect to the NA48/2 experiment, to collect at lower intensity (by a factor of 10) kaon decays in two leptons. The nominal beam momentum was 74 GeV and the trigger was designed to maximize the number of decays in electron and neutrino collected.

The NA62 collaboration developed a completely new apparatus with respect to the previous experiments. The main goal is to measure the very rare decay $K^+ \to \pi^+\nu\bar{\nu}$ with 10% precision. The first physics run started in 2016, after a period of detectors commissioning. The proton beam extracted from the SPS is used to generate a secondary hadron beam selected in charge (positive) and momentum (75 GeV/c) through a system of dipoles and collimators. The nominal intensity is of 750 MHz, but only a small fraction of the hadrons in the beam are kaons (about $\sim 6\%$). The kaons are positively identified by a Cherenkov beam detector (KTAG). The momentum of each particle in the beam is measured by a beam spectrometer composed by three stations of silicon pixels detector (Gigatracker) exposed at full intensity. An about 150 m long decay region is followed by a straw tubes spectrometer in vacuum in order to minimize the material present along the path of the photons produced in kaon decays (mainly $K^+ \to \pi^+\pi^0$). These photons are detected by the LKr calorimeter and a system of vetoes placed along the fiducial volume (12 rings of lead glass) and around the beam direction. The photon veto system guarantees a coverage up to 50 mrad. The particle identification is done by using a Ring Imaging Cherenkov (RICH), electromagnetic and hadronic calorimeters, and Muon Veto (MUV). A detailed description of the NA62 beam line and detectors can be found in [5].

3 Strategies for Heavy Neutrino searches

Two different strategies can be used to look for HNL in kaons. The first approach directly searches for decay of HNL in a pair of opposite charge particle, depending on the assumption made on the new particle properties. The second approach searches for HNL produced in kaon and pion decays by looking for peaks in the missing mass spectrum in two body decay.

The huge sample of three tracks events collected by the NA48/2 experiments allows to use the first approach to search for HNL decay. On the other hand the search in production is performed in the NA62-RK and NA62 experiments where the single charged particle missing mass is used to look for rare processes.

3.1 NA48/2: HNL search in $K^\pm \to \pi\mu\mu$

Short-living sterile neutrinos decaying promptly to $\pi\mu$ have been searched in the data sample $K^\pm \to \pi\mu\mu$ by NA48/2 in data collected in 2003-2004. The heavy neutrino is produced in $K^\pm \to \mu^\pm N_4$ decay where $N_4 \to \pi^\pm\mu^\mp$. Since the new state decays in the same primary vertex, the $\pi\mu$ pair would produce a narrow peak in the $M_{\pi\mu}$ spectrum (the invariant mass distribution). In the νMSM framework the branching ratio of this decay is proportional to mass and mixing mass matrix element introduced by the new particle:

$$BR(N_4 \to \pi\mu) \sim |U_{\mu 4}| \cdot m_N^3$$

Two different samples were collected by NA48/2: lepton number violating decays $K^\pm \to \pi^\mp\mu^\pm\mu^\pm$ ($K^{LNV}_{\pi\mu\mu}$) where the muons have the same sing and lepton number conserving candidates $K^\pm \to \pi^\pm\mu^\pm\mu^\mp$ ($K^{LNC}_{\pi\mu\mu}$) where the muons have opposite sign. Since the lifetime of the possible new particle is supposed to be shorter than $\sim 10\ ps$, the topology of the event, K^{LNV} and K^{LNC}, is indistinguishable from a three track event, both at trigger and preliminary offline selection stage. The frequent $K^\pm \to \pi^\pm\pi^+\pi^-$ ($K_{3\pi}$) decay is used as normalization channel in order to evaluate the number of kaons decaying in the 98 m fiducial region. The similar geometrical acceptance and trigger efficiency between signal and normalization allow a first order cancellation of systematic effects. For each track a momentum between 5 and 55 GeV/c is required, while the total longitudinal momentum must be compatible with the beam momentum and the total transverse momentum must be less than 10 MeV/c. The pion and muon identification is done by using the MUV detector and the energy deposited in the LKr with respect to the momentum as measured by the DCH. For each event with vertex in the decay region the invariant mass $M_{\pi\mu\mu}$ is computed.

The signal region is defined with $|M_{\pi^\mp\mu^\pm\mu^\pm} - M_K| < 5\ MeV/c$ and $|M_{\pi^\pm\mu^\pm\mu^\mp} - M_K| < 8\ MeV/c$ (M_K is the nominal kaon mass) for LNV and LNC respectively. The different cuts are due to different background contamination. In Fig. 1 the invariant mass of all events passing the selection are shown, for both data and Monte Carlo. For the LNV selection one event is observed in the signal region with a background expectation of $N_{bkg} = 1.16 \pm 0.87_{stat} \pm 0.12_{syst}$, evaluated with Monte Carlo simulation, due to $K_{3\pi}$ with two pions decaying in muons. An upper limit (90 % confidence level) is set to $BR(K^\pm \to \pi^\mp\mu^\pm\mu^\pm)$ with the Rolke-Lopez method [6]:

$$BR(K^\pm \to \pi^\mp\mu^\pm\mu^\pm) < 8.6 \times 10^{-11}$$

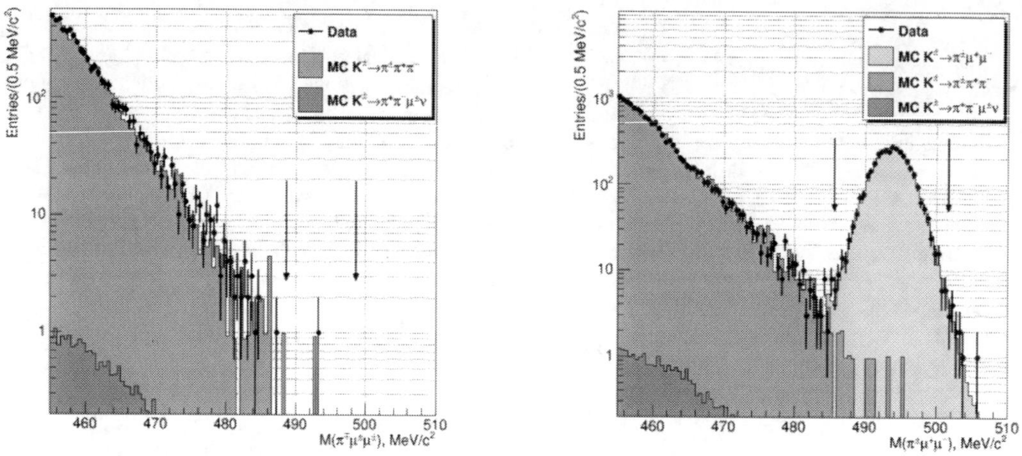

Figure 1. Invariant $M_{\pi\mu\mu}$ mass distribution for SM forbidden LNV (left) and SM allowed LNC (right) decays.

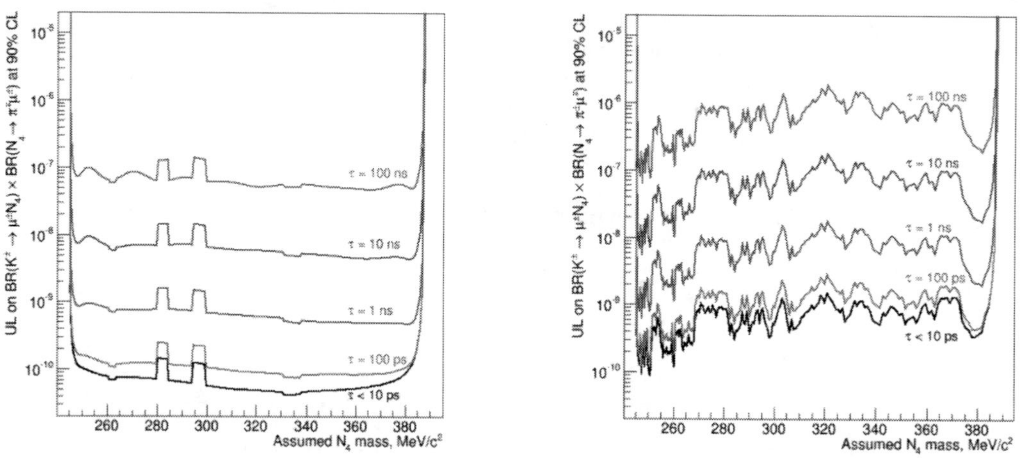

Figure 2. Upper limits at 90% CL for the $BR(K^{\pm} \to \mu^{\pm}N_4)BR(N_4 \to \pi^{\mp}\mu^{\pm})$ (left) and $BR(K^{\pm} \to \mu^{\pm}N_4)BR(N_4 \to \pi^{\pm}\mu^{\mp})$ (right).

For all the events survived to the selection, both for LNV and LNC, a scan of the invariant $M_{\pi\mu}$ mass is performed in order to search for two-body resonances. The mass steps are given by $\sigma(M_{\pi\mu})/2$, where $\sigma(M_{\pi\mu})$ is the mass resolution. In total 284 and 267 bins are considered for LNV and LNC respectively. In each mass bin the significance of the signals never exceed 3 standard deviations. The Upper Limits on the product $BR(K^{\pm} \to \mu^{\pm}N_4)BR(N_4 \to \pi^{\mp}\mu^{\pm})$ and $BR(K^{\pm} \to \mu^{\pm}N_4)BR(N_4 \to \pi^{\pm}\mu^{\mp})$ are evaluated as a function of the mass hypothesis (M_i) and the lifetime of the possible resonance (τ), by using the Rolke-Lopez method. The results are shown in Fig. 2.

The upper limit on $BR(K^{\pm} \to \mu^{\pm}N_4)BR(N_4 \to \pi\mu)$ can be used to define the upper limit on the matrix element $|U_{\mu 4}|$ according to [7].

3.2 NA62-RK and NA62: HNL search in $K^+ \to l^+\nu$ $(l = \mu, e)$

NA62-RK collected data in 2007, with a trigger optimized to collect leptonic kaon decays, with electrons and muons. This allows to search for $K^+ \to \mu^+N_4$. Assuming heavy neutrinos decaying only in standard model particles with a lifetime $\tau > 1\mu s$ (which corresponds to $|U_{\mu 4}|^2 < 10^{-4}$), the mean free path in the mass range considered is longer than 10 Km. For this reason the only expected signature is a narrow peak in the muon missing mass. The events are selected by requiring a single track identified as a muon, coming from the decay region and with momentum compatible with a kaon decay. The missing mass is defined as $m^2_{miss} = (P_K - P_{track})^2$ where P_{track} represents the 4-momentum as reconstructed by the spectrometer and P_K the nominal kaon 4-momentum evaluated as the average of the 3π total momentum in the $K^+ \to \pi^+\pi^+\pi^-$ decay. The signal region is defined as $300 < m_{miss} < 375(MeV/c^2)$. A total of $N_{\mu\nu} = 5.977 \times 10^7$ events are selected in the data

 https://doi.org/10.1051/epjconf/201817901009

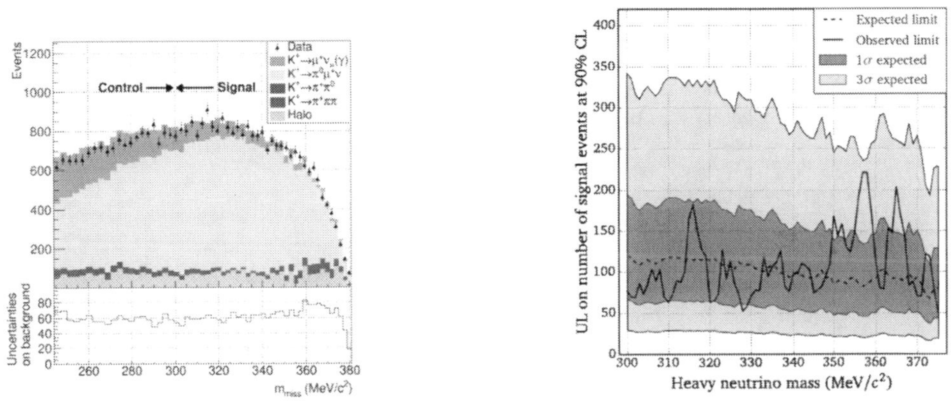

Figure 3. Reconstructed missing mass in $K \to \mu^+ N_4$ hypothesis (2007 data) (left); Upper limit on number of signal events in $K \to \mu^+ N_4$ hypothesis (2007 data) (right).

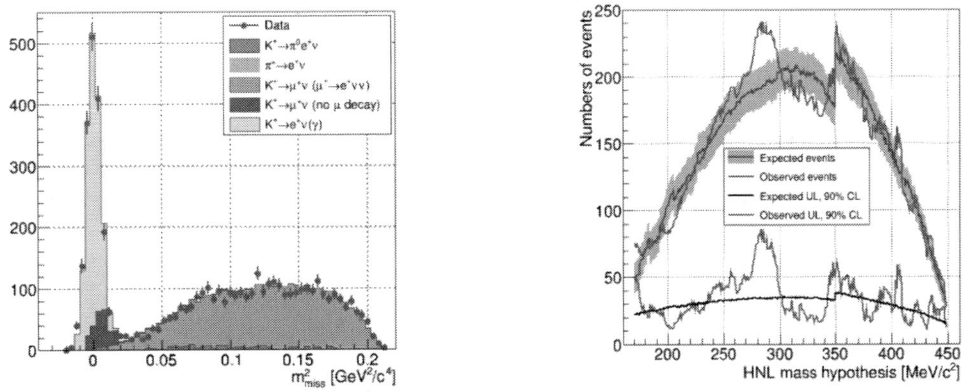

Figure 4. Reconstructed missing mass in $K \to e^+ N_4$ hypothesis (2015 data) (left); Upper limit on number of signal events in $K \to e^+ N_4$ hypothesis (2015 data) (right).

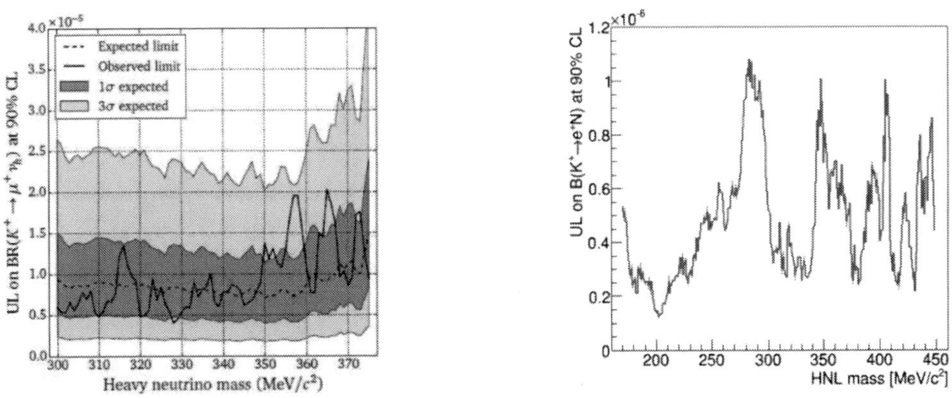

Figure 5. Upper limit on BR($K^+ \to \mu^+ N_4$) (2007 data) (left); Upper limit on BR($K^+ \to e^+ N_4$) (2015 data) (right).

sample considered for this analysis. Possible contribution to the background from other kaon decays is evaluated by Monte Carlo, while the main background from the beam halo muon components is studied using a control sample, obtained with the K^- beam only. In Fig. 3 (left) the missing mass spectrum is shown together with the background contribution.

The peak search is performed in steps of 1 MeV/c^2 in the signal region. For each bin of missing mass the Rolke-Lopez method for the case of a Poisson process in presence of gaussian background is applied to find the 90% confidence intervals. No significance above three sigma is found in the whole region considered (Fig. 3 (right)). The upper limit on the number of observed events ($N_{\mu N}$) is converted in an upper limit on the BR($K^+ \to \mu^+ N_4$) using the number of kaon decays (N_K) and the acceptance given by the Monte Carlo ($A_{\mu N}$) for each mass hypothesis:

$$BR(K^{\pm} \to \mu^+ N_4) = \frac{N_{\mu N}}{N_K \cdot A_{\mu N}}$$

In Fig. 5 (left) the limits on the BR is shown. These limits can be used to constrain the $|U_{l4}|^2$ matrix element by using the following formula:

$$|U_{\mu 4}|^2 = \frac{BR(K^+ \to \mu^+ N_4)}{BR(K^+ \to \mu^+ \nu) \cdot \rho(M_{N_4})}$$

where $\rho(M_{N_4})$ is a kinematical factor to account for the heavy neutrino mass. Similar search in the $K^+ \to e^+ N_4$ channel, is performed on the 2015 data collected by the completely new NA62 apparatus during 5 days of special minimum bias run. In Fig. 4 (left) the missing mass is shown. A total of $N_{ev} = 3.01 \times 10^8$ events are selected with background coming from kaon and pion decays. The mass scan, in step of 1 MeV/c^2, is performed in the signal region of $170 < m_{miss} < 448(MeV/c^2)$. The upper limit on the expected events is shown in Fig. 4 (right), while the corresponding upper limit on the BR($K^+ \to e^+ N_4$) is shown in Fig. 5 (right).

4 Status of the $K^+ \to \pi^+ \nu\bar{\nu}$ analysis in NA62.

NA62 collected data for $\pi\nu\bar{\nu}$ analysis in 2016 with reduced intensity (40%) with respect to the nominal intensity of 3×10^{12} pps. About 5% of these data, corresponding to 2.3×10^{10} kaons, have been preliminary analyzed in order to study signal sensitivity and background.

A reduction factor of $O(10^{12})$ on background rejection must be obtained exploiting both kinematical differences between signal and other most frequent kaon decays and active vetoes by detectors designed on purpose. The pion missing mass, the main kinematic variable used for signal to background separation, is defined as $m^2_{miss} = (P_K - P_\pi)^2$, where P_K is the kaon 4-momentum as measured by the GTK and P_π is the pion momentum as measured by the STRAWS spectrometer. This variable allows to define two regions with reduced background contribution. The remaining background contamination in the two regions,

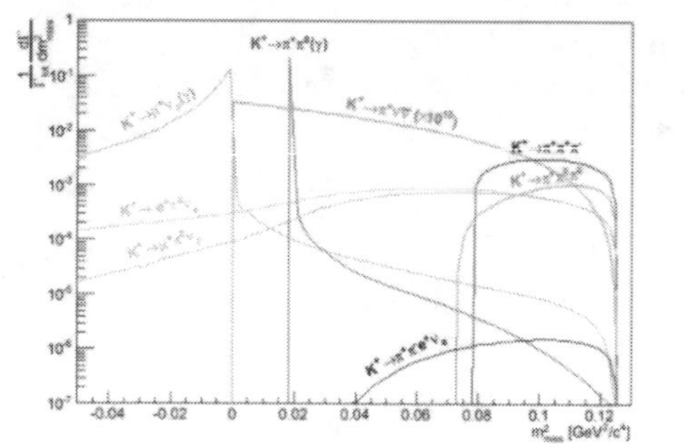

Figure 6. Missing mass distribution for signal and background. The signal is multiply by a factor 10.

shown in Fig. 6, is due to non Gaussian tails in $K_{2\pi}$ ($K^+ \to \pi^+\pi^0$) and $K_{\mu 2}$ ($K^+ \to \mu^+\nu$) decays, kaon decays with neutrinos and undetected or mis-identified particles and beam related background. The signal selection is done by requiring a good single downstream track resolution with no extra in-time activity in the veto detectors (LKr, LAV, IRC, SAC and MUV3) and a good measurement, by the GTK, of the momentum of the kaon identified by the KTAG. The pion identification is done by using RICH and calorimeters (LKr, MUV1 and MUV2) in order to further suppress muons and electrons contribution. The kaon mistagging is at level of 2% for 40% of nominal intensity, while the signal efficiency due to the kaon identification is about 75%. The signal region is limited to 50 m, starting 10 m after the last collimator, in order to suppress the contribution to background from pions originated from beam particles interacting in GTK or from kaons decaying before the fiducial region. In Fig. 7 the selected events are shown in a missing mass versus momentum 2D space. The two red boxes represent the two signal regions.

The level of background in the signal regions, measured from data, is $\sim 6 \times 10^{-4}$ for $K_{2\pi}$ and $\sim 3 \times 10^{-4}$ for $K_{\mu 2}$. The use of the γ-vetoes allows to obtain the level

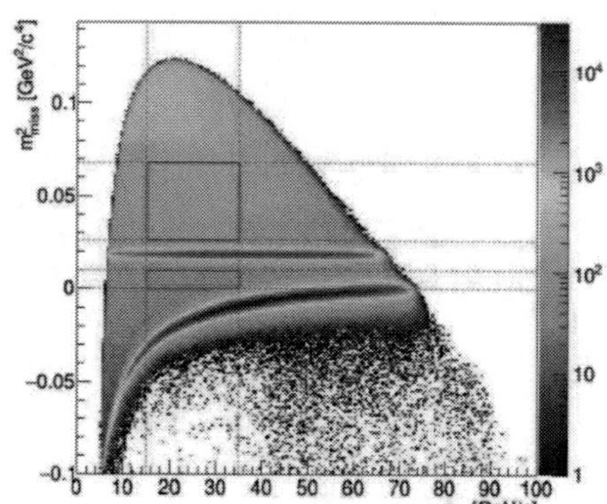

Figure 7. Distribution of m^2_{miss} versus pion momentum in minimum bias sample. The signal regions are identified by red boxes.

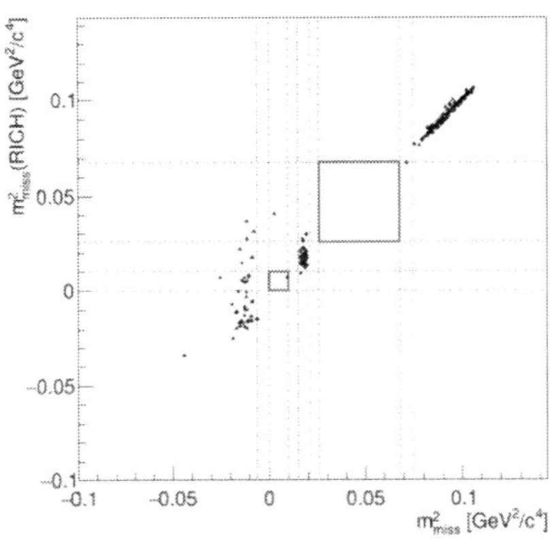

Figure 8. Events passing the full selection (see the text for the axis variable definition). The red lines identify the signal regions, while the black dashed lines the background control regions. The single event present in region 1 is discarded if the missing mass is reconstructed using the nominal beam momentum (to control the GTK non Gaussian tails).

of 1.2×10^{-7} on π^0 reduction, as measured on minimum bias triggered events. The identification of the muon with RICH, MUV3 and calorimeters allows to reach the level of 10^{-7} keeping a $\sim$ 65% of efficiency on positive pion identification.

In Fig. 8 the remaining events are shown: the missing mass is defined both using the pion momentum measured by the STRAWS (X axis) and by the RICH (Y axis) (in pion hypothesis) in order to reduce the contribution of the non Gaussian tails. The kaon flux is measured from a $K^+ \to \pi^+\pi^0$ from a minimum bias trigger. Extrapolating from the control regions, the background in the signal regions is estimated as 0.024, 0.017 and 0.011 events from $K^+ \to \pi^+\pi^0$, $K^+ \to \pi^+\pi^+\pi^-$ and $K^+ \to \mu^+\nu$ respectively. The expected number of SM $K^+ \to \pi^+\nu\bar{\nu}$, in the data sample considered, is about 0.064. No events are observed in the signal region.

5 Conclusions

The kaon sector allows to study both short and long-living HNL in the mass range of few hundreds of MeV. NA48/2 sets the world best limit on the BR of a Majorana sterile neutrino from the LNV $K^\pm \to \pi^\mp\mu^\pm\mu^\pm$, improving by an order of magnitude the previous result.

For heavy neutrino lifetime shorter than 100 ps limits in the range $10^{-10} - 10^{-9}$ are set. For stable (or long-living) HNL, the preliminary results in the peak search of HNL by the NA62-RK and the NA62 experiments, both in muon and positron channel, improve the existing limits in $|U_{l4}|$. In Fig. 9 a comparison with the previous HNL searches in π^+ and K^+ is shown. The full analysis of the 2015 minimum bias run is on-going. The background level in the heavy neutrino signal region is a factor 100 lower than in 2007, for both positron and muon channel. This will allow to finally reach a sensitivity of $10^{-6} - 10^{-7}$ on both channels.

The status of the NA62 experiment is presented. A total of 5% of the statistics collected in the 2016 run at 40% of the final beam intensity has been analyzed to study signal acceptance and background. No SM events are expected and observed in the sample considered. The blind analysis of the full 2016 data set is on-going.

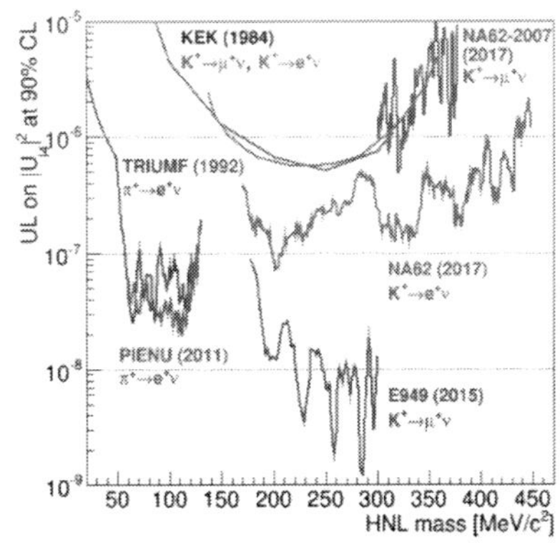

Figure 9. Upper limits on $|U_{l4}|^2$ at 90% CL from NA62 compared to limits coming from previous analysis.

References

[1] T. Asaka and M. Shaposhnikov, Phys. Lett. B **620** (2005) 17 doi:10.1016/j.physletb.2005.06.020 [hep-ph/0505013].
[2] J. R. Batley *et al.* [NA48/2 Collaboration], Eur. Phys. J. C **52** (2007) 875 doi:10.1140/epjc/s10052-007-0456-7 [arXiv:0707.0697 [hep-ex]].
[3] V. Fanti *et al.* [NA48 Collaboration], Nucl. Instrum. Meth. A **574** (2007) 433. doi:10.1016/j.nima.2007.01.178
[4] C. Lazzeroni *et al.* [NA62 Collaboration], Phys. Lett. B **698** (2011) 105 doi:10.1016/j.physletb.2011.02.064 [arXiv:1101.4805 [hep-ex]].
[5] E. Cortina Gil *et al.* [NA62 Collaboration], JINST **12** (2017) no.05, P05025 doi:10.1088/1748-0221/12/05/P05025 [arXiv:1703.08501 [physics.ins-det]].
[6] J. R. Batley *et al.* [NA48/2 Collaboration], Phys. Lett. B **769** (2017) 67 doi:10.1016/j.physletb.2017.03.029 [arXiv:1612.04723 [hep-ex]].
[7] G. Cvetic, C. Dib, S. K. Kang and C. S. Kim, Phys. Rev. D **82** (2010) 053010 doi:10.1103/PhysRevD.82.053010 [arXiv:1005.4282 [hep-ph]].

QED contributions to electron g-2

Stefano Laporta[1,2,a,b]

[1] Dipartimento di Fisica e Astronomia, Università di Padova, Via Marzolo 8, I-35131 Padova, Italy
[2] Istituto Nazionale Fisica Nucleare, Sezione di Padova, Via Marzolo 8, I-35131 Padova, Italy

Abstract. In this paper I briefly describe the results of the numerical evaluation of the mass-independent 4-loop contribution to the electron g-2 in QED with 1100 digits of precision. In particular I also show the semi-analytical fit to the numerical value, which contains harmonic polylogarithms of $e^{\frac{i\pi}{3}}$, $e^{\frac{2i\pi}{3}}$ and $e^{\frac{i\pi}{2}}$, one-dimensional integrals of products of complete elliptic integrals and six finite parts of master integrals, evaluated up to 4800 digits. I give also some information about the methods and the program used.

Electron g-2

An electron of mass m and spin s has a magnetic moment μ

$$\mu = g\frac{e\hbar}{2mc}s \qquad (1)$$

where g is the gyromagnetic ratio. According to the Dirac's theory[1], the electron has $g = 2$. The interaction with photons shifts slightly g; we define the anomaly $a_e = (g-2)/2$.

Experimental measurements

The current measurements of a_e are based on the Penning trap method, developed at the University of Washington. The anomaly is expressed as the ratio of two frequencies which can measured to a very high precision. For the development of this technique, the Noble prize in Physics 1989 was awarded to H. Dehmelt. Their final results were [2]:

$$a_{e^-}^{exp} = 1\ 159\ 652\ 188.4(4.3)\times 10^{-12} \quad (4.3\ \text{ppb})\,, \qquad (2)$$

$$a_{e^+}^{exp} = 1\ 159\ 652\ 187.9(4.3)\times 10^{-12} \quad (4.3\ \text{ppb})\,. \qquad (3)$$

The most recent value obtained with the same technique by the Harvard group is [3, 4]:

$$a_{e^-}^{exp} = 1\ 159\ 652\ 180.73(.28)\times 10^{-12} \quad (0.24\ \text{ppb})\,. \qquad (4)$$

Theoretical expression

In the standard model

$$a_e^{SM} = a_e^{QED} + a_e^{\text{weak}} + a_e^{\text{hadr}}\,. \qquad (5)$$

The QED contribution can be split up in mass-independent and mass-dependent parts:

$$a_e^{QED} = A_1 + A_2\left(\frac{m_e}{m_\mu}\right) + A_2\left(\frac{m_e}{m_\tau}\right) + A_3\left(\frac{m_e}{m_\mu}, \frac{m_e}{m_\tau}\right)\,. \qquad (6)$$

The functions $A^{(i)}$ can be expanded in power series

$$A_i = A_i^{(2)}\left(\frac{\alpha}{\pi}\right) + A_i^{(4)}\left(\frac{\alpha}{\pi}\right)^2 + A_i^{(6)}\left(\frac{\alpha}{\pi}\right)^3 + A_i^{(8)}\left(\frac{\alpha}{\pi}\right)^4 + A_i^{(10)}\left(\frac{\alpha}{\pi}\right)^5 + \ldots. \qquad (7)$$

The mass-independent coefficients at 1, 2 and 3 loop are known in analytical form [5–7].

$$A_1^{(2)} = \frac{1}{2}; \qquad (8)$$

$$A_1^{(4)} = \frac{197}{144} + \frac{1}{12}\pi^2 - \frac{1}{2}\pi^2\ln 2 + \frac{3}{4}\zeta(3) \qquad (9)$$

$$= -0.328478965579\ldots, \qquad (10)$$

$$A_1^{(6)} = \frac{83}{72}\pi^2\zeta(3) - \frac{215}{24}\zeta(5) \qquad (11)$$

$$+\frac{100}{3}\left[\left(a_4 + \frac{1}{24}\ln^4 2\right) - \frac{1}{24}\pi^2\ln^2 2\right] - \frac{239}{2160}\pi^4 + \frac{139}{18}\zeta(3) - \frac{298}{9}\pi^2\ln 2 \qquad (12)$$

$$+\frac{17101}{810}\pi^2 + \frac{28259}{5184} = 1.181241456\ldots,$$

where $\zeta(n) = \sum_{i=0}^{\infty} i^{-n}$, $a_n = \sum_{i=0}^{\infty} 2^{-i} i^{-n}$. In table 1 we list some older theoretical evaluations of the two, three and four loop coefficients.

I have evaluated the 4-loop contribution $A_1^{(8)}$ up to 1100 digits of precision [8], finalizing a twenty-year effort [9–15] begun after the completion of the calculation

[a] e-mail: stefano.laporta@pd.infn.it

[b] Talk given at "Flavour changing and conserving processes" 2017 (FCCP2017), 7-9 September 2017, Villa Orlandi, Anacapri, Capri, Italy. The author thanks the organizers for financially supporting his participation in the workshop.

$A_1^{(4)}$	$A_1^{(6)}$
−2.973... [26]	1.49(20) [27]
−0.328478965579... [6]	1.195(26) [28]
	1.17611(42) [29]
	1.181259(40) [30]
	1.181241456587... [7]

$A_1^{(8)}$	$A_1^{(10)}$
−1.434(138) [20]	7.795(336) [25]
−1.5098(384) [31]	6.599(223) [25]
−1.7283(35) [21]	
−1.9144(35) [22]	
−1.9106(20) [24]	
−1.91298(84) [25]	
−1.912245764926... [8]	

Table 1. Numerical results of various evaluations of $A_1^{(4)}$, $A_1^{(6)}$, $A_1^{(8)}$ and $A_1^{(10)}$.

contribution	value in units of 10^{-12}
$A_1^{(2)}(\alpha/\pi)$	1 161 409 733.631(720)
$A_1^{(4)}(\alpha/\pi)^2$	− 1 772 305.065(3)
$A_1^{(6)}(\alpha/\pi)^3$	14 804.203
$A_1^{(8)}(\alpha/\pi)^4$	− 55.667
$A_1^{(10)}(\alpha/\pi)^5$	0.446(15)
$A_2^{(4)}(m_e/m_\mu)(\alpha/\pi)^2$	2.804
$A_2^{(6)}(m_e/m_\mu)(\alpha/\pi)^3$	− 0.092
$A_2^{(8)}(m_e/m_\mu)(\alpha/\pi)^4$	0.026
$A_2^{(10)}(m_e/m_\mu)(\alpha/\pi)^5$	− .0002
$A_2^{(4)}(m_e/m_\tau)(\alpha/\pi)^2$	0.010
$A_2^{(6)}(m_e/m_\tau)(\alpha/\pi)^3$	− 0.0008
a_e(hadronic v.p.)	1.866(11)
a_e(hadronic v.p.,NLO)	− 0.223(1)
a_e(hadronic v.p.,NNLO)	0.028(1)
a_e(hadronic l-l)	0.035(10)
a_e(weak)	0.0297(5)

Table 2. Contributions to a_e.

of $A_1^{(6)}$ [7]. The first digits of the result are

$$A_a^{(4)} = -1.912245764926445574152647167439830054\ldots \tag{13}$$

The full-precision result is shown in table 3. The result (13) is in excellent agreement (0.9σ) with the numerical value

$$A_1^{(8)}(\text{Ref. [25]}) = -1.91298(84)\,, \tag{14}$$

latest result of a really impressive pluridecennial effort [16–25], and with the independent value

$$A_1^{(8)}(\text{Ref. [32]}) = -1.87(12)\,. \tag{15}$$

There are 12672 diagrams which contribute at 5-loop level. They have been computed through a gigantic effort [25]; the most recent value is

$$A_1^{(10)}(\text{Ref. [25]}) = 6.599(223)\,. \tag{16}$$

```
-1.91224576492644557415264716743983005406087339065872534517132
98480060384439806517061427608927000036315837558415331473270056
37851491285453919028043270502738223043455789570455627293099412
96699760277782211578472033906415190816652709797086743811501215
51479722743221642734319279759586074050057837384960701874328314
02483802519224946074229855893046350614049225266343109442400023
56356881280620645494013224977594300429288836761748899236915180
87808698970526357853375377696411702453619601349757449436126848
61751626068323871867473038315059627418780153055148794005369777
98369464278684326918431175889581159743566950433048349073613426
58649953116387811743475385423488364085558444188223721745670687 1
04182330743051744305573945961171550858961148995261266061246994
07311840392747234002346496953173548258481799822409737371077365
74046451352112309124252811113721530215445372101481112115984897
08842232798797204842014451228284515165852365617865945926009917
33031721302865467212345340500349104700728924487200616044261325
44906900043191519823004748818149431103849537829940629675867875
38524978194698979313216219797575067670114290489796208505078...
```

Table 3. First 1100 digits of $A_1^{(8)}$.

We also note the MonteCarlo approach of Ref. [36] which might in future provide an independent value. Concerning the mass-dependent part $A_2(r)$, $A_2^{(4)}(r)$ is known in analytical form [37], as well as $A_2^{(6)}(r)$ [38–42]; the first terms of the expansion for small r of the 4-loop coefficient $A_2^{(8)}(r)$ are known analytically [43, 44] . $A_2^{(10)}(m_e/m_\mu)$ and $A_2^{(10)}(m_e/m_\tau)$ have been calculated numerically [25]; the first terms of the expansion for small mass ratios of $A_3^{(6)}(m_e/m_\mu, m_e/m_\tau)$ and $A_3^{(8)}(m_e/m_\mu, m_e/m_\tau)$ are known analytically [44]. The values of the hadronic and weak contributions [45–48] are shown in table 2, where the most precise determination of the fine structure constant [49, 50]

$$\alpha^{-1} = 137.035\,998\,996(85) \quad (0.62\ \text{ppb})$$

is used. Inserting all the contributions (see Eq.s(8-16) and table 2) into Eq.s(5-7) one finds

$$a_e^{\text{th}} = 1\,159\,652\,182.031(15)(15)(720) \times 10^{-12}\,, \tag{17}$$

where the first error comes from $A_1^{(10)}$, the second one from the hadronic and electroweak corrections, the last one from α. Conversely, assuming the validity of the theory and using the experimental measurement (4), one finds

$$\alpha^{-1}(a_e) = 137.035\,999\,1500(18)(18)(330) \quad (0.25\ \text{ppb})\,,$$

where the errors come respectively from $A_1^{(10)}$, hadronic and electroweak corrections, and the experimental measurement. This determination of α is more precise than the experimental one and is that used in the periodic least-squares adjustements of fundamental constants[48].

Gauge-invariant sets

The contributions to g-2 of single Feynman diagrams may be I.R. or U.V. divergent, and depends on the gauge chosen. Therefore, it is convenient to regroup the diagrams in gauge invariant sets, whose contributions are finite. The number of diagrams at 2, 3 and 4 loops are respectively 7, 72 and 891; they can be arranged in, respectively 3, 9 (see Ref. [51]) and 25 gauge invariant sets. Typical elements of the sets at four loops are shown in Fig.1; the values of the contributions of each gauge set are shown in Ref. [8] with 40 digits of precision.

https://doi.org/10.1051/epjconf/201817901010

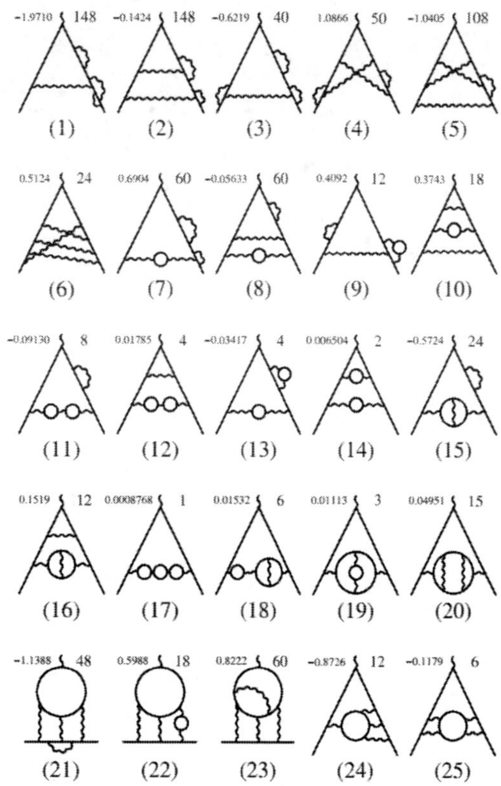

Figure 1. Typical representative diagrams of gauge-invariant sets. For each set only one diagrams is shown. Top left: contribution of the set to g-2; top right: number of diagrams of the set

The numerical fit

By using the integration-by-parts identities[9, 52, 53] the contributions of all 4-loop vertex diagrams can be reduced to a linear combination of 334 master integrals belonging to 220 topologies. Due to the expected analytical complexity of the result, a completely analytical calculation of all the master integrals (for example similar to that at three-loop) has seemed out of reach. So I have used a different approach: compute very high-precision numerical values of a sufficient number of terms of the expansions of the master integrals in Laurent series of $\epsilon = (4-D)/2$ dimensions; make the right analytical ansatz on the analytical form; fit the rational coefficients of the ansatz by using the PSLQ algorithm [54, 55].

As a result, one finds that the analytical expressions of the coefficients of the ϵ–expansions of the master integrals contain values of harmonic polylogarithms [57] with argument 1, $\frac{1}{2}$, $e^{\frac{i\pi}{3}}$, $e^{\frac{2i\pi}{3}}$ and $e^{\frac{i\pi}{2}}$, a family of one-dimensional integrals of products of elliptic integrals, and the finite terms of the ϵ–expansions of six master integrals. Work is still in progress to fit analytically these six unknown elliptical constants. The result of the analytical fit for $A_1^{(8)}$ can be written as follows:

$$A_1^{(8)} = T + \sqrt{3}V_a + V_b + W_b + \sqrt{3}E_a + E_b + U\,, \tag{18}$$

$$\begin{aligned}
T =& \frac{1243127611}{130636800} + \frac{30180451}{25920}\zeta(2) - \frac{255842141}{2721600}\zeta(3) - \frac{8873}{3}\zeta(2)\ln 2 \\
&+ \frac{6768227}{2160}\zeta(4) + \frac{19063}{360}\zeta(2)\ln^2 2 + \frac{12097}{90}\left(a_4 + \frac{1}{24}\ln^4 2\right) \\
&- \frac{2862857}{6480}\zeta(5) - \frac{12720907}{64800}\zeta(3)\zeta(2) - \frac{221581}{2160}\zeta(4)\ln 2 \\
&+ \frac{9656}{27}\left(a_5 + \frac{1}{12}\zeta(2)\ln^3 2 - \frac{1}{120}\ln^5 2\right) + \frac{191490607}{46656}\zeta(6) \\
&+ \frac{10358551}{43200}\zeta^2(3) - \frac{40136}{27}a_6 + \frac{26404}{27}b_6 - \frac{700706}{675}a_4\zeta(2) \\
&- \frac{26404}{27}a_5\ln 2 + \frac{26404}{27}\zeta(5)\ln 2 - \frac{63749}{50}\zeta(3)\zeta(2)\ln 2 \\
&- \frac{40723}{135}\zeta(4)\ln^2 2 + \frac{13202}{81}\zeta(3)\ln^3 2 - \frac{253201}{2700}\zeta(2)\ln^4 2 \\
&+ \frac{7657}{1620}\ln^6 2 + \frac{2895304273}{435456}\zeta(7) + \frac{670276309}{193536}\zeta(4)\zeta(3) \\
&+ \frac{85933}{63}a_4\zeta(3) - \frac{142793}{18}a_5\zeta(2) + \frac{7121162687}{967680}\zeta(5)\zeta(2) \\
&- \frac{195848}{21}a_7 + \frac{195848}{63}b_7 - \frac{116506}{189}d_7 - \frac{4136495}{384}\zeta(6)\ln 2 \\
&- \frac{1053568}{189}a_6\ln 2 + \frac{233012}{189}b_6\ln 2 + \frac{407771}{432}\zeta^2(3)\ln 2 \\
&- \frac{8937}{2}a_4\zeta(2)\ln 2 + \frac{833683}{3024}\zeta(5)\ln^2 2 - \frac{3995099}{6048}\zeta(3)\zeta(2)\ln^2 2 \\
&- \frac{233012}{189}a_5\ln^2 2 + \frac{1705273}{1512}\zeta(4)\ln^3 2 + \frac{602303}{4536}\zeta(3)\ln^4 2 \\
&- \frac{1650461}{11340}\zeta(2)\ln^5 2 + \frac{52177}{15876}\ln^7 2\,,
\end{aligned} \tag{19}$$

$$\begin{aligned}
V_a =& -\frac{14101}{480}\mathrm{Cl}_4\left(\frac{\pi}{3}\right) - \frac{169703}{1440}\zeta(2)\mathrm{Cl}_2\left(\frac{\pi}{3}\right) \\
&+ \frac{494}{27}\mathrm{Im}H_{0,0,0,1,-1,-1}\left(e^{i\frac{\pi}{3}}\right) + \frac{494}{27}\mathrm{Im}H_{0,0,0,1,-1,1}\left(e^{i\frac{2\pi}{3}}\right) \\
&+ \frac{494}{27}\mathrm{Im}H_{0,0,0,1,1,-1}\left(e^{i\frac{2\pi}{3}}\right) + 19\mathrm{Im}H_{0,0,1,0,1,1}\left(e^{i\frac{2\pi}{3}}\right) \\
&+ \frac{437}{12}\mathrm{Im}H_{0,0,0,1,1,1}\left(e^{i\frac{2\pi}{3}}\right) + \frac{29812}{297}\mathrm{Cl}_6\left(\frac{\pi}{3}\right) \\
&+ \frac{4940}{81}a_4\mathrm{Cl}_2\left(\frac{\pi}{3}\right) - \frac{520847}{69984}\zeta(5)\pi - \frac{129251}{81}\zeta(4)\mathrm{Cl}_2\left(\frac{\pi}{3}\right) \\
&- \frac{892}{15}\mathrm{Im}H_{0,1,1,-1}\left(e^{i\frac{2\pi}{3}}\right)\zeta(2) - \frac{1784}{45}\mathrm{Im}H_{0,1,1,-1}\left(e^{i\frac{\pi}{3}}\right)\zeta(2) \\
&+ \frac{1729}{54}\zeta(3)\mathrm{Im}H_{0,1,-1}\left(e^{i\frac{\pi}{3}}\right) + \frac{1729}{36}\zeta(3)\mathrm{Im}H_{0,1,1}\left(e^{i\frac{2\pi}{3}}\right) \\
&+ \frac{837190}{729}\mathrm{Cl}_4\left(\frac{\pi}{3}\right)\zeta(2) + \frac{25937}{4860}\zeta(3)\zeta(2)\pi - \frac{223}{243}\zeta(4)\pi\ln 2 \\
&+ \frac{892}{9}\mathrm{Im}H_{0,1,-1}\left(e^{i\frac{\pi}{3}}\right)\zeta(2)\ln 2 + \frac{446}{3}\mathrm{Im}H_{0,1,1}\left(e^{i\frac{2\pi}{3}}\right)\zeta(2)\ln 2 \\
&- \frac{7925}{81}\mathrm{Cl}_2\left(\frac{\pi}{3}\right)\zeta(2)\ln^2 2 + \frac{1235}{486}\mathrm{Cl}_2\left(\frac{\pi}{3}\right)\ln^4 2\,,
\end{aligned} \tag{20}$$

$$\begin{aligned}
V_b =& \frac{13487}{60}\mathrm{Re}H_{0,0,0,1,0,1}\left(e^{i\frac{\pi}{3}}\right) + \frac{13487}{60}\mathrm{Cl}_4\left(\frac{\pi}{3}\right)\mathrm{Cl}_2\left(\frac{\pi}{3}\right) \\
&+ \frac{136781}{360}\mathrm{Cl}_2^2\left(\frac{\pi}{3}\right)\zeta(2) + \frac{651}{4}\mathrm{Re}H_{0,0,0,1,0,1,-1}\left(e^{i\frac{\pi}{3}}\right) \\
&+ 651\mathrm{Re}H_{0,0,0,0,1,1,-1}\left(e^{i\frac{\pi}{3}}\right) - \frac{17577}{32}\mathrm{Re}H_{0,0,1,0,0,1,1}\left(e^{i\frac{2\pi}{3}}\right) \\
&- \frac{87885}{64}\mathrm{Re}H_{0,0,0,1,0,1,1}\left(e^{i\frac{2\pi}{3}}\right) - \frac{17577}{8}\mathrm{Re}H_{0,0,0,0,1,1,1}\left(e^{i\frac{2\pi}{3}}\right) \\
&+ \frac{651}{4}\mathrm{Cl}_4\left(\frac{\pi}{3}\right)\mathrm{Im}H_{0,1,-1}\left(e^{i\frac{\pi}{3}}\right) + \frac{1953}{8}\mathrm{Cl}_4\left(\frac{\pi}{3}\right)\mathrm{Im}H_{0,1,1}\left(e^{i\frac{2\pi}{3}}\right) \\
&+ \frac{31465}{176}\mathrm{Cl}_6\left(\frac{\pi}{3}\right)\pi + \frac{211}{4}\mathrm{Re}H_{0,1,0,1,-1}\left(e^{i\frac{\pi}{3}}\right)\zeta(2) \\
&+ \frac{211}{2}\mathrm{Re}H_{0,0,1,1,-1}\left(e^{i\frac{\pi}{3}}\right)\zeta(2) + \frac{1899}{16}\mathrm{Re}H_{0,1,0,1,1}\left(e^{i\frac{2\pi}{3}}\right)\zeta(2) \\
&+ \frac{1899}{8}\mathrm{Re}H_{0,0,1,1,1}\left(e^{i\frac{2\pi}{3}}\right)\zeta(2) + \frac{211}{4}\mathrm{Im}H_{0,1,-1}\left(e^{i\frac{\pi}{3}}\right)\mathrm{Cl}_2\left(\frac{\pi}{3}\right)\zeta(2) \\
&+ \frac{633}{8}\mathrm{Im}H_{0,1,1}\left(e^{i\frac{2\pi}{3}}\right)\mathrm{Cl}_2\left(\frac{\pi}{3}\right)\zeta(2)\,,
\end{aligned} \tag{21}$$

$$\begin{aligned}
W_b =& \zeta(2)\Bigg(-\frac{28276}{25}\mathrm{Cl}_2\left(\frac{\pi}{2}\right)^2 + 104\Bigg(4\mathrm{Re}H_{0,1,0,1,1}(i) \\
&+ 4\mathrm{Im}H_{0,1,1}(i)\mathrm{Cl}_2\left(\frac{\pi}{2}\right) - 2\mathrm{Cl}_4\left(\frac{\pi}{2}\right)\pi + \mathrm{Cl}_2^2\left(\frac{\pi}{2}\right)\ln 2\Bigg)\Bigg)\,,
\end{aligned} \tag{22}$$

$$E_a = \pi\left(-\frac{28458503}{691200}B_3 + \frac{250077961}{18662400}C_3\right) + \frac{483913}{77760}\pi f_2(0,0,1)$$
$$+\pi\Big(\frac{4715}{1944}\ln 2\; f_2(0,0,1) + \frac{270433}{10935}f_2(0,2,0) - \frac{188147}{4860}f_2(0,1,1)$$
$$+\frac{188147}{12960}f_2(0,0,2)\Big) + \pi\Big(\frac{826595}{248832}\zeta(2)f_2(0,0,1) \tag{23}$$
$$-\frac{5525}{432}\ln 2\; f_2(0,0,2) + \frac{5525}{162}\ln 2\; f_2(0,1,1)$$
$$-\frac{5525}{243}\ln 2\; f_2(0,2,0) + \frac{526015}{248832}f_2(0,0,3) - \frac{4675}{768}f_2(0,1,2)$$
$$+\frac{1805965}{248832}f_2(0,2,1) - \frac{3710675}{1119744}f_2(0,3,0) - \frac{75145}{124416}f_2(1,0,2)$$
$$-\frac{213635}{124416}f_2(1,1,1) + \frac{168455}{62208}f_2(1,2,0) + \frac{69245}{124416}f_2(2,1,0)\Big),$$

$$E_b = \zeta(2)\Big(\frac{2541575}{82944}f_1(0,0,2) - \frac{556445}{6912}f_1(0,1,1) + \frac{54515}{972}f_1(0,2,0)$$
$$-\frac{75145}{20736}f_1(1,0,1)\Big) - \frac{4715}{1458}\zeta(2)f_1(0,0,1),$$

$$U = -\frac{541}{300}C_{81a} - \frac{629}{60}C_{81b} + \frac{49}{3}C_{81c} - \frac{327}{160}C_{83a} + \frac{49}{36}C_{83b} + \frac{37}{6}C_{83c}. \tag{24}$$

C_{8xy} are the ϵ^0 coefficients of the ϵ−expansion of six master integrals (see f, f', f'', g, g', g'' of Fig.2); they appear only in the gauge-invariant sets 24 and 25. In the above expressions $b_6 = H_{0,0,0,0,1,1}\left(\frac{1}{2}\right)$, $b_7 = H_{0,0,0,0,0,1,1}\left(\frac{1}{2}\right)$, $d_7 = H_{0,0,0,0,1,-1,-1}(1)$, $\mathrm{Cl}_n(\theta) = \mathrm{Im}\mathrm{Li}_n(e^{i\theta})$. $H_{i_1,i_2,...}(x)$ are the harmonic polylogarithms. The integrals f_j are defined as follows:

$$f_m(i,j,k) = \int_1^9 ds\, D_1(s)\mathrm{Re}\left(\sqrt{3^{m-1}}D_m(s)\right)\times \tag{25}$$

$$\left(s - \frac{9}{5}\right)\ln^i(9-s)\ln^j(s-1)\ln^k(s),$$

$$D_m(s) = \frac{2}{\sqrt{(\sqrt{s}+3)(\sqrt{s}-1)^3}}\times \tag{26}$$

$$K\left(m - 1 - (2m-3)\frac{(\sqrt{s}-3)(\sqrt{s}+1)^3}{(\sqrt{s}+3)(\sqrt{s}-1)^3}\right);$$

$K(x)$ is the complete elliptic integral of the first kind. The constants B_3 and C_3, defined in Ref. [14], admit the hypergeometric representations:

$$B_3 = \frac{4\pi^{\frac{3}{2}}}{3}\left(\frac{\Gamma^2(\frac{7}{6})\Gamma(\frac{1}{3})}{\Gamma^2(\frac{2}{3})\Gamma(\frac{5}{6})}\,{}_4F_3\left({\frac{1}{6}\ \frac{1}{3}\ \frac{1}{3}\ \frac{1}{2} \atop \frac{5}{6}\ \frac{5}{6}\ \frac{2}{3}};1\right)\right. \tag{27}$$

$$\left.+\frac{\Gamma^2(\frac{5}{6})\Gamma(-\frac{1}{3})}{\Gamma^2(\frac{1}{3})\Gamma(\frac{1}{6})}\,{}_4F_3\left({\frac{1}{2}\ \frac{2}{3}\ \frac{2}{3}\ \frac{5}{6} \atop \frac{7}{6}\ \frac{7}{6}\ \frac{4}{3}};1\right)\right), \tag{28}$$

$$C_3 = \frac{486\pi^2}{1925}\,{}_7F_6\left({\frac{7}{4}\ -\frac{1}{3}\ \frac{1}{3}\ \frac{2}{3}\ \frac{4}{3}\ \frac{3}{2}\ \frac{3}{2} \atop \frac{3}{4}\ 1\ \frac{7}{6}\ \frac{11}{6}\ \frac{13}{6}\ \frac{17}{6}};1\right), \tag{29}$$

Fig.2 shows the fundamental elliptic master integrals which contain irreducible combinations of B_3, C_3 and $f_m(i,j,k)$.

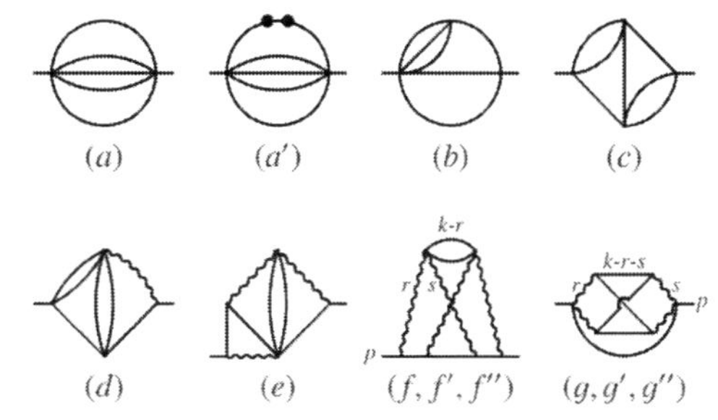

Figure 2. Minimal set of master integrals which contain all the elliptic constants. The double dot in (a') means that denominator is raised to the power three. (f, f', f'') and (g, g', g'') have numerators respectively equal to $(1, p.k, (p.k)^2)$.

More in detail, the ϵ-expansions of the master integrals were first calculated with a precision of 200-300 digits, then recalculated with a precision of 4096 bit (1232 digits). A first run of PSLQ on the coefficients with a trial basis identified which coefficients are linearly independent (that is, which contain new analytical terms). For each of these "unknown constants", the simplest master integral which contains it was identified; these special master integral were recalculated with a precision of 8,16 or 32 kbits (2464, 4932 or 9864 digits). The structure the master integrals was analysed to guess the analytical form of the new elements; several runs of PSLQ with basis containing (all or selections of) these new elements were performed until the PSLQ fit was successful and the correct basis was found.

To give an example, the PSLQ basis needed to fit T of Eq.19 contains $F_{10} - 1 = 54$ terms (F_i are the Fibonacci numbers, $F_1 = F_2 = 1$). Due to the presence of harmonic polylogarithms of complex argument, the general basis needed to perform the fit of V_a and V_b contains $F_{17}-1 = 1596$ terms. The practical maximum size of basis usable in PSLQ fits turned out to be ~ 500, so some guessing has been necessary to identify which elements are really needed in the fits (for example, only 93 elements for V_a and V_b). The precision of the calculation of elements of the basis is up to 9864 digits. We note that the multi-pair parallel version [55] of the PSLQ algorithm has been essential to work out these difficult analytical fits in reasonable times.

Method of calculation

All the 104 4-loop self-mass diagrams are generated with a C program; for each self-mass diagram, the corresponding vertex diagrams are generated by inserting a photon in all possible ways. For the sake of subsequent checks, one keeps also the diagrams which give no contribution because of Furry's theorem. The contribution to g-2 of the diagram is extracted from the unrenormalized amplitude of each vertex diagram using suitable projectors, with a FORM[56] program. The results are expressions which typically contain 100-30000 different Feynman integrals. The total number of topologically distinct master integrals is 334. There are 82 topologies which have more than one master integral. For these topologies, products of scalar products are chosen as numerators, the choice depending on the momentum flow of the diagram. For the sake of subsequent checks, one processes *separately* the contributions from each one of the 104 self-mass diagrams. The master integrals are calculated separately for each self-mass diagram; therefore the total number of the master integrals actually calculated is 4607. Because of differences in momentum routing and therefore in the numerators, topologies with more than one master integral will have a slightly different set of master integrals in different self-mass diagrams. This provides a *lot of* useful checks. The reduction is exact in D, and is performed by generating large systems of integration-by-parts identities and solving them with the algorithm [9] implemented in SYS. These large systems contain $4 \cdot 10^6 - 50 \cdot 10^6$ identities, with physical disk space ranging from 4GB to 100GB. The reduction to master integrals has been repeated with 32 and 64 bit versions of SYS. In addition, the principal self-mass diagrams were reprocessed using a different momentum flow, checking that reduction to master integrals remained the same (after converting different sets of master integrals).

For the numerical calculation of master integrals I have used combinations of the the difference equation method [9, 10] and the differential equation method [59–61]. An approach consists in calculating with difference equations (inserting the exponent n of the first electron propagator) the integral obtained putting the photon mass λ equal to the electron mass m, and integrating a differential equation in λ from $\lambda = m$ to $\lambda = 0$. A second approach consists in calculating with difference equations the diagram obtained by putting the the external momentum $p = 0$, and integrating a differential equation in p^2 from $p^2 = 0$ to $p^2 = -m^2$. The systems of difference and differential equations for master integrals are obtained by building systems of integration-by-parts identities and solving them by using the algorithm [9, 10] using rational arithmetic in D. The sizes of the systems of difference or differential equations to be numerically solved are in the range 1MB-3GB. Difference equations are solved using the Laplace transformation method (integral representation of solutions and differential equation of the integrand). Differential equations are solved numerically, by using series expansion with truncated expansions in $\epsilon = (4 - D)/2$ as coefficients. The minimum number of terms of the expansion in ϵ is 9 ($\epsilon^{-4} \ldots\ \epsilon^4$). There are cancellations of ϵ terms in intermediate steps; no care is used to avoid cancellations, as the corresponding numerical zeroes provide extremely useful checks. In the worst case 37 terms of expansions are needed. The standard precision of calculations is 4 kbit (1232 digits). About 130 digits are lost due to cancellations, so that the result has 1100 digits exact. Note that the analytical fit of some selected master integrals required a precision much higher, up to 16 kbit (9864 digits).

The renormalization counterterms are generated with two procedures written in *C* and FORM. Their expressions are reduced to master integrals already known in analytical form. I have chosen the Feynman gauge. I checked explicitly the (internal) gauge invariance of the contributions for arbitrary gauge of the photon line going into 1-,2- or 3-loop vacuum polarization diagrams.

In order to perform this calculation, in 1995 I begun writing from scratch a comprehensive *C* program, SYS [9], now 24000 lines long, containing all the parts necessary to the calculations, like a simplified fast algebraic symbolic manipulator, a numerical solver of systems of difference and differential equations, and a library of arbitrary precision mathematical routines, for integer and floating point numbers, independent of any other existing in the literature.. Another important feature is the thorough protection of large buffers and I/O with checksums. Throughout the calculation, I have discovered several subtle corruptions, due to different causes, like marginal coupling of non-ECC RAM modules (error-rate: 1 bit per week), failing RAID systems (error-rate: one corrupted 64KB block every 100GB of data), etc...). The prompt identification and elimination of the source of errors has resulted in an highly reliable result.

References

[1] P. A. M. Dirac, Proc. Roy. Soc. Lond. A **117** 610 (1928).
[2] R. S. Van Dyck, P. B. Schwinberg and H. G. Dehmelt, Phys. Rev. Lett. **59** 26 (1987).
[3] D. Hanneke, S. Fogwell and G. Gabrielse, Phys. Rev. Lett. **100** 120801 (2008).
[4] D. Hanneke, S. F. Hoogerheide and G. Gabrielse, Phys. Rev. A **83** 052122 (2011).
[5] J. Schwinger, Phys. Rev. **73** 416 (1948).
[6] A. Petermann, *Helv. Phys. Acta* **30** 407 (1957); C.M. Sommerfield, *Phys. Rev.* **107** 328 (1957).
[7] S. Laporta and E. Remiddi, Phys. Lett. B **379** 283 (1996).
[8] S. Laporta, Phys. Lett. B **772** 232 (2017).
[9] S. Laporta, Int. J. Mod. Phys. A **15** 5087 (2000).
[10] S. Laporta, Phys. Lett. B **504** 188 (2001).
[11] S. Laporta, Phys. Lett. B **523** 95 (2001).
[12] S. Laporta, Acta Phys. Polon. B **34** 5323 (2003).
[13] S. Laporta, P. Mastrolia and E. Remiddi, Nucl. Phys. B **688** 165 (2004).
[14] S. Laporta, Int. J. Mod. Phys. A **23** 5007 (2008).
[15] S. Laporta, Subnucl. Ser. **45** 409 (2009).
[16] T. Kinoshita and W. B. Lindquist, Phys. Rev. D **27** 867 (1983).
[17] T. Kinoshita and W. B. Lindquist, Phys. Rev. D **27** 877 (1983).
[18] T. Kinoshita and W. B. Lindquist, Phys. Rev. D **27** 886 (1983).
[19] T. Kinoshita and W. B. Lindquist, Phys. Rev. D **39** 2407 (1989).

[20] T. Kinoshita and W. B. Lindquist, Phys. Rev. D **42** 636 (1990).
[21] T. Kinoshita and M. Nio, Phys. Rev. D **73** 013003 (2006).
[22] T. Aoyama, M. Hayakawa, T. Kinoshita and M. Nio, Phys. Rev. Lett. **99** 110406 (2007).
[23] T. Aoyama, M. Hayakawa, T. Kinoshita and M. Nio, Phys. Rev. D **77**, 053012 (2008).
[24] T. Aoyama, M. Hayakawa, T. Kinoshita and M. Nio, Phys. Rev. Lett. **109** 111807 (2012).
[25] T. Aoyama, M. Hayakawa, T. Kinoshita and M. Nio, Phys. Rev. D **91** 033006 (2015) ; Erratum: Phys. Rev. D **96** 019901 (2017).
[26] R. Karplus and N. M. Kroll, Phys. Rev. **77** 536 (1950).
[27] M. J. Levine and J. Wright, Phys. Rev. Lett. **26** 1351 (1971).
[28] P. Cvitanovic and T. Kinoshita, Phys. Rev. D **10** 4007 (1974).
[29] T. Kinoshita, in *Quantum Electrodynamics*, edited by T. Kinoshita (World Scientific, Singapore,1990), pp. 218-321.
[30] T. Kinoshita, Phys. Rev. Lett. **75** 4728 (1995).
[31] V. W. Hughes and T. Kinoshita, Rev. Mod. Phys. **71**, S133 (1999).
[32] P. Marquard, A. V. Smirnov, V. A. Smirnov, M. Steinhauser and D. Wellmann, arXiv:1708.07138 [hep-ph]
[33] T. Kinoshita and M. Nio, Phys. Rev. D **73** (2006) 053007
[34] T. Kinoshita, T. Aoyama, M. Hayakawa and M. Nio, Nucl. Phys. Proc. Suppl. **160** (2006) 235.
[35] M. Nio, T. Aoyama, M. Hayakawa and T. Kinoshita, Nucl. Phys. Proc. Suppl. **169** (2007) 238.
[36] S. A. Volkov, J. Exp. Theor. Phys. 122 1008 (2016); Zh. Eksp. Teor. Fiz. 149 1164 (2016).
[37] H. H. Elend, Phys. Lett. **20** 682 (1966).
[38] M. A. Samuel and G. w. Li, Phys. Rev. D **44** 3935 (1991) ; Erratum: Phys. Rev. D **48** 1879 (1993).
[39] G. Li, R. Mendel and M. A. Samuel, Phys. Rev. D **47** 1723 (1993).
[40] S. Laporta and E. Remiddi, Phys. Lett. B **301** 440 (1993).
[41] S. Laporta, Nuovo Cim. A **106** 675 (1993).
[42] M. Passera, Phys. Rev. D **75** 013002 (2007).
[43] A. L. Kataev, Phys. Rev. D **86** 013010 (2012).
[44] A. Kurz, T. Liu, P. Marquard and M. Steinhauser, Nucl. Phys. B **879** 1 (2014).
[45] D. Nomura and T. Teubner, Nucl. Phys. B **867** 236 (2013).
[46] A. Kurz, T. Liu, P. Marquard and M. Steinhauser, Phys. Lett. B **734** 144 (2014).
[47] J. Prades, E. de Rafael, and A Vainshtein, in *Lepton Dipole Moments*, edited by B. L. Roberts and W. J. Marciano (World Scientific, Singapore, 2009), 303.
[48] P. J. Mohr, B. N. Taylor and D. B. Newell, Rev. Mod. Phys. **84** 1527 (2012).
[49] R. Bouchendira, P. Clade, S. Guellati-Khelifa, F. Nez and F. Biraben, Phys. Rev. Lett. **106** 080801 (2011).
[50] P. J. Mohr, D. B. Newell and B. N. Taylor, Rev. Mod. Phys. **88** 035009 (2016).
[51] P. Cvitanovic, Nucl. Phys. B **127** 176 (1977).
[52] K. G. Chetyrkin and F. V. Tkachov, Nucl. Phys. B **192** (1981) 159.
[53] F. V. Tkachov, Phys. Lett. **100B** (1981) 65.
[54] H. R. P. Ferguson and D. H. Bailey, RNR Technical Report RNR-91-032.
[55] D. H. Bailey and D. J. Broadhurst, Math. Comput. **70** 1719 (2001).
[56] J. A. M. Vermaseren, math-ph/0010025
[57] E. Remiddi and J. A. M. Vermaseren, *Int. J. Mod. Phys.* **A15** 725 (2000).
[58] S. Laporta and E. Remiddi, Nucl. Phys. B **704** 349 (2005).
[59] A. V. Kotikov, Phys. Lett. B **254** 158 (1991).
[60] E. Remiddi, Nuovo Cim. A **110** 1435 (1997).
[61] T. Gehrmann and E. Remiddi, Nucl. Phys. B **580** 485 (2000).

Status of the COMET experiment

R. P. Litchfield[1,a]

[1] Imperial College, London
On behalf of the COMET collaboration

Abstract. The COMET experiment, currently under construction at J-PARC, is designed to search for the the neutrinoless flavour-changing charged-lepton process $\mu^- + A \to e^- + A$, which is a promising channel to look for new physics. The experiment will make use of the high-power proton beam available at the J-PARC Main Ring to achieve a sensitivity orders of magnitude better than the current limits. The experiment will be staged, with 'Phase-I' scheduled to begin commissioning soon while the 'Phase-II' design is refined in parallel. A description the design of both Phases is presented, along with the latest developments.

1 Muon to electron conversion

Lepton flavour is conserved in the Standard Model without massive neutrinos, but it is not protected by any symmetry and is now known to be violated by non-zero neutrino masses with mixing. Neutrino mixing also gives rise to the charged lepton conversion process, but the rate of such processes can be calculated and is hugely suppressed, down to the level of 1 in 10^{54}, because the neutrino masses are very much smaller than the W boson mass. Thus the charged-lepton channel is an ideal place to look for new physics, as there is no known reason for it to be forbidden, it is only 'accidentally' (at least as far as we know) not present in the Standard Model. Physics scenarios enhancing leptonic flavour violation processes are a central theme of this workshop, so details are not repeated here. Suffice to say that the process is predicted in a wide range of new physics models, unless parameters are tuned specifically to suppress it.

The COMET (COherent Muon to Electron Transition) experiment is currently under construction at J-PARC in Tōkai, Japan. [1] It will search for evidence of the conversion of muons to electrons without the production of neutrinos, $\mu^- + A \to e^- + A$, a process that conserves total lepton number, but not lepton flavour.

As noted in the name, COMET will look specifically for the coherent conversion process: Muons that are stopped and bound in the $1s_\mu$ orbit of an atomic nucleus convert to electrons, and the 4-momentum transfer to the nucleus is no more than required to conserve energy and momentum. The electron is therefore mono-energetic at around 105 MeV, which is the muon rest mass minus the binding and nuclear recoil energies.

[a] e-mail: p.litchfield@ic.ac.uk

2 Experimental concept

The experimental signature that COMET will use is the single mono-energetic electron produced by conversion of stopped muons. The clean sample of stopped muons is provided by a delayed window technique, making use of the relatively long lifetime of bound muons to 'wait out' all the other activity associated with providing an intense muon source, as illustrated in Fig. 1. The delayed window technique requires that the primary beam for the muon source is pulsed, with a relatively large time gap between pulses. The lifetime of stopped muons decreases with increasing atomic number, as the muon is bound more tightly to the nucleus. So although the rate of the process is enhanced for heavier nuclei, the choice of stopping target is mostly driven by requiring a good match to the size of the gap between pulses provided by the accelerator. The J-PARC Main Ring will provide proton bunches at 1.2 µs intervals, and this makes aluminium the preferred stopping target, although titanium is also a good choice and could be used at a later date.

The ultimate goal of COMET is to discover (or constrain) the μ–e process with a sensitivity of four orders of magnitude better than the current (90% C.L.) limit of 7×10^{-13}, from the SINDRUM-II experiment [2]. Although there is no intrinsic background, this requires excellent control over accidental backgrounds that may give rise to electrons near the signal energy. It also requires an intense muon source: even with perfect efficiency, an experiment aiming for a 10^{-17} sensitivity would need to use 10^{17} muons. Since this is a new approach that provides a marked improvement in sensitvity, the experiment will be staged into two Phases, with Phase-I intended to provide input for the detailed design of Phase-II, as well as producing an intermediate result in a timely fashion.

Because of the delayed window, prompt backgrounds induced by the beam should be relatively easy to control,

 https://doi.org/10.1051/epjconf/201817901011

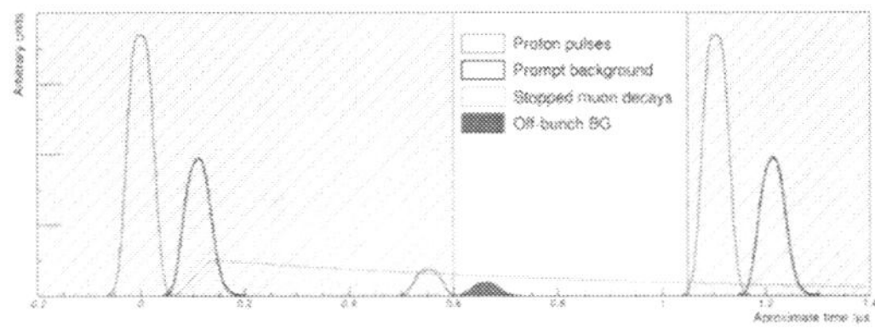

Figure 1. Cartoon of the delayed window technique. The prompt backgrounds (blue) should all occur a few hundred nanoseconds after the primary beam time, while the signal occurs according to the lifetime of the muonic atoms, which is matched to the bunch frequency. An important aspect is whether there are non-negligible off-time bunches (grey fill). In reality, such a background is not expected to be a problem, as described in the next section.

but provide the main constraint on the how soon the delayed window can be used after the beam pulse[1]. More problematic are backgrounds originating from the long-lived muons. The most obvious is from decay-in-orbit, $\mu^- + A \rightarrow e^- + \nu + \bar{\nu} + A$, which is the fate of around 39% of muons bound on aluminium. This process is very similar to free muon decay, but since the leptonic system can recoil against the heavy nucleus is is possible for the the decay energy to be carried almost entirely by the electron. The theoretical maximum energy is therefore very close to that of the mono-energetic line of the conversion signal. This is heavily suppressed, and the vast majority of electrons have much lower energies, but the fraction is large enough to makes it necessary for the experiment to have good (100 keV-scale) momentum resolution. In COMET this will be provided by using large magnetic fields and low-mass tracking detectors.

Another muon-related background is from decay-in-flight. If the muon has momentum of at least 77 MeV/c, it is possible, depending on the decay angle, for the electron to end up in the 105 MeV signal region. Thus while collection of lower energy muons is a priority, it is also important to avoid allowing higher-energy muons into the vicinity of the stopping target. This drives the design of the muon collection system.

The three key design points of the experiment are then:

- An intense *pulsed* muon source, emphasising less-energetic muons that will stop in a thin aluminium target.
- A muon beamline that can efficiently collect and transport low-energy negative muons (and their parent pions) from the point of production to the stopping target, but filter out high-energy and positive muons and pions.
- Detectors that can make accurate momentum measurements of 105 MeV electrons originating in the stopping target.

[1] A very cautious window—starting later than naively expected—should be used because of the possibility of stragglers, and to allow for the detector readout to go quiet.

3 COMET Design

The protons of the COMET primary beam will be resonant slow extracted from the J-PARC Main Ring at a beam momentum of 8 GeV/c. In order to get a longer delayed window between bunches only four of the nine accelerator buckets will filled in each cycle. In order to eliminate 'off-pulse' prompt backgrounds the ratio of protons in 'empty' to filled buckets should be no higher than 10^{-9}. Initial tests indicate that levels of 10^{-11} and lower are already achievable. The experiment will monitor the primary beam using highly radiation-tolerant diamond-based detectors.

The beam is directed towards a graphite production target that sits in a 5 T solenoid field. Secondary particles produced at the target travel in helical paths in the solenoid field, and are thus captured in the secondary beamline. A peculiarity of the set-up it that the solenoid system is designed to collect pions emerging in the *backward* direction, as forward going secondaries are more energetic (so will not produce muons that can be stopped) and contain a larger fraction of other undesirable hadrons, such as anti-protons.

The field of this *Capture* Solenoid decreases smoothly from 5 T at the target position down to 3 T at the rear section. As secondaries travel through this gradient field, their helix angle decreases, effectively increasing the angular acceptance of the secondary beamline around the target.

The lower field at the exit of the capture solenoid is matched to a *Transport* solenoid which serves several purposes: The first and simplest is that it provides propagation space in which the pions can decay to produce the muons required by the experiment. The transport solenoid is also curved, which results in the helical trajectories drifting either up or down depending on the charge and momentum of the particle. Since the aperture of the transport solenoid is finite, this acts as a filter of high-momentum and wrong-sign particles, an effect which is enhanced by including a collimator at the end of the transport section. A dipole field is applied and tuned so that pions and muons with appropriate momenta (between about 20 and 50 MeV/c) are transmitted with high efficiency, and others are removed.

The subsequent sections of the experiment differ significantly between Phase-I and Phase-II, so are described separately. Both Phases are illustrated and compared in Fig. 2.

3.1 Phase-I

In Phase-I, the transport solenoid is around 4 m long and bends through 90°, with a collimator (again inside a solenoid) at the end, to provide a narrower band of selected momenta. After the collimator solenoid sits a much larger (1.07 m radius) *Detector* solenoid, inside which is the aluminium stopping target, which consists of a number of thin aluminium foils on solenoid axis. In Phase-I the outer part of the detector solenoid will be instrumented with a Cylindrical Drift Chamber (CDC). The drift chamber uses an all-stereo-wire design with 18 layers of sense wires. The chamber was completed in 2016 and is currently (at time

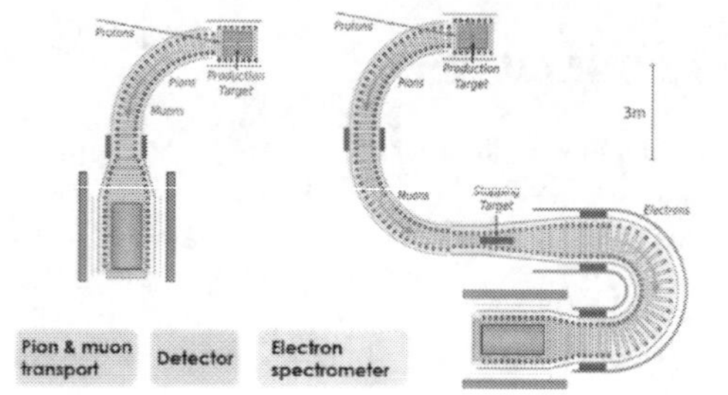

Figure 2. Plan view of Phase-I [left] and Phase-II [right] of COMET. The Capture section and first 90° are common to both Phases.

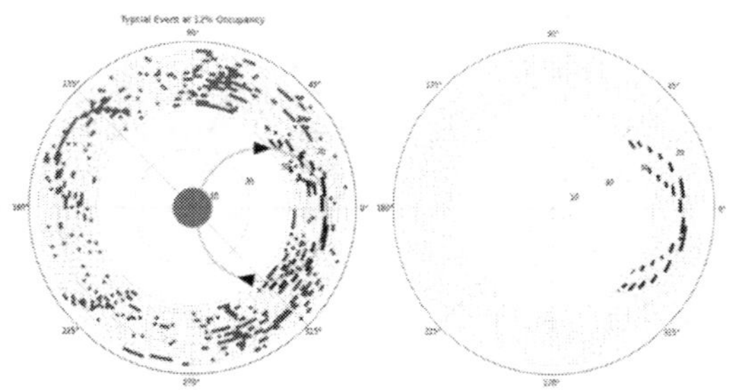

Figure 3. Reconstruction of simulated tracks. In the first step [left] hits are classified as signal-like (blue) or background-like (red). In the second step [right], a Hough transform based on the signal-like hits is used to select hits belonging to a track, which can be reconstructed in 3D using the two stereo projections.

of the workshop) taking cosmic ray data using an external scintillator trigger at KEK. The fully installed detector will trigger from scintillator and Cherenkov hodoscope rings place near the upstream and downstream ends of the CDC.

Reconstruction of events recorded by the CDC is under development using Monte-Carlo simulations. Even with the filtering afforded by the transport solenoid, the number of particles passing through the centre of the Detector Solenoid would make it impossible to operate a drift chamber at small radius. The 50 cm minimum radius of the CDC means that it escapes this direct beam flash, and the majority of decay-in-orbit electrons. But there is still expected to be a lot of activity causing many hits unrelated to the triggering track. Reconstruction first focusses on identifying noise hits based on local features, then finds tracks hits based on the global pattern of the signal-like hits. These tracks are then fit to obtain the transverse and longitudinal components of the track momentum (using the two stereo projections) and indicate whether the tracked particle originated in the target foils. Figure 3 illustrates the high detector occupancy, and how effective this this two-step track finding algorithm is.

3.2 Phase-II

In Phase-II, the transport solenoid will be extended with a second 90° section to provide better momentum filtering, which will be necessary at higher primary beam power. The stopping target will sit in a straight solenoid section, and downstream of this there will be a second (180°) set of curved *Spectrometer* Solenoids. This second set of curved solenoids will have a much larger inner radius and are designed to collect electrons near 105 MeV/c. In this way they act a filter for signal electrons, reducing backgrounds and enabling the use of higher beam intensities. The COMET software framework is designed to be used by both phases, so detailed simulation of Phase-II is ongoing in parallel with preparations for Phase-I. This commonality also allows for rapid inclusion of tuned simulations from Phase-I into the Phase-II design optimisation.

Because the majority of decay-in-orbit electrons are removed by this curved electron spectrometer, the detector solenoid (which is reused and sits after the spectrometer) can be instrumented with full-plane detectors. The baseline design uses several planes of straw tubes as a tracker to reconstruct the helical trajectory, followed by a LYSO-crystal EM calorimeter. The design is being evaluated in Phase-I with a partial plane of crystals and 5 tracking stations, each containing of 4 planes of straws in an YYXX configuration. During Phase-I, this detector can be exchanged with the CDC, and will be used for evaluation of the detector technology and to characterise the secondary beam that reaches the detector solenoid. This can be used to check the simulations for both Phase-I and Phase-II, and to optimise collimator designs for the Phase-II spectrometer.

A small ($\sim 10 \times 10\,\mathrm{cm}^2$) prototype using straws arranged as a single station of the Straw Tube Tracker and an 8×8 crystal section of the Calorimeter was taken to a electron test-beam at ELPH in Sendai in March 2017 (see Fig. 4). This was a fully integrated test of the system, and included the front-end readout cards for both detectors and a DAQ using the same MIDAS framework planned for the real experiment. The test was very successful, meeting all of the planned goals with spare time to try out some extra configurations and trial the trigger system. Analysis of the data is currently (as of Sept '17) ongoing.

4 Sensitivity projections and timeline

The quantity measured by a μ–e conversion experiment is usually expressed as ratio

$$R = \frac{N(\mu A \to eA)}{N(\mu A \to \nu A')} \tag{1}$$

which is to say the relative rates of the conversion process and the (flavour-conserving) weak Nuclear Muon Capture (NMC) process. Because of the short range of the weak interaction, the rate of NMC depends on how tightly the muon is bound to the nucleus. Thus as well as the lifetime of the bound muon, the relative rates of NMC and decay-in-orbit depend on the nucleus. For ^{27}Al the NMC

Figure 4. The test-beam prototype Phase-II detector, during the run at ELPH. The Straw Tube Tracker and Calorimeter pieces are indicated in green and red respectively.

fraction (f) is 61%. The ratio R is easier to calculate theoretically than the fraction of muons converting to all those captured, and it makes it easier to compare measurements using different nuclei.

Experimental sensitivity is expressed using the a quantity called Single Event Sensitivity (SES). This is the value of R such that the number of events seen by an experiment has an expectation value of one, and is essentially just the reciprocal of the number of events that would be expected if $R = 1$. This can therefore be expressed using the total number of stopped muons produced and the various efficiencies:

$$\mathrm{SES} = \frac{1}{N_\mu \cdot g \cdot f \cdot A}, \qquad (2)$$

where $g = 0.9$ is the probability for the nucleus to remain in the ground state after the conversion. There are two experiment-dependent quantities: N_μ, the number of stopped muons; and A the experimental acceptance. In Phase-I, the latter is dominated by the geometric acceptance of the detector around the target (18%), and time acceptance of the delayed window (30%). Overall, for COMET Phase-I, with a selection tuned for a background expectation of 0.03 events, projections from simulations give $A = 4.1\%$ and $N_\mu = 1.5 \times 10^{16}$, assuming a goal of 5 months of running. This gives a single event sensitivity of 3×10^{-15}. For COMET Phase-II, which is designed to run at a primary beam power of 56 kW (c.f. 3.2 kW for Phase-I) with better acceptance, the projected single event sensitivity is 2.6×10^{-17} after one year of running.

The schedule for operation of Phase-I is dependent on progress of facility construction, which (since J-PARC is a running facility) must take into account operation of the J-PARC accelerators for existing experiments. The most recent assessment projects that commissioning of the Phase-I configuration could begin as early as the end of Japanese Financial Year (April~March) 2018, with physics data-taking primarily in JFY 2019 and 2020. Fabrication of the extended solenoids and final development of detectors for Phase-II would precede in parallel, so that it could commence relatively soon after the completion of Phase-I.

Acknowledgements

On behalf of the COMET collaboration, I thank KEK and J-PARC, Japan for hosting the COMET experiment and providing the necessary infrastructure.

Additionally, the work is supported in part by: Japan Society of the Promotion of Science (JSPS) KAKENHI Grant No. 25000004; JSPS KAKENHI Grant No.JP17H06135; National Natural Science Foundation of China (NSFC) under Contracts No. 11335009 and 11475208; Research program of the Institute of High Energy Physics (IHEP) under Contract No. Y3545111U2; the State Key Laboratory of Particle Detection and Electronics of IHEP, China, under Contract No.H929420BTD; IN2P3-CNRS, France; Shota Rustaveli National Science Fund (SRNSF), Georgia; Institute for Basic Science of the Republic of Korea, Project No. IBS-R017-D1-2017-a00; Russian Foundation for Basic Research Grant No. RFBR-17-02-01073; Science and Technology Facilities Council, United Kingdom; JSPS London Short Term Predoctoral Fellowship program; Daiwa Anglo-Japanese Foundation Small Grant; and Royal Society International Joint Projects Grant. Crucial computing support from all partners is gratefully acknowledged, in particular from CC-IN2P3, France.

References

[1] COMET Review Document (TDR), and other documents available at `https://comet.kek.jp`

[2] W. H. Bertl *et al.* [SINDRUM II Collaboration], Eur. Phys. J. C **47**, 337 (2006).

Measuring the leading hadronic contribution to the muon g-2 via the μ-e elastic scattering

Umberto Marconi[1,a] and Fulvio Piccinini[2,b]

[1] INFN Sezione di Bologna, Via Irnerio, 46 - 40126 Bologna (BO) Italy.
[2] INFN Sezione di Pavia, Via Agostino Bassi, 6 - 27100 Pavia (PV) Italy.

Abstract. The precision measurement of the anomalous magnetic moment g-2 of the muon presently exhibits a 3.5 σ deviation between theory and experiments. In the next few years the anomalous magnetic moment will be measured to higher precisions at Fermilab and J-PARC. The theoretical prediction can be improved by reducing the uncertainty on the leading hadronic correction a_μ^{HLO} to the g-2. Here we present a novel approach to determine a_μ^{HLO} with space-like data, by means of precise measurement of the hadronic shift of the effective electromagnetic coupling α exploiting the elastic scattering of 150 GeV muons (currently available at CERN North area) on atomic electrons of a low-Z target. The direct measurement of a_μ^{HLO} in the space-like region will provide a new independent determination competitive with the time-like dispersive approach, and will consolidate the theoretical prediction of the muon g-2 in the Standard Model. It will allow therefore a firmer interpretation of the measurements of the future muon g-2 experiments at Fermilab and J-PARC

1 Introduction

The discrepancy between the experimental value of the muon anomalous magnetic moment $a_\mu = (g-2)/2$ and the Standard Model (SM) prediction, $\Delta a_\mu \sim (28 \pm 8) \times 10^{-10}$ is a long standing issue in particle physics [1, 2]. The current accuracy of the SM predictions, $\sim 5 \times 10^{-10}$, are limited by strong interaction effects, at the low energy scale implied. However by using analyticity and unitarity, it was shown [3] that the leading-order (LO) hadronic contribution to the muon g-2, $a_\mu^{\rm HLO}$, can be computed via a dispersion integral of the hadron production cross section in e^+e^- annihilation at low-energy. The present error on $a_\mu^{\rm HLO}$, $\sim 4 \times 10^{-10}$ or slightly better, with a relative accuracy of 0.6%, constitutes the main uncertainty of the SM prediction [4]. Alternative evaluations of $a_\mu^{\rm HLO}$ can be obtained with QCD lattice calculations [5]. The current lattice QCD results are not yet competitive with the dispersive approach via time-like data, with errors that are expected to decrease significantly in the next few years [6]. The $O(\alpha^3)$ hadronic light-by-light contribution, $a_\mu^{\rm HLbL}$, which has the second largest error in the theoretical evaluation, contributing with an uncertainty of (2.5–4)$\times 10^{-10}$, cannot at present be determined from data and its calculation relies on the use of specific models [7–10].

The error achieved by the BNL E821 experiment [11], $\delta a_\mu^{\rm Exp} = 6.3 \times 10^{-10}$, corresponding to 0.54 ppm, is dominated by the available statistics. New experiments at Fermilab and J-PARC, aiming at measuring the muon g-2 to a precision of 1.6×10^{-10} (0.14 ppm), are in preparation [12–15].

Together with the experimental plans, an improvement on the value of the LO hadronic contribution is highly desirable. The proposal described here is to determine $a_\mu^{\rm HLO}$ from a measurement of the effective electromagnetic coupling α in the space-like region, where the vacuum polarization is expectd to be a smooth function of the squared momentum transfer. In this approach the hadronic contribution to the running of α can be measured by means of the t-channel $\mu - e$ elastic scattering process, from which $a_\mu^{\rm HLO}$ can be determined directly [17] [1].

2 Measuring the Hadronic Leading contribution with space-like data.

For the calculation of the hadronic leading contribution $a_\mu^{\rm HLO}$ with the t-channel approach an alternative formula to the dispersion integral [3, 21] can be exploited [18, 22], namely:

$$a_\mu^{\rm HLO} = \frac{\alpha}{\pi} \int_0^1 dx\,(1-x)\,\Delta\alpha_{\rm had}[t(x)]\,, \qquad (1)$$

where $\Delta\alpha_{\rm had}(t)$ is the hadronic contribution to the running of the fine-structure constant, evaluated at

$$t(x) = \frac{x^2 m_\mu^2}{x-1} < 0, \qquad (2)$$

[a] e-mail: umberto.marconi@bo.infn.it
[b] e-mail: fulvio.piccinini@pv.infn.it

[1] The method has been originally proposed [18] by using Bhabha scattering data. A method to determine the running of α by using small-angle Bhabha scattering was proposed in [19] and applied to LEP data in [20].

that is the space-like (negative) squared four-momentum transfer of the process. In contrast with the dispersive integral, the integrand of Eq. (1) is a smooth function and free of resonance poles. Fig. 1 (left) shows $\Delta\alpha_{\rm had}$, and for comparison $\Delta\alpha_{\rm lep}$, as a function of the variables x and t. The range $x \in (0,1)$ corresponds to $t \in (-\infty, 0)$, with $x = 0$ for $t = 0$. The expected integrand of Eq. (1), calculated with the routine `hadr5n12` [23], which uses time-like hadroproduction data and perturbative QCD, is plotted in Fig. 1 (right). The peak of the integrand occurs at $x_{\rm peak} \simeq 0.914$ (corresponding to $t_{\rm peak} \simeq -0.108$ GeV^2) and $\Delta\alpha_{\rm had}(t_{\rm peak}) \simeq 7.86 \times 10^{-4}$ (see Fig. 1 (right)).

We propose to use Eq. (1) to determine $a_\mu^{\rm HLO}$ by measuring the running of α using the CERN muon beam of energy E_μ = 150 GeV, colliding on electron at rest of a fixed target. This technique is similar to the one used for the measurement of the pion form factor, as described in [24].

It looks very appealing for the following reasons:

- It is a t-channel process, making the dependence on t of the differential cross section proportional to $|\alpha(t)/\alpha(0)|^2$:

$$\frac{d\sigma}{dt} = \frac{d\sigma_0}{dt}\left|\frac{\alpha(t)}{\alpha(0)}\right|^2, \tag{3}$$

  where $d\sigma_0/dt$ is the effective Born cross section, including virtual and soft photons, analogously to Ref. [25], where small-angle Bhabha scattering at high energy was considered. The vacuum polarization effect, in the leading photon t-channel exchange, is incorporated in the running of α and gives rise to the factor $|\alpha(t)/\alpha(0)|^2$.

- Given the incoming muon energy E_μ the t variable is related to the energy of the scattered electron E'_e or its angle θ_e through:

$$t = (p_\mu - p'_\mu)^2 = (p_e - p'_e)^2 = 2m_e^2 - 2m_e E'_e, \tag{4}$$

$$s = (p'_\mu + p'_e)^2 = (p_\mu + p_e)^2 = m_\mu^2 + m_e^2 + 2m_e E_\mu, \tag{5}$$

$$E'_e = m_e \frac{1 + r^2\cos^2\theta_e}{1 - r^2\cos^2\theta_e}, \quad \theta_e = \arccos\left(\frac{1}{r}\sqrt{\frac{E'_e - m_e}{E'_e + m_e}}\right), \tag{6}$$

  where the angle θ_e spans the range (0–31.85) mrad for the electron energy E'_e in the range (1–139.8) GeV.

- For E_μ = 150 GeV, it turns out that $s \simeq 0.164$ GeV^2 and -0.143 GeV$^2 < t < 0$ GeV^2. It implies that the region of x extends up to 0.93, while the peak of the integrand function of Eq. (1) is at x_{peak} = 0.914, corresponding to an electron scattering angle of 1.5 mrad, as visible in Fig. 1 (right).

- The angles of the scattered electron and muon are correlated as shown in the Fig. 2, drawn for incoming muon energy of 150 GeV. This constraint is extremely important to select elastic scattering events, rejecting background events from radiative or inelastic processes and to minimize systematic effects in the determination of t. Note that for scattering angles of (2–3) mrad there is an ambiguity between the outgoing electron and muon, as their angles and momenta are similar, to be resolved by means of μ/e discrimination.

- The boosted kinematics allows the same detector to cover the whole acceptance. Many systematic errors, *e.g.* on the efficiency, will cancel out (at least at first order) in the relative ratios of event counts in the high and low q^2 regions (signal and normalization regions).

Assuming to use a muon beam of 150 GeV with an average intensity of $\sim 1.3 \times 10^7$ muon/s, with a running time of 2×10^7 s/yr, and using 30 experimental points in x (supplemented with large $|t|$ contributions that can be derived from pQCD), we estimate the statistical sensitivity of this experiment on the value of $a_\mu^{\rm HLO}$ to be $\sim 0.3\%$. Such a beam is available at the CERN North Area.

3 Detection technique

The CERN muon beam M2 presents ideal characteristics to perform for the measurement. The beam intensity, of more than 10^7 muon/s, can provide the required event yield and its time structure allows to tag the incident muon. The target must be of low-Z material, and the whole target thick such (order of 60 cm) to get enough electron scattering centres. The low-Z requirement is in order to minimize multiple scattering and to have high radiation length. The idea is them to use a segmented target of 1 cm thin layers, distributed in 60 identical modules. Each detection module has the size of 50 cm and consists of a thin target of Be (or C), coupled to three Si planes for tracking (no magnetic field applied). Fig. 3 shows the basic layout. Downstream the tracking modules, as particle identifiers, we plan to use a calorimeter for the electrons. They are required in order to solve the muon-electron ambiguity for electron scattering angles around (2–3) mrad *cf*. Fig. 2. The preliminary studies of such an apparatus, performed by using GEANT4, indicate an angular resolution of ~ 0.02 mrad for the outgoing particles. The detector acceptance covers both the region of the signal, with the electron emitted at extremely forward angles and high energies, and the normalization region, where the electron has much lower energy (around 1 GeV) and an emission angle of some tens of mrad. Due to the boosted kinematics of the collisions, the detector covers most of the acceptance, and let all the scattering angles in the laboratory system to be accessed by a single detector element.

4 Considerations on systematic uncertainties

Significant contributions of the hadronic vacuum polarization to the $\mu e \to \mu e$ differential cross section are essentially restricted to electron scattering angles below 10 mrad, corresponding to electron energies above 10 GeV. The net effect of these contributions is to increase the cross section by a few per mille: a precise determination of $a_\mu^{\rm HLO}$ requires not only high statistics, but also a high level of control of systematic uncertainties, both on the theory as well as the experimental side, as the final goal of the experiment is equivalent to a determination of the differential cross section with ~10 ppm systematic uncertainty at the peak of the integrand function (*cf*. Fig. 1).

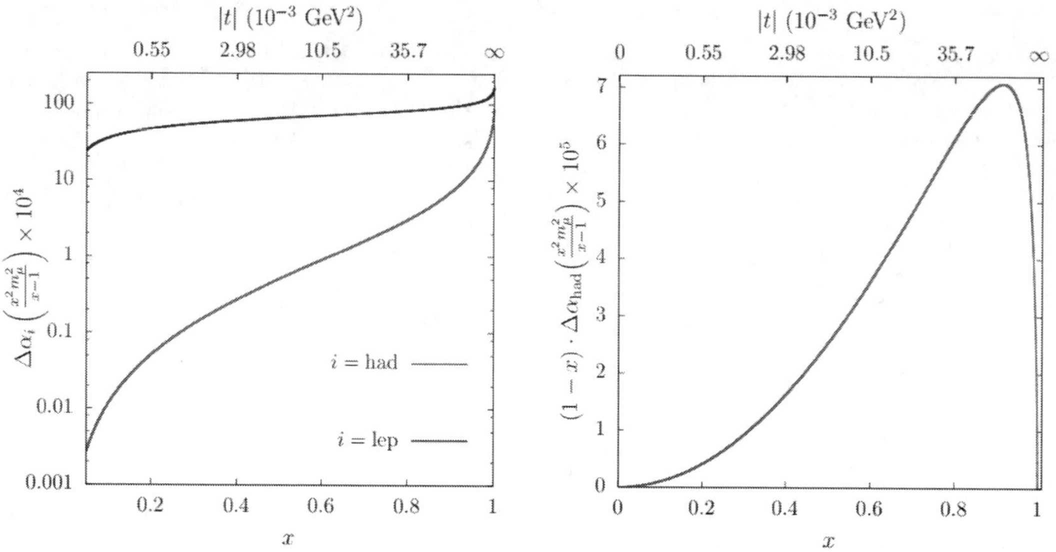

Figure 1. Left: $\Delta\alpha_{\rm had}[t(x)] \times 10^4$ (red) and, for comparison, $\Delta\alpha_{\rm lep}[t(x)] \times 10^4$ (blue), as a function of x and t (upper scale). Right: the integrand $(1-x)\Delta\alpha_{\rm had}[t(x)] \times 10^5$ as a function of x and t. The peak value is at $x_{\rm peak} \simeq 0.914$, corresponding to $t_{\rm peak} \simeq -0.108$ GeV^2.

4.1 Experimental systematic uncertainties

A crucial requirement on the experimental side is to keep the efficiency highly uniform over the entire q^2 range, including the normalization region, and over all the detector components. This motivates the choice of a purely angular measurement: an acceptance of tens of mrad can be covered with a single sensor of modern silicon detectors, positioned at a distance of ~ 1 m from the target. It has to be stressed that particle identification (electromagnetic calorimeter and muon filter) is necessary to solve the electron-muon ambiguity in the region below 5 mrad. The wrong assignment probability can be measured with the data by using the rate of muon-muon and electron-electron events.

Another requirement for reaching very high accuracy is to measure all the relevant contributions to systematic uncertainties from the data themselves. An important effect, which distinguishes the normalization from the signal region, is multiple scattering, as the electron energy in the normalization region is as low as 1 GeV. Multiple scattering breaks the muon-electron two-body angular correlation, moving events out of the kinematic line in the 2D plot of Fig. 2. In addition, multiple scattering in general causes acoplanarity, while two-body events are planar, within the resolution. These facts allow effects to be modelled and measured by using data. An additional handle on multiple scattering could be the inclusion of a thin layer in the apparatus, made of the same material as the main target modules. This possibility will be studied in detail with simulation.

In high-precision experiments several systematic effects can be explored within the experiment itself. In this respect the proposed modularity of the apparatus will help. A test with a single module could provide a proof-of-concept of the proposed method.

4.2 Theoretical systematic uncertainties

Since the final goal of the experiment is a high precision measurement of a differential cross section, the latter has to be calculated with all the radiative corrections relevant at the precision scale of 0.01%. Such an accuracy has never been achieved in a collision experiment. The most precise experiments has been conducted at leptonic colliders such as LEP and flavour factories [26], where the theoretical precision at the level of 0.1% (or even better, as for small angle Bhabha scattering at LEP) was achieved. Previous quantitative considerations of Eq. (3) and Eq. (6) are based on the QED Leading Order (LO) expression of the differential cross section for $\mu e \to \mu e$

$$\frac{d\sigma}{dt} = \frac{4\pi\alpha^2}{\lambda(s, m_\mu^2, m_e^2)}\left[\frac{(s-m_\mu^2-m_e^2)^2}{t^2} + \frac{s}{t} + \frac{1}{2}\right]. \quad (7)$$

The effect of running of α introduced in Eq. (3) is obtained by resumming to all orders the gauge invariant subset of fermionic corrections at $O(\alpha)$. At the same next-to-leading order (NLO), the class of QED corrections due to the insertion of a photon line, virtual and real, contribute. These corrections are known in the literature for μe scattering [27] but need to be implemented in a Monte Carlo event generator [28] taylored to simulate the present energy and kinematics conditions, in order to design the optimal detector configuration and the event selection for data analysis. The typical theoretical uncertainty of a QED NLO calculation is of the order of 1%. The techniques to match NLO fixed order calculation with all-orders resummed calculations of leading logarithmic contributions

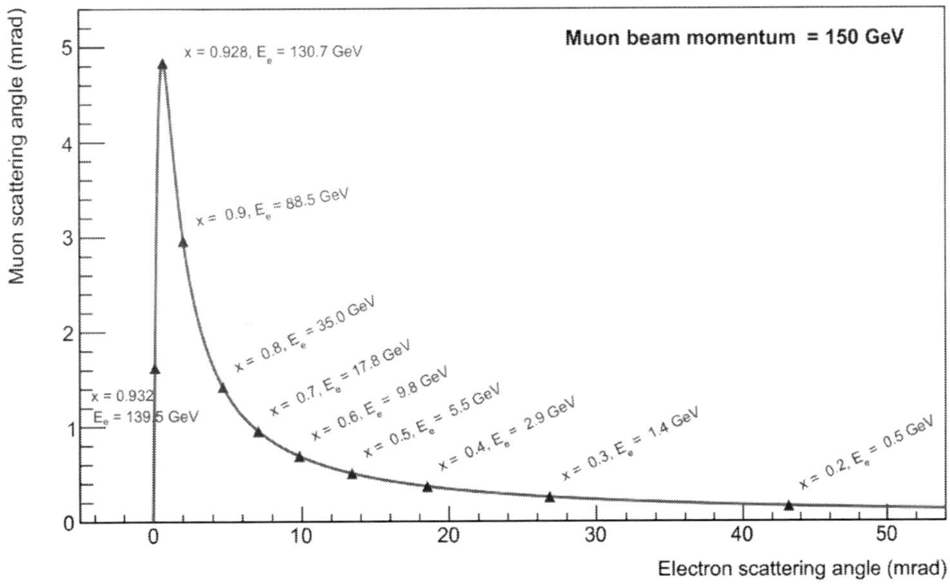

Figure 2. The relation between the muon and electron scattering angles for 150 GeV incident muon beam momentum. Blue triangles indicate reference values of the Feynman's x and electron energy.

are well known in QED [29] and allow to reach accuracies at the 0.1% level. In order to go beyond this limit, the NNLO QED corrections are required. A first step in this direction has been undertaken through the evaluation of the planar master integrals related to the two-loop photonic box diagrams of the virtual QED corrections to the $\mu e \to \mu e$ process, in the approximation of massless electron [30]. Once the calculation of the non-planar contributions will be carried out, all the building blocks for the complete photonic QED NNLO calculation will be available. At two loop level, also fermionic one-loop insertions on top of QED NLO diagrams have to be evaluated. Such contributions have already been evaluated for Bhabha scattering [31]. The virtual contribution involves also the effect of hadronic vacuum polarization, which requires a dedicated study [32]. The real pair emission is also particularly important since one of the leptons of the additional pair can produce by itself a signal track. The evaluation of such contributions requires, therefore, a detailed Monte Carlo simulation wih realistic geometric acceptance. Such kind of studies have been performed already for LEP physics [33].

Another class of contributions is given by the Z^0 tree-level exchange diagram and the full NLO corrections in the electroweak standard model. Even if they are expected to be small, it is worth to quantify their size. These corrections have been calculated in the literature [34]. A complete evaluation and Monte Carlo implementation is in progress [28].

5 Conclusions

The experiment MUonE presented to determine the leading hadronic contribution to the muon g-2, by scattering high-energy muons on atomic electrons of a low-Z target through the process $\mu e \to \mu e$, is primarily based on a precise measurement of the scattering angles of the two outgoing particles as the q^2 of the muon-electron interaction can be directly determined by the electron (or muon) scattering angle. An advantage of the muon beam is the possibility of employing a modular apparatus, with the target subdivided in subsequent layers. A low-Z solid target is preferred in order to provide the required event rate, limiting at the same time the effect of multiple scattering as well as of other types of muon interactions (pair production, bremsstrahlung and nuclear interactions). The normalization of the cross section is provided by the very same $\mu e \to \mu e$ process in the low-q^2 region, where the effect of the hadronic corrections on $\alpha(t)$ is negligible. Such a simple and robust technique has the potential to keep systematic effects under control, aiming at reaching a systematic uncertainty of the same order as the statistical one. For this purpose a preliminary detector layout has been described. On the theoretical side, the radiative corrections at NNLO accuracy have to be considered in full detail and a Monte Carlo event generator able to match fixed order NNLO predictions with resummation given by the Parton Shower technique is needed. By considering a beam of 150 GeV muons with an average intensity of $\sim 1.3 \times 10^7$ muon/s, currently available at the CERN North Area, a statistical uncertainty of $\sim 0.3\%$ can be achieved on a_μ^{HLO} in

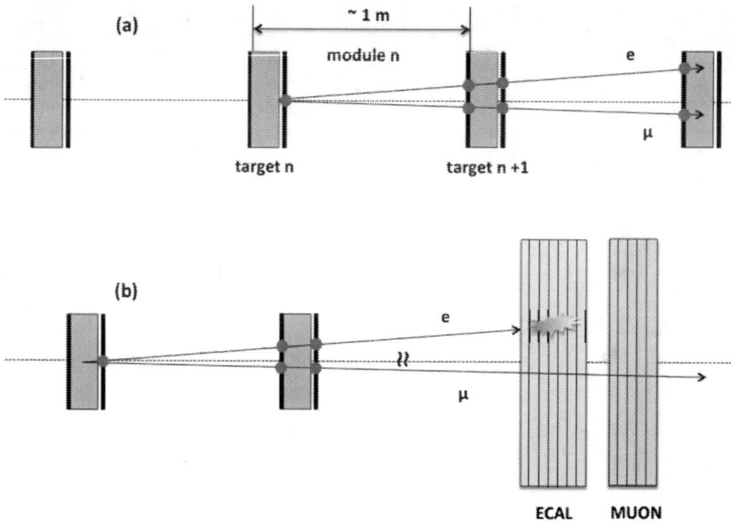

Figure 3. Scheme of a possible detector layout. (a) The detector is a modular system. Each module consists of a low-Z target (Be or C) and three silicon tracking stations within the distance of 50 cm. (b) To perform the μ-e discrimination in the case of small scattering angles (with both θ_μ and θ_e below 5 mrad) the detector is equipped with an electromagnetic calorimeter and a muon detector.

two years of data taking. A test performed using a single detector module, exploiting the muon beam facility, could provide a validation of the proposed method.

Acknowledgements

We thank G. Abbiendi, C.M. Carloni Calame, C. Matteuzzi, G. Montagna, O. Nicrosini, M. Passera, R. Tenchini, L. Trentadue and G. Venanzoni for fruiful and pleasant collaboration. We thank the organizers of the FCCP2017 Workshop for their kind invitation.

References

[1] T. Blum, A. Denig, I. Logashenko, E. de Rafael, B. Lee Roberts, T. Teubner and G. Venanzoni, arXiv:1311.2198 [hep-ph].

[2] F. Jegerlehner, EPJ Web Conf. **118** (2016) 01016.

[3] C. Bouchiat, L. Michel, J. Phys. Radium 22 (1961) 121;
L. Durand, Phys. Rev. **128** (1962) 441 [Erratum-ibid. **129** (1963) 2835];
M. Gourdin, E. De Rafael, Nucl. Phys. B **10** (1969) 667.

[4] D. Nomura, these proceedings.

[5] C. Aubin, T. Blum, Phys. Rev. D **75** (2007) 114502;
P. Boyle, L. Del Debbio, E. Kerrane, J. Zanotti, Phys. Rev. D **85** (2012) 074504;
X. Feng, K. Jansen, M. Petschlies, D.B. Renner, Phys. Rev. Lett. **107** (2011) 081802;
M. Della Morte, B. Jager, A. Juttner, H. Wittig, JHEP **1203** (2012) 055;
T. Blum *et al.*, Phys. Rev. Lett. **116** (2016) no.23, 232002;
B. Chakraborty, C. T. H. Davies, P. G. de Oliviera, J. Koponen and G. P. Lepage, arXiv:1601.03071 [hep-lat].

[6] T. Blum, M. Hayakawa, T. Izubuchi, PoS LATTICE **2012** (2012) 022;
B. Chakraborty, C. Lehner, R. Van de Water, talks at ICHEP 2016, https://indico.cern.ch/event/432527/contributions/speakers

[7] M. Knecht, A. Nyffeler, Phys. Rev. D **65** (2002) 073034;
K. Melnikov, A. Vainshtein, Phys. Rev. D **70** (2004) 113006;
J. Prades, E. de Rafael, A. Vainshtein, arXiv:0901.0306 [hep-ph].
F. Jegerlehner, A. Nyffeler, The Muon g-2, Phys. Rept. **477** (2009) 1;
A. Nyffeler, Phys. Rev. D **79** (2009) 073012.

[8] F. Jegerlehner, A. Nyffeler, Phys. Rept. **477** (2009) 1.

[9] C. Lehner, these proceedings.

[10] A. Nyffeler, these proceedings.

[11] G. W. Bennett et al. (The g-2 Collaboration), Phys. Rev. **D73** (2006) 072003.

[12] J. Grange *et al.* [Muon g-2 Collaboration], arXiv:1501.06858 [physics.ins-det].

[13] N. Saito [J-PARC g-2/EDM Collaboration], AIP Conf. Proc. **1467** (2012) 45.

[14] T. Gorringe, these proceedings.

[15] G. Marshall, these proceedings.

[16] M. Davier, A. Hoecker, B. Malaescu and Z. Zhang, Eur. Phys. J. **C71** (2011) 1515, Erratum-ibid. **C72** (2012) 1874.

[17] G. Abbiendi, C. M. Carloni Calame, U. Marconi, C. Matteuzzi, G. Montagna, O. Nicrosini, M. Passera, F. Piccinini, R. Tenchini, L. Trentadue and G. Venanzoni, Eur. Phys. J. C (2017) 77 139.

[18] C. M. Carloni Calame, M. Passera, L. Trentadue and G. Venanzoni, Phys. Lett. B **746** (2015) 325

[19] A.B. Arbuzov, D. Haidt, C. Matteuzzi, M. Paganoni and L. Trentadue, Eur. Phys. J. C **34** (2004) 267.

[20] G. Abbiendi *et al.* [OPAL Collaboration], Eur. Phys. J. C **45** (2006) 1.

[21] F. Jegerlehner, Springer Tracts Mod. Phys. **226**, 2008.

[22] B.E. Lautrup, A. Peterman, E. de Rafael, Phys. Rept. **3** (1972) 193.

[23] S. Eidelman and F. Jegerlehner, Z. Phys. C **67** (1995) 585;
F. Jegerlehner, Nucl. Phys. Proc. Suppl. **181-182** (2008) 135

[24] S. R. Amendolia *et al.*, Phys. Lett. B **146** (1984) 116; S. R. Amendolia *et al.* [NA7 Collaboration], Nucl. Phys. B **277** (1986) 168.

[25] A.B. Arbuzov, V.S. Fadin, E.A. Kuraev, L.N. Lipatov, N.P. Merenkov, L. Trentadue, Nucl. Phys. B **485** (1997) 457;

[26] B.F.L. Ward et al., Phys. Lett. **B450** (1999) 262; S. Actis et al., Eur. Phys. J. C66 (2010) 585-686.

[27] K.E. Eriksson, Nuovo Cimento **19** (1961) 1029; A.I. Nikishow, Sov. Phys. JETP (1961) 1929; K.E. Eriksson, B. Larsson and G.A. Rinander, Nuovo Cimento **30** (1963) 1434; P. Van Nieuwenhuizen, Nucl. Phys. B **28** (1971) 429; T. V. Kukhto, N. M. Shumeiko and S. I. Timoshin, J. Phys. G **13** (1987) 725; D. Y. Bardin and L. Kalinovskaya, DESY-97-230, hep-ph/9712310; N. Kaiser, J. Phys. G **37** (2010) 115005.

[28] M. Alacevich et al., work in progress.

[29] G. Balossini et al., Nucl.Phys. B758 (2006) 227-253.

[30] P. Mastrolia et al., JHEP 1711 (2017) 198.

[31] C.M. Carloni Calame et al., JHEP 1107 (2011) 126; Nucl. Phys. Proc. Suppl. 225-227 (2012) 293

[32] M. Fael and M. Passera, work in progress.

[33] G. Montagna et al., Nucl. Phys. **B547** (1999) 39; Phys. Lett. **B459** (1999) 649.

[34] E. Derman and W.J. Marciano, Ann. Phys. **121** (1979) 147; G. D'Ambrosio, Lett. Nuovo Cim. **38** (1983) 593; J.C. Montero, V. Pleitez and M.C. Rodriguez, Phys. Rev. **D58** (1998) 097505.

A strange program for LHCb

Diego Martínez Santos[1,a], on behalf of LHCb

[1]Instituto Galego de Física de Altas Enerxías (IGFAE) Rúa de Xoaquín Díaz de Rábago, s/n Campus Vida Universidade de Santiago de Compostela E-15782 Santiago de Compostela, Spain

Abstract. The LHCb experiment is expanding its physics program towards studies of rare decays of strange particles. In this talk, we reviewed the published results by LHCb in $K_S^0 \to \mu^+\mu^-$ and $\Sigma^+ \to p\mu^+\mu^-$ decays, as well as sensitivity studies and prospects for other strangeness decays.

1 Introduction

The Standard Model (SM) of particle physics is very successful at describing accelerator data, but it is known to be incomplete as it fails to explain gravity, dark matter, or neutrino masses. For this reason, dynamics Beyond the Standard Model (BSM) is being searched for. Direct searches at LHC are limited to few TeV by the maximum energy of the accelerator, while precise measurements of flavour decays can a priori access higher energy scales. This is particularly true for models with new sources of flavour violation, in which the CKM matrix is not the only source of flavour and CP violation. In such context, $s \to d$ quark transitions are central, because they have the strongest CKM suppression. Examples of these are kaon oscillations, rare decays such as $K \to \pi\nu\bar{\nu}$, being searched for in NA62, or $K_S^0 \to \mu^+\mu^-$, which is being searched for at LHCb. Here we review the status and prospects for strangeness decays at the LHCb.

2 The LHCb experiment and strangeness trigger

The LHCb detector at the LHC [1] is a single-arm forward spectrometer covering the pseudorapidity range $2 < \eta < 5$, designed for the study of particles containing b or c quarks. It has collected an integrated luminosity of 1fb^{-1} at $\sqrt{s} = 7\,TeV$ in 2011 and 2 fb^{-1} at $\sqrt{s} = 7\,TeV$ in 2012. In 2015 it started its second Run (hereafter Run-II) of data taking, and expects to collect up to 8 fb^{-1} $\sqrt{s} = 13\,TeV$ by the end of 2018. The LHCb Upgrade, scheduled to start by 2021, is expected to collect 50 fb^{-1} at $\sqrt{s} = 14\,TeV$ by the end of 2029. Finally, an *Expression of Interest* for a Phase-II Upgrade, aiming to collect ~300 fb^{-1} has been submitted to the LHC Council [2].

Due to its optimization for b and c decays, the trigger efficiency for strangeness decays in its first years of data taking has been very low (1-2% for kaons with dimuon final-states, to be compared with 80-90% for dimuon B-decays). Yet, the strangeness corssection at LHC is very large, with ~10^{13} K_S^0 produced in the LHCb acceptance per fb^{-1}, and it allowed the publication of world best results in rare decays of strange particles. The flexibility of the software trigger (HLT) allowed increasing significantly the trigger efficiency for strangeness decays to dimuons for Run-II data [3], as shown in Table 1. Since in the LHCb upgrade there will be no hardware trigger (L0) and all the trigger steps will be software based, the trigger efficiency is expected to be as high as for B decays.

Table 1. Approximate trigger efficiencies for benchmark $s \to d\mu^+\mu^-$ transitions at LHCb for 2012 and Run-II. Numbers obtained from [3]. The efficiency in 2011 data is lower than in 2012 data, being approximately ~1%. The maximum efficiency that can be achieved, taking into account L0 constraints, is 30%.

Channel	ε(2012)%	ε(Run-II)%
$K_S^0 \to \mu^+\mu^-$	2.9	25
$\Sigma^+ \to p\mu^+\mu^-$	0.83	11
$K_S^0 \to \pi^0\mu^+\mu^-$	2.6	24

3 Results

3.1. $K_S^0 \to \mu^+\mu^-$

In the SM, the $K_S^0 \to \mu^+\mu^-$ decay is long-distance dominated, with subdominant short-distance

[a] Corresponding author: diego.martinez.santos@cern.ch

contributions from Z-penguin and W-box. Yet, the long-distance contribution is still very small in absolute terms, and the decay rate is very supressed [4] [5] [6]:

$$BR(K_S^0 \to \mu^+\mu^-)_{SM} = \\ = (5.18 \pm 1.50_{LD} \pm 0.18_{SD}) \times 10^{-12}.$$

Therefore, small BSM contributions and BSM/SM interferences can sizably modify the branching fraction. This has been proven to be the case in leptoquark models [7] [8] as well as supersymmetric models [9], despite of other experimental constraints.

The 2011 LHCb data allowed improving by a factor of thirty [10] the world best upper limit prior to LHC [11]. Recently, the analysis of the full Run-I data has been published [12], corresponding to an integrated luminosity of 3fb^{-1} and improved reconstruction, trigger and selection. The analysis uses a Boosted Decision Tree (BDT) to separate signal from combinatorial background and dedicated particle identification to separate $K_S^0 \to \mu^+\mu^-$ decays from misidentified $K_S^0 \to \pi^+\pi^-$ decays. The mass shift on the K_S^0 peak caused in the $\pi^\pm \to \mu^\pm$ misidentification is also crucial for the analysis, since it separates the $K_S^0 \to \pi^+\pi^-$ peak away from the signal, and only the tail of the distribution leaks into the region in which genuine $K_S^0 \to \mu^+\mu^-$ decays are expected. This can be seen in Figure 1 (top), where the invariant mass distribution of the dimuon candidates is shown, for one of the most sensitive BDT bins. A simultaneous maximum likelihood fit of the dimuon invariant mass spectrum of all the BDT bins is performed, yielding no significant signal. A Bayesian upper limit (see Figure 1 bottom) is derived by integrating the likelihood times a flat prior on the BR, resulting in an upper limit

$$BR(K_S^0 \to \mu^+\mu^-)_{EXP} < 8(10) \times 10^{-10}$$

The extrapolated sensitivity as a function of the integrated luminosity times trigger efficiency is shown in Figure 2. The trigger efficiency for Run-II is estimated to be 25%, while for the LHCb upgrade is expected to be close to 100%. It can be seen that the LHCb upgrade can explore a good part of the 10^{-11} region, and a Phase-II upgrade could get very close to the SM prediction.

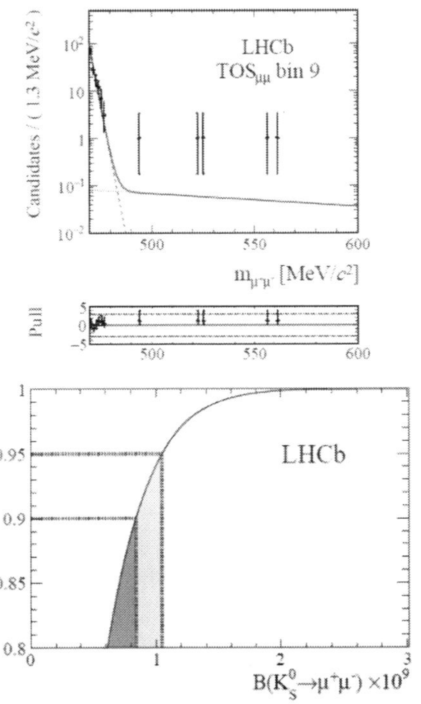

Figure 1. *Top:* Invariant mass distribution of the dimuon candidates in one of the most sensitive BDT bins. The green dashed line shows the contribution from misidentified $K_S^0 \to \pi^+\pi^-$ decays. The red dotted line shows the contribution from combinatorial background. No significant signal (which would have shown as a nearly-Gaussian peak at ~500 MeV/c^2) has been found. *Bottom:* cumulative function of the posterior probability of the branching fraction. The intersections with (upper limit) 90% and 95% credible intervals are shown with dashed red lines.

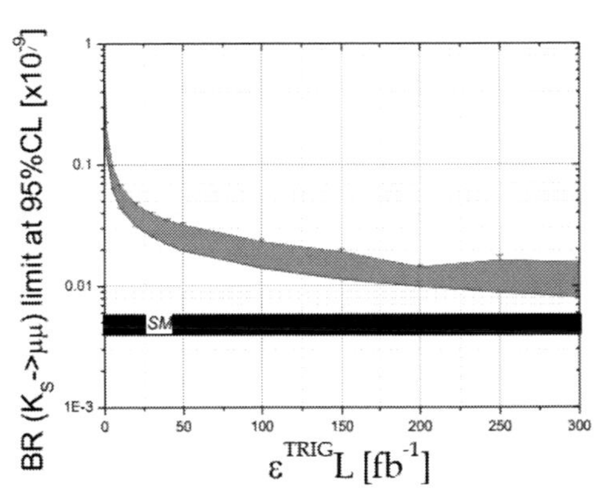

Figure 2. Expected LHCb sensitivity to $BR(K_S^0 \to \mu^+\mu^-)$ as a function of the integrated luminosity (times trigger efficiency). The trigger efficiency for Run-II is estimated to be ~25%, while for LHCb upgrade is expected to be close to

https://doi.org/10.1051/epjconf/201817901013

the 100%. The black band shows the SM prediction with its uncertainty. The width of the green band shows an estimated uncertainty on the sensitivity prediction, including model systematics and statistical subtraction of the $K_L^0 \to \mu^+\mu^-$contribution.

3.2 $\Sigma^+ \to p\mu^+\mu^-$

The $\Sigma^+ \to p\mu^+\mu^-$ decay is a flavour-changing neutral-current process, allowed only at loop level in the SM. The process is dominated by long-distance contributions for a predicted branching fraction of [13]

$$1.6 \times 10^{-8} < BR(\Sigma^+ \to p\mu^+\mu^-)_{SM} < 9.0 \times 10^{-8} ,$$

while the short-distance SM contributions are suppressed at a branching fraction of about 10^{-12}. Evidence for this decay was seen by the HyperCP experiment [14] with a measured branching fraction

$$BR(\Sigma^+ \to p\mu^+\mu^-)^{HyperCP}_{EXP} = (8.6^{+6.6}_{-5.4} \pm 5.5) \times 10^{-8},$$

which is compatible with the SM prediction. Remarkably, the three observed decays have almost the same dimuon invariant mass of m_X^0 = 214.3 ± 0.5 MeV/c^2, which lies close to the kinematic limit. This result is sometimes referred to as "the Hyper-CP anomaly", since it suggests a potential BSM particle at a mass of ~214 MeV/c^2.

The LHCb collaboration has recently published [15] a preliminary search for the $\Sigma^+ \to p\mu^+\mu^-$ decay, showing a strong evidence for this decay with a significance of four standard deviations (see Figure 3, top). The dimuon invariant mass distribution does not show evidence for a low mass resonance, disproving the HyperCP-anomaly (see Figure 3, bottom). This measurement was based on Run-I data with no trigger path dedicated to this channel. In the time between the FCCP workshop and the write-up of the proceedings, the search has been published into a paper [16], providing in addition a branching fraction measurement

$$BR(\Sigma^+ \to p\mu^+\mu^-)^{LHCb}_{EXP} = (2.1^{+1.6}_{-1.2}) \times 10^{-8}.$$

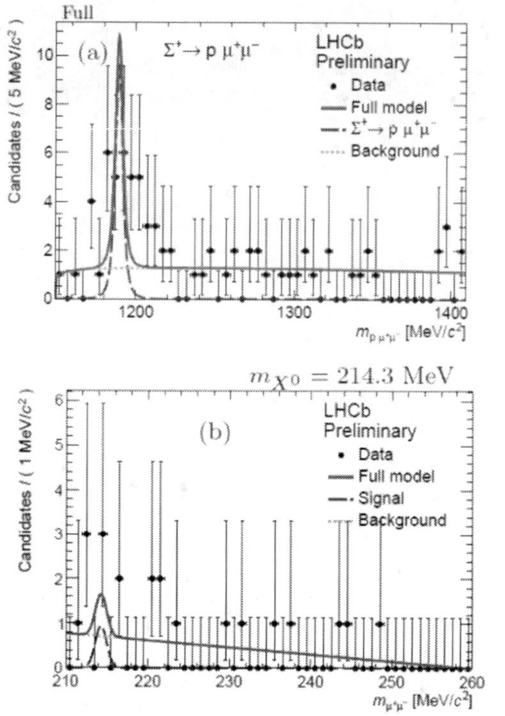

Figure 3. *Top:* Invariant mass distribution of the $\Sigma^+ \to p\mu^+\mu^-$ candidates. The green dotted line represents the background, and the red dashed line the $\Sigma^+ \to p\mu^+\mu^-$signal, with a significance of four standard deviations. *Bottom:* dimuon mass distribution of the candidates. The green dotted line represents an exponential continuum, and the red dashed line a potential $X^0 \to \mu^+\mu^-$signal, which is found to be not significant.

4 Sensitivity studies

4.1 $K_S^0 \to \pi^0\mu^+\mu^-$

The $K_S^0 \to \pi^0\mu^+\mu^-$ decay is another $s \to d\mu^+\mu^-$ transition with a very low branching fraction, predicted at the 10^{-9} level in the SM. The closely related $K_L^0 \to \pi^0\mu^+\mu^-$ decay has been shown to have sensitivity to BSM dynamics, such as Extra Dimensions [17] or leptoquarks [8]. Its sensitivity is limited by the uncertainty on the SM contribution, which is dominated by the experimental precision on the K_S^0 mode, the NA48 measurement of $BR(K_S^0 \to \pi^0\mu^+\mu^-)$ [18]:

$$BR(K_S^0 \to \pi^0\mu^+\mu^-)_{EXP} = (2.9^{+1.5}_{-1.2} \pm 0.2) \times 10^{-9}$$

Hence, improvements on the experimental measurement of $BR(K_S^0 \to \pi^0\mu^+\mu^-)$ translate into improved BSM search potential of $K_L^0 \to \pi^0\mu^+\mu^-$. The LHCb experiment has evaluated its sensitivity to $BR(K_S^0 \to \pi^0\mu^+\mu^-)$ [19] with two different reconstruction strategies: explicit requirement of the

$\pi^0 \rightarrow \gamma\gamma$ decay (FULL), and reconstruction based on the dimuon information only (PARTIAL). In both cases the LHCb upgrade is expected to overtake the precision of NA48, as shown in Figure 4. The sensitivity as a function of the dimuon mass has not been evaluated yet.

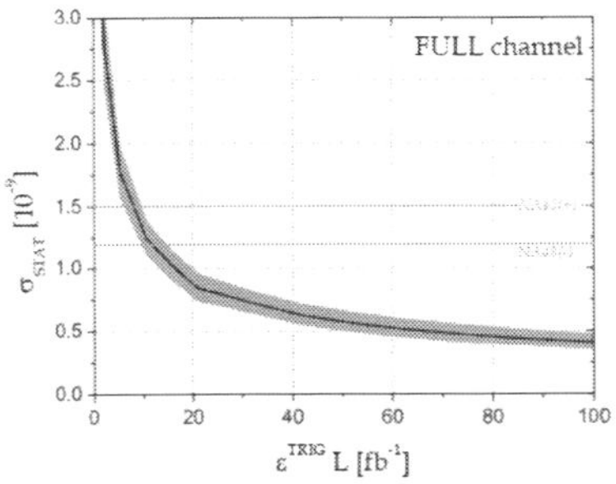

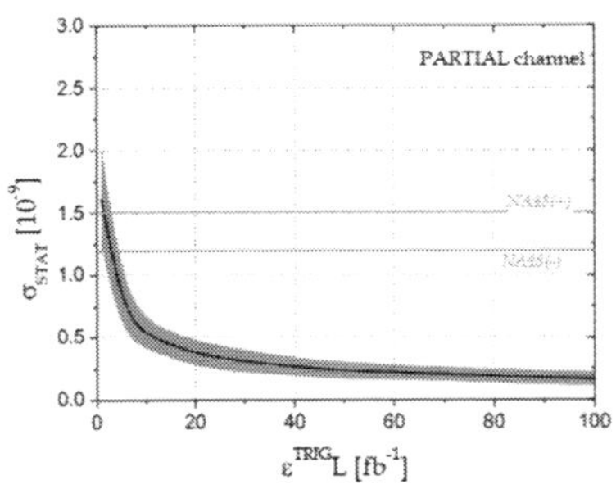

Figure 4. Expected LHCb sensitivity to $BR(K_S^0 \rightarrow \pi^0\mu^+\mu^-)$ as a function of the integrated luminosity (times trigger efficiency). The trigger efficiency for Run-II is estimated to be ~24%, while for LHCb upgrade is expected to be close to 100%.

4.2 $K_S^0 \rightarrow \pi^+\pi^-e^+e^-$

Although most of the efforts have been put in dimuon modes, sensitivities to final states with electrons, such as $K_S^0 \rightarrow \pi^+\pi^-e^+e^-$, [20] are also under study, showing that an LHCb strangeness program with electrons is also possible.

4.3 Other analyses

It must be noted that the LHCb capabilities are not restricted to the four channels discussed above. Indeed, other measurements such as K^+ mass, $K_S^0 \rightarrow \gamma\mu^+\mu^-$, kaons from Φ decays, Lepton Flavour Violation, Lepton Universality Violation, and semileptonic decays of hyperons could be feasible and are under study.

4 Conclusions

Results and prospects of the LHCb experiment in rare decays of strange particles have been discussed. These include updated limits in $BR(K_S^0 \rightarrow \mu^+\mu^-)$ as well as the disproval of the Hyper-CP anomaly in $\Sigma^+ \rightarrow p\mu^+\mu^-$ decays. Future prospects have also been discussed, showing the potential to reach $BR(K_S^0 \rightarrow \mu^+\mu^-)$ at nearly the SM level, as well as to overtake NA48 in the determination of $BR(K_S^0 \rightarrow \pi^0\mu^+\mu^-)$.

References

1. LHCb Collab., "The LHCb detector at the LHC", JINST 3 (2008) S08005.
2. LHCb Collab., "Expression of Interest for a Phase-II LHCb Upgrade: Opportunities in flavour physics, and beyond, in the HL-LHC era", CERN-LHCC-2017-003.
3. F. Dettori et al.," Low-p T dimuon triggers at LHCb in Run 2", LHCb-PUB-2017-023 .
4. G. Ecker and A. Pich, "The Longitudinal muon polarization in KL->μμ," Nucl. Phys.B366 (1991) 189–205.
5. G. Isidori and R. Unterdorfer, "On the short-distance constraints from K_{L,S} -> mu^+ mu^- ",JHEP 0401 (2004) 009, arXiv:hep-ph/0311084.
6. G. D'Ambrosio and T. Kitahara, "Direct CP Violation in K0->μμ", Phys. Rev. Lett. 119 no. 20, (2017) 201802, arXiv:1707.06999 [hep-ph].
7. I.Dorsner et al., "Limits on scalar leptoquark interactions and consequences for GUTs", JHEP 1111 (2011) 002, arXiv:1107.5393 [hep-ph]].
8. C. Bobeth, A. J. Buras, "Leptoquarks meet ε ′/ε ", arXiv:1712.01295 [hep-ph].
9. V. Chobanova et al., "Probing SUSY effects with KS->μμ", arXiv:1711.11030v1 [hep-ph]. Submitted to JHEP.
10. LHCb Collaboration, JHEP 01 (2013) 090, [arXiv:1209.4029 [hep-ex]].
11. S. Gjesdal et al., "Search for the decay K0S->2μ", Phys. Lett. B 44 (1973) 217.

12. LHCb Collab., "Improved limit on the branching fraction of the rare decay K 0 S →μ + μ -", Eur. Phys. J. C, 77 10 (2017) 678.
13. X.-G. He, J. Tandean, and G. Valencia," The Decay Σ -> pμμ within the standard model", Phys. Rev. D72 (2005) 074003, arXiv:hep-ph/0506067.
14. HyperCP collaboration, H. Park et al.," Evidence for the decay Σ->pμμ", Phys.Rev. Lett. 94 (2005) 021801, arXiv:hep-ex/0501014.
15. LHCb Collab.,"Evidence for the decay Σ->pμμ", LHCb-CONF-2016-013.
16. LHCb Collab., "Evidence for the decay Σ->pμμ", arXiv:1712.08606 [hep-ex] .
17. M. Bauer et al, JHEP 1009:017,2010, arXiv:0912.1625 [hep-ph].
18. NA48 Collab.,"Observation of the rare decay K_S -> pi^0mu^+mu^-", Phys. Lett. B599 197-211,2004. [arXiv:hep-ex/0409011].
19. V. Chobanoba et al.," Sensitivity of LHCb and its upgrade in the measurement of B(K 0 S →π 0 μ + μ −)", LHCb-PUB-2018-017.
20. C. Marin Benito et al.," Feasibility study of KS-->ππee in LHCb", LHCb-PUB-2016-016.

On μe-scattering at NNLO in QED

P. Mastrolia[1,2,*], M. Passera[2], A. Primo[3], U. Schubert[4], and W. J. Torres Bobadilla[5]

[1]Dipartimento di Fisica ed Astronomia, Università di Padova, Via Marzolo 8, 35131 Padova, Italy
[2]INFN, Sezione di Padova, Via Marzolo 8, 35131 Padova, Italy
[3]Physik-Institut, Universität Zürich, Winterthurerstrasse 190, CH-8057 Zürich, Switzerland
[4]High Energy Physics Division, Argonne National Laboratory, Argonne, IL 60439, USA
[5]Instituto de Física Corpuscular, Universitat de València - Consejo Superior de Investigaciones Científicas, Parc Científic, E-46980 Paterna, Valencia, Spain

Abstract. We report on the current status of the analytic evaluation of the two-loop corrections to the μe-scattering in Quantum Electrodynamics, presenting state-of-the art techniques which have been developed to address this challenging task.

1 Introduction

The elastic scattering of muons and electrons is one of the simplest and cleanest processes in particle physics. In spite of this simplicity, μe scattering measurements are scarse. In the 60s, experiments at CERN and Brookhaven measured this scattering cross section using accelerator-produced muons [1–4]. At the same time, μe collisions were measured by cosmic-ray experiments [5–8]. The scattering of muons off polarized electrons was then proposed as a polarimeter for high-energy muon beams in the late 80s [9] and measured by the SMC collaboration at CERN a few years later [10].

Recently, a new experiment, MUonE, has been proposed at CERN to measure the differential cross section of the elastic scattering of high-energy muons on atomic electrons as a function of the spacelike (negative) squared momentum transfer [11]. This measurement will provide the running of the effective electromagnetic coupling in the spacelike region and, as a result, a new and independent determination of the leading hadronic contribution to the muon g-2 [11, 12]. In order for this new determination to be competitive with the present dispersive one, which is obtained via timelike data, the μe differential cross section must be measured with statistical and systematic uncertainties of the order of 10ppm. This high experimental precision demands an analogous accuracy in the theoretical prediction.

Until recently, the process $\mu e \to \mu e$ had received little attention also on the theory side. The few existing theoretical studies mainly focused on its QED corrections at next-to-leading order (NLO) [13–19] and tests of the Standard Model (SM) [20–22]. The QED corrections at next-to-next-to-leading order (NNLO), crucial to interpret the high-precision data of future experiments like MUonE, are not known, although some of the two-loop corrections which were computed for Bhabha scattering in QED [23, 24], for the heavy-to-light quark decay [25–29] and the $t\bar{t}$ production [30–33] in QCD can be applied to elastic μe scattering as well.

*e-mail: pierpaolo.mastrolia@pd.infn.it

In [34], we took a first step towards the calculation of the full NNLO QED corrections to μe scattering. In particular, we considered the evaluation of the master integrals (MIs) occurring in the decomposition of the genuine two-loop $2 \to 2$ planar box-diagrams, namely all the two-loop four-point topologies for μe scattering except for the crossed double box diagram. Given the small value of the electron mass m_e when compared to the muon one m, we worked in the approximation $m_e = 0$. In this case, integration-by-parts identities [35–37] yielded the identification of a set of 65 MIs, which we computed analytically by means of the differential equation method [38–40]. Elaborating on recent ideas to simplify the system-solving strategy [41, 42], we chose a set of MIs obeying a system of first-order differential equations (DEQs) in the kinematical variables s/m^2 and t/m^2 which is linear in the space-time dimension d, and, by means of Magnus exponential matrix [42], we derived an equivalent system of equations in *canonical form* [41], where the d-dependence of the associated matrices is factorized from the kinematics. Let us emphasize that the use of Magnus exponential matrix to identify a canonical basis of master integrals turned out to be very effective in the context of multi-loop integrals involving several scales [42–45]. We found that the matrices associated with the canonical systems admit a logarithmic-differential ($d$log) form, whose entries are rational functions of the kinematics; therefore, the canonical MIs can be cast in a Taylor series around $d = 4$, with coefficients written as combinations of generalised polylogarithms (GPLs) [46–49]. The final determination of the MIs was achieved after imposing the boundary conditions, implemented by requiring the regularity of the solutions at special kinematics points, and by using simpler integrals as independent input.

The analytic expressions of the MIs were numerically evaluated with the help of `GiNaC` [50] and were successfully tested against the values provided by the computer code `SecDec` [51]. The package `Reduze` [52] was used throughout the calculations.

It is important to observe that the MIs of the QED corrections to $\mu e \to \mu e$ scattering are related by crossing to the MIs of the QCD corrections to the $t\bar{t}$-pair production at hadron colliders. The analytic evaluation of the MIs for the leading-color corrections to $pp \to t\bar{t}$, due to planar diagrams only, was already considered in refs. [30–33]. They correspond to the MIs appearing in the evaluation of the Feynman graphs associated to the topologies T_i with $i \in \{1,2,3,7,8,9,10\}$ in figure 1, which we (re)computed in [34], independently. The MIs for the planar topology T_4 and T_5, instead, would correspond to the MIs of subleading-color contributions to $t\bar{t}$-pair production, and were not considered previously.

For certain classes of MIs, like the ones of the processes $\mu e \to \mu e$ and $pp \to t\bar{t}$, the choice of the boundary conditions may still constitute a challenging problem. In some cases considered in refs. [30–33], the direct integration of the MIs in special kinematic configurations was addressed by using techniques based on Mellin-Barnes representations [53, 54]. Alternatively, our approach exploited either the regularity conditions at pseudo-thresholds or the expression of the integrals at well-behaved kinematic points. The latter were obtained by solving simpler auxiliary systems of differential equations, hence limiting the use of direct integration only to a simple set of input integrals. Our preliminary studies make us believe that the strategy we adopted for the determination of the considered integrals is not only limited to the planar contributions, but it can be applied to the non-planar graphs as well. In particular, in [34], we showed its application for the determination of the MIs for the non-planar vertex graph [25–29]. Moreover, due to the similarity of the cases, we are confident that it can be very helpful for completing the analytic evaluation of the MIs needed for the two-loop QCD corrections to $pp \to t\bar{t}$, which are currently known only numerically [55–59].

The evaluation of the MIs is only a first step towards the complete evaluation of the two-loop amplitudes. By means of the *adaptive integrand decomposition* [60–62] and the integration-by-parts identities, we can decompose the whole amplitude in terms of MIs. To achieve this task, we have been developing a general framework for the automatic evaluation of two-loop amplitudes, called AIDA (Adaptive Integrand Decomposition Algorithm), and we present its first application to the case of μe-scattering.

In the following, we report on our findings.

2 LO cross section and NLO QED corrections

Let us consider the elastic scattering

$$\mu^+(p_1) + e^-(p_2) \to e^-(p_3) + \mu^+(p_4), \tag{1}$$

and define the Mandelstam variables

$$s = (p_1+p_2)^2, \quad t = (p_2-p_3)^2, \quad u = (p_1-p_3)^2, \tag{2}$$

satisfying $s+t+u = 2m^2+2m_e^2$, with the physical requirements $s > (m_e+m)^2$, $-\lambda(s,m^2,m_e^2)/s < t < 0$, and $\lambda(x,y,z) = x^2+y^2+z^2-2xy-2xz-2yz$ is the Källen function.

The LO QED prediction for the differential cross section of the scattering in (1) is

$$\frac{d\sigma_0}{dt} = 4\pi\alpha^2 \frac{\left(m^2+m_e^2\right)^2 - su + t^2/2}{t^2\lambda\left(s,m^2,m_e^2\right)}, \tag{3}$$

where α is the fine-structure constant. The NLO QED corrections to this cross section were computed long time ago [13–18] and revisited more recently [19]. As a first check, we recalculated these corrections and found perfect agreement with ref. [19], both for the virtual corrections and the soft photon emissions. We note that some of the pioneering publications, like [14, 16], contain typos or errors, so that they cannot be directly employed.

In the rest of this paper we will work in the approximation of vanishing electron mass, $m_e = 0$, i.e. with the kinematics specified by $p_1^2 = p_4^2 = m^2$ and $p_2^2 = p_3^2 = 0$. The master integrals will be conveniently evaluated in the non-physical region $s < 0, t < 0$.

3 Four-point topologies

We focus on the evaluation of the master integrals (MIs) of the planar two-loop four-point functions contributing to μe scattering, drawn in figure 1. For completeness, we will discuss also the evaluation of the MIs of the one-loop four-point function in figure 2.

We consider ℓ-loop m-denominator Feynman integrals in $d = 4-2\epsilon$ dimensions of the type

$$\int \prod_{i=1}^{\ell} \widetilde{\mathrm{d}^d k_i} \frac{1}{D_1^{n_1}\ldots D_m^{n_m}}, \quad n_i \in \mathbb{Z}. \tag{4}$$

In our conventions, the integration measure is defined as

$$\widetilde{\mathrm{d}^d k_i} = \frac{\mathrm{d}^d k_i}{(2\pi)^d}\left(\frac{i S_\epsilon}{16\pi^2}\right)^{-1}\left(\frac{m^2}{\mu^2}\right)^{\epsilon}, \tag{5}$$

with μ being the 't Hooft scale of dimensional regularization and

$$S_\epsilon = (4\pi)^\epsilon \Gamma(1+\epsilon). \tag{6}$$

We display the relevant planar four-point topologies at one- and two-loop in families:

- the one-loop integral family, depicted in figure 2;
- the first two-loop integral family, which includes the topologies T_1, T_2, T_3, T_7 and T_8 of figure 1;
- the second two-loop family, which contains topologies T_4, T_5, T_9 and T_{10} shown in figure 1;

For all families, k_1 and k_2 denote the loop momenta. In the following sections, MIs will be represented by diagrams where thick lines stand for massive particles (muon), whereas thin lines stand for massless ones (electron, photon).

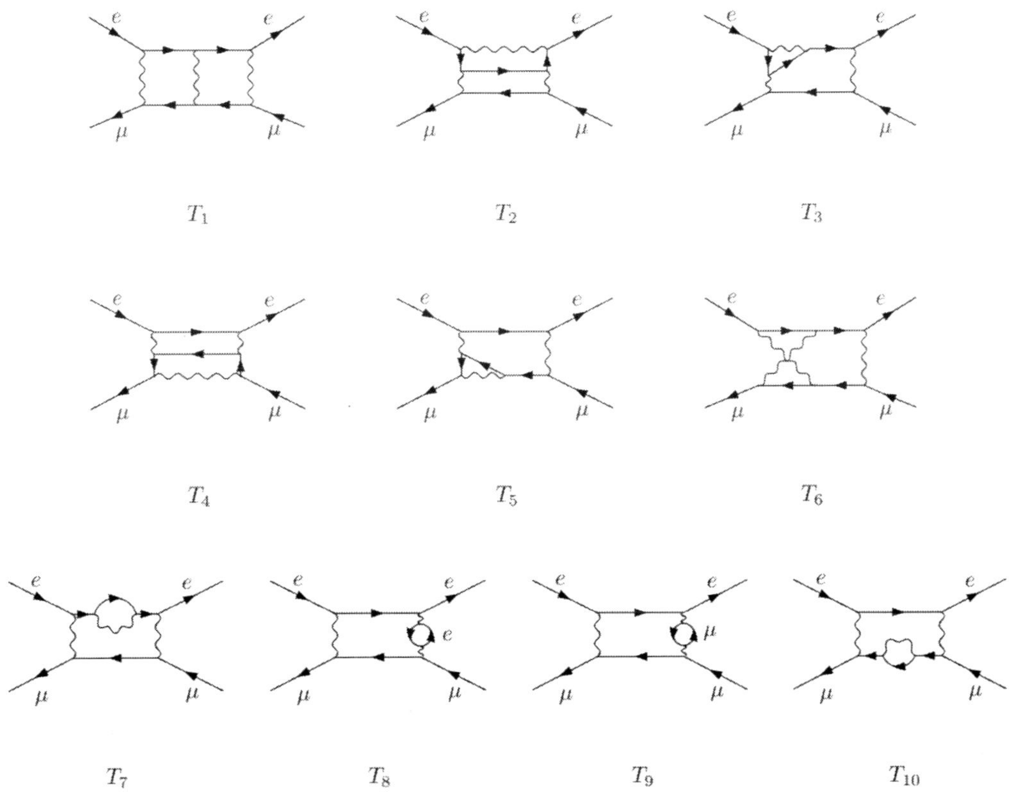

Figure 1: Two-loop four-point topologies for μe scattering

4 System of differential equations

In order to determine all MIs appearing in the three integral families defined above, we initially derive their DEQs in the dimensionless variables $-s/m^2$ and $-t/m^2$. Upon the change of variable,

$$-\frac{s}{m^2} = x, \quad -\frac{t}{m^2} = \frac{(1-y)^2}{y}, \tag{7}$$

the coefficients of the DEQs are rational functions of x and y. According to our system solving strategy, by means of integration-by-parts identities (IBPs), we identify an initial set of MIs $\mathbf{F}$ that fulfills a system of DEQs

$$\frac{\partial \mathbf{F}}{\partial x} = \mathbb{A}_x(\epsilon, x, y)\mathbf{F}, \qquad \frac{\partial \mathbf{F}}{\partial y} = \mathbb{A}_y(\epsilon, x, y)\mathbf{F}, \tag{8}$$

where the matrices $\mathbb{A}_x(\epsilon, x, y)$ and $\mathbb{A}_y(\epsilon, x, y)$ are *linear* in the dimensional regularization parameter $\epsilon = (4-d)/2$, being d the number of space-time dimensions. According to the algorithm described in [42–45], by means of Magnus exponential matrix, we identify a set of MIs $\mathbf{I}$ obeying canonical systems of DEQs [41], where the dependence on ϵ is factorized from the kinematics,

$$\frac{\partial \mathbf{I}}{\partial x} = \epsilon\hat{\mathbb{A}}_x(x, y)\mathbf{I}, \qquad \frac{\partial \mathbf{I}}{\partial y} = \epsilon\hat{\mathbb{A}}_y(x, y)\mathbf{I}. \tag{9}$$

After combining both systems of DEQs into a single total differential, we arrive at the following canonical form

$$d\mathbf{I} = \epsilon d\mathbb{A}\mathbf{I}, \qquad d\mathbb{A} \equiv \hat{\mathbb{A}}_x dx + \hat{\mathbb{A}}_y dy, \tag{10}$$

where the generic form of the total differential matrix for the considered MIs reads as,

$$d\mathbb{A} = \sum_{i=1}^{9} \mathbb{M}_i \, d\log(\eta_i), \tag{11}$$

with $\mathbb{M}_i$ being constant matrices. The arguments η_i of this $d$log-form, which contain all the dependence of the DEQ on the kinematics, are referred to as the *alphabet* and they consist in the following 9 letters:

$$\begin{aligned} &\eta_1 = x, && \eta_2 = 1 + x, \\ &\eta_3 = 1 - x, && \eta_4 = y, \\ &\eta_5 = 1 + y, && \eta_6 = 1 - y,, \\ &\eta_7 = x + y, && \eta_8 = 1 + xy, \\ &\eta_9 = 1 - y(1 - x - y). \end{aligned} \tag{12}$$

The MIs presented in this paper are computed in the kinematic region where all letters are real and positive,

$$x > 0, \quad 0 < y < 1, \tag{13}$$

which corresponds to the Euclidean region $s < 0$, $t < 0$. All MIs are chosen to be finite in the $\epsilon \to 0$ limit, in such a way that $\mathbf{I}(x, y)$ admits a Taylor expansion in ϵ,

$$\mathbf{I}(\epsilon, x, y) = \mathbf{I}^{(0)}(x, y) + \epsilon\,\mathbf{I}^{(1)}(x, y) + \epsilon^2 \mathbf{I}^{(2)}(x, y) + \ldots\,, \tag{14}$$

with the n-th order coefficient given by

$$\mathbf{I}^{(n)}(x, y) = \sum_{i=0}^{n} \Delta^{(n-i)}(x, y; x_0, y_0) \mathbf{I}^{(i)}(x_0, y_0), \tag{15}$$

where $\mathbf{I}^{(i)}(x_0, y_0)$ is a vector of boundary constants and $\Delta^{(k)}$ the weight-k operator

$$\Delta^{(k)}(x, y; x_0, y_0) = \int_\gamma \underbrace{d\mathbb{A} \ldots d\mathbb{A}}_{\text{k times}} \qquad \Delta^{(0)}(x, y; x_0, y_0) = 1\,, \tag{16}$$

which iterates k ordered integrations of the matrix-valued 1-form $d\mathbb{A}$ along a piecewise-smooth path γ in the xy-plane. Since the alphabet given in eq. (12) is rational and has only algebraic roots, the iterated integrals (16) can be directly expressed in terms of GPLs, which are defined as

$$G(\vec{w}_n; x) \equiv G(w_1, \vec{w}_{n-1}; x) \equiv \int_0^x dt \frac{1}{t - w_1} G(\vec{w}_{n-1}; t), \tag{17}$$

$$G(\vec{0}_n; x) \equiv \frac{1}{n!} \log^n(x), \tag{18}$$

with $\vec{w}_n$ being a vector of n arguments. The number n is referred to as the *weight* of $G(\vec{w}_n; x)$ and amounts to the number of iterated integrations needed to define it. Equivalently one has

$$\frac{\partial}{\partial x} G(\vec{w}_n; x) = \frac{\partial}{\partial x} G(w_1, \vec{w}_{n-1}; x) = \frac{1}{x - w_1} G(\vec{w}_{n-1}; x). \tag{19}$$

GPLs fulfill shuffle algebra relations of the form

$$G(\vec{m}; x)\, G(\vec{n}; x) = G(\vec{m}; x) ⧢ G(\vec{n}; x) = \sum_{\vec{p} = \vec{m} ⧢ \vec{n}} G(\vec{p}; x), \tag{20}$$

where the shuffle product $\vec{m} ⧢ \vec{n}$ denotes all possible merges of $\vec{m}$ and $\vec{n}$ while preserving their respective orderings.

The analytic continuation of the MIs to the physical region defined in sec. 2 can be obtained through by-now standard techniques.

5 One-loop master integrals

In this section we briefly discuss the computation of the master integrals of the one-loop four-point graph shown in figure 2. We choose the following set of MIs, which satisfy an ϵ-linear DEQ,

$$\mathrm{F}_1 = \epsilon\,\mathcal{T}_1\,, \qquad \mathrm{F}_2 = \epsilon\,\mathcal{T}_2\,, \qquad \mathrm{F}_3 = \epsilon\,\mathcal{T}_3\,,$$
$$\mathrm{F}_4 = \epsilon^2\,\mathcal{T}_4\,, \qquad \mathrm{F}_5 = \epsilon^2\,\mathcal{T}_5\,, \tag{21}$$

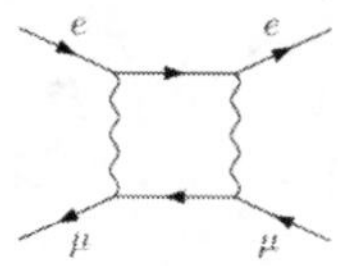

Figure 2: One-loop four-point topology for μe scattering

where the $\mathcal{T}_i$ are depicted in figure 3. With the help of the Magnus algorithm we can identify the corresponding canonical basis

$$\mathrm{I}_1 = \mathrm{F}_1\,, \qquad \mathrm{I}_2 = -s\,\mathrm{F}_2\,,$$
$$\mathrm{I}_3 = -t\mathrm{F}_3\,, \qquad \mathrm{I}_4 = \lambda_t\,\mathrm{F}_4\,,$$
$$\mathrm{I}_5 = (s - m^2)\mathrm{F}_5\,. \tag{22}$$

with $\lambda_t = \sqrt{-t}\,\sqrt{4m^2 - t}$.

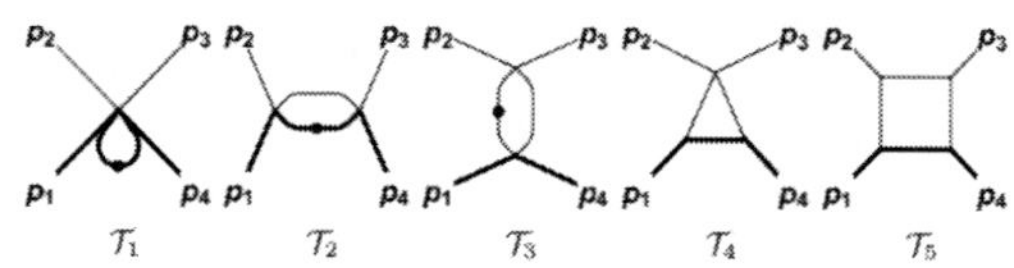

Figure 3: One-loop MIs $\mathcal{T}_{1,\ldots,5}$.

This set of MIs satisfies a canonical DEQ of the form given in eq. (10). The integration of the DEQ in terms of GPLs as well as the fixing of boundary constants is straightforward. $\mathrm{I}_{1,3}$ are obtained by direct integration and, by using the normalization of eq.(5), are given by

$$\mathrm{I}_1(\epsilon) = 1, \mathrm{I}_3(\epsilon, y) = \left(\frac{(1-y)^2}{y}\right)^{-\epsilon} \left(1 - \zeta_2 \epsilon^2 - 2\zeta_3 \epsilon^3 + O\left(\epsilon^4\right)\right). \tag{23}$$

The boundary constants for I_2, I_4 and I_5 can be fixed by respectively demanding regularity at pseudothresholds $s \to 0$, at $t \to 4m^2$, and at $s = -t \to m^2/2$. The final expression of the other MIs are,

$$\mathrm{I}_i(\epsilon, x, y) = \sum_{k=0}^{2} \mathrm{I}_i^{(k)}(x, y)\epsilon^k + O(\epsilon^3)\,, \tag{24}$$

with

$$\begin{aligned}
\mathrm{I}_2^{(0)}(x) &= 0\,,\\
\mathrm{I}_2^{(1)}(x) &= -\,G(-1; x)\,,\\
\mathrm{I}_2^{(2)}(x) &= 2G(-1, -1; x) - G(0, -1; x)\,,
\end{aligned} \tag{25}$$

$$\begin{aligned}
\mathrm{I}_4^{(0)}(y) &= 0\,,\\
\mathrm{I}_4^{(1)}(y) &= 0\,,\\
\mathrm{I}_4^{(2)}(y) &= -\,4\zeta_2 - G(0, 0; y) + 2G(0, 1; y)\,,
\end{aligned} \tag{26}$$

$$\begin{aligned}
\mathrm{I}_5^{(0)}(x, y) &= \;\; 2\,,\\
\mathrm{I}_5^{(1)}(x, y) &= -\,2G(-1; x) + G(0; y) - 2G(1; y)\,,\\
\mathrm{I}_5^{(2)}(x, y) &= -\,5\zeta_2 + 2G(-1; x)\,(2G(1; y) - G(0; y))\,.
\end{aligned} \tag{27}$$

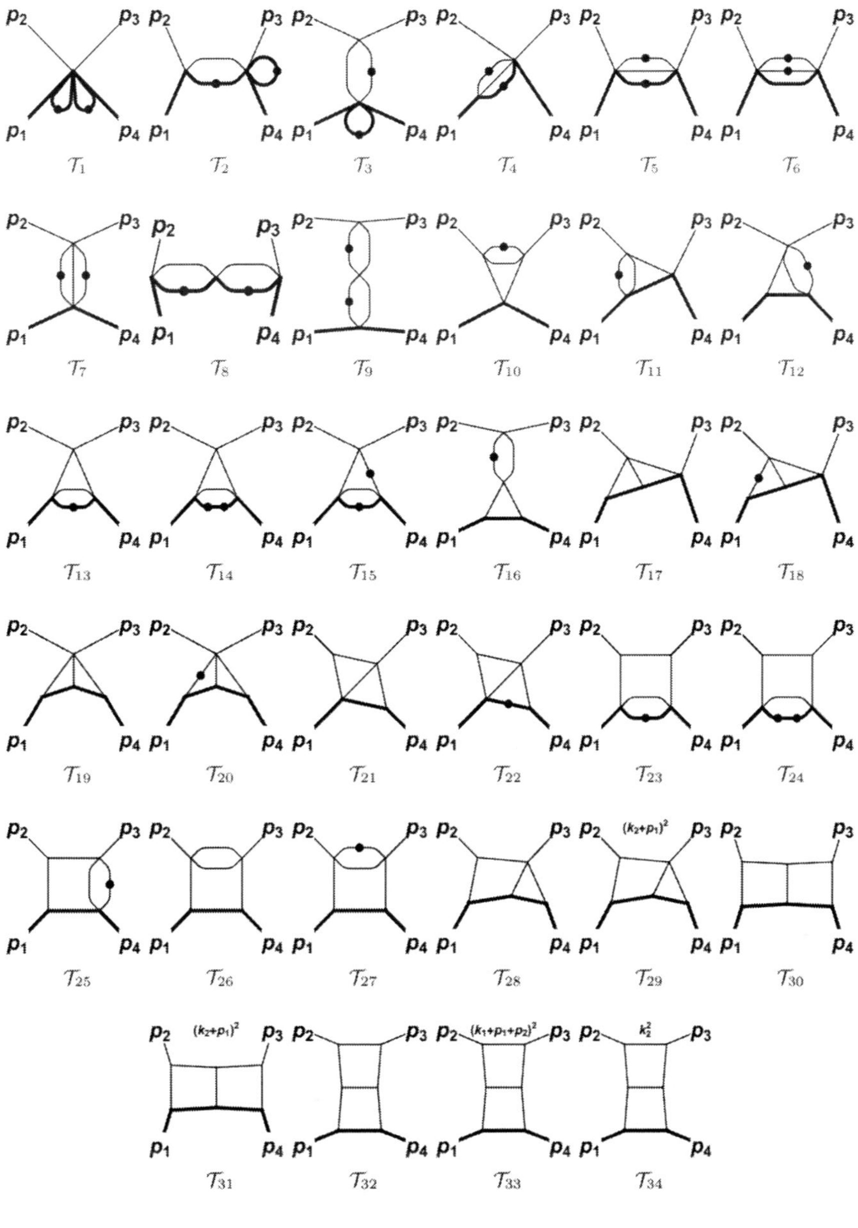

Figure 4: Two-loop MIs $\mathcal{T}_{1,\ldots,34}$ for the first integral family.

6 Two-loop master integrals

In this section we present the results for the planar two-loop MIs contributing to the NNLO virtual QED corrections to μe scattering.

6.1 The first integral family

For the two-loop first integral family associated to the topologies T_1, T_2, T_3, T_7 and T_8 of figure 1, the following set of 34 MIs fulfill an ϵ-linear system of DEQs,

$$
\begin{array}{lll}
F_1 = \epsilon^2 \mathcal{T}_1\,, & F_2 = \epsilon^2 \mathcal{T}_2\,, & F_3 = \epsilon^2 \mathcal{T}_3\,, \\
F_4 = \epsilon^2 \mathcal{T}_4\,, & F_5 = \epsilon^2 \mathcal{T}_5\,, & F_6 = \epsilon^2 \mathcal{T}_6\,, \\
F_7 = \epsilon^2 \mathcal{T}_7\,, & F_8 = \epsilon^2 \mathcal{T}_8\,, & F_9 = \epsilon^2 \mathcal{T}_9\,, \\
F_{10} = \epsilon^3 \mathcal{T}_{10}\,, & F_{11} = \epsilon^3 \mathcal{T}_{11}\,, & F_{12} = \epsilon^3 \mathcal{T}_{12}\,, \\
F_{13} = \epsilon^3 \mathcal{T}_{13}\,, & F_{14} = \epsilon^2 \mathcal{T}_{14}\,, & F_{15} = \epsilon^2 \mathcal{T}_{15}\,, \\
F_{16} = \epsilon^3 \mathcal{T}_{16}\,, & F_{17} = \epsilon^4 \mathcal{T}_{17}\,, & F_{18} = \epsilon^3 \mathcal{T}_{18}\,,
\end{array}
$$

$$\begin{aligned}
&F_{19} = \epsilon^4 \mathcal{T}_{19}, \quad F_{20} = \epsilon^2(1+2\epsilon)\mathcal{T}_{20}, \quad F_{21} = \epsilon^4 \mathcal{T}_{21},\\
&F_{22} = \epsilon^3 \mathcal{T}_{22}, \quad F_{23} = \epsilon^3 \mathcal{T}_{23}, \quad F_{24} = \epsilon^2 \mathcal{T}_{24},\\
&F_{25} = \epsilon^3 \mathcal{T}_{25}, \quad F_{26} = \epsilon^3(1-2\epsilon)\mathcal{T}_{26}, \quad F_{27} = \epsilon^3 \mathcal{T}_{27},\\
&F_{28} = \epsilon^4 \mathcal{T}_{28}, \quad F_{29} = \epsilon^3(1-2\epsilon)\mathcal{T}_{29}, \quad F_{30} = \epsilon^4 \mathcal{T}_{30},\\
&F_{31} = \epsilon^4 \mathcal{T}_{31}, \quad F_{32} = \epsilon^4 \mathcal{T}_{32}, \quad F_{33} = \epsilon^4 \mathcal{T}_{33},\\
&F_{34} = \epsilon^4 \mathcal{T}_{34}, \qquad (28)
\end{aligned}$$

where the $\mathcal{T}_i$ are depicted in figure 4.

Through the Magnus exponential, we rotate this set of integrals to the canonical basis

$$\begin{aligned}
I_1 &= F_1,\\
I_2 &= -s\,F_2,\\
I_3 &= -tF_3,\\
I_4 &= m^2\,F_4,\\
I_5 &= -sF_5,\\
I_6 &= 2m^2\,F_5 + (m^2-s)\,F_6,\\
I_7 &= -tF_7,\\
I_8 &= s^2\,F_8,\\
I_9 &= t^2F_9,\\
I_{10} &= -t\,F_{10},\\
I_{11} &= (m^2-s)\,F_{11},\\
I_{12} &= \lambda_t\,F_{12},\\
I_{13} &= \lambda_t\,F_{13},\\
I_{14} &= \lambda_t\,m^2\,F_{14},\\
I_{15} &= (t-\lambda_t)\left(\frac{3}{2}F_{13} + m^2F_{14}\right) - m^2\,t\,F_{15},\\
I_{16} &= -t\,\lambda_t\,F_{16},\\
I_{17} &= (m^2-s)\,F_{17},\\
I_{18} &= m^2(m^2-s)\,F_{18},\\
I_{19} &= \lambda_t\,F_{19},\\
I_{20} &= \frac{\lambda_t - t}{2}(F_{12} - 4\,F_{19}) - m^2t\,F_{20},\\
I_{21} &= (m^2-s-t)\,F_{21},\\
I_{22} &= -m^2\,t\,F_{22},\\
I_{23} &= s\,t\,F_{23},\\
I_{24} &= -m^2\,t\,F_{23} + (s-m^2)\,m^2\,t\,F_{24},\\
I_{25} &= -(m^2-s)\,t\,F_{25},\\
I_{26} &= \lambda_t\,F_{26},\\
I_{27} &= -(m^2-s)\,t\,F_{27},\\
I_{28} &= (m^2-s)\,\lambda_t\,F_{28},\\
I_{29} &= -2t\,F_{21} - (m^2-s)(2(\lambda_t - t)F_{28} - F_{29}),\\
I_{30} &= -(m^2-s)^2t\,F_{30},\\
I_{31} &= (m^2-s)^2\,F_{31},\\
I_{32} &= (m^2-s)\,t^2\,F_{32},\\
I_{33} &= -\lambda_t\,t\,F_{33},\\
I_{34} &= -m^2\,t^2\,F_{32} + t^2\,F_{34}. \qquad (29)
\end{aligned}$$

This set of MIs **I** satisfies a system of DEQ of the form given in eq.(10), which can easily be integrated in terms of GPLs.

6.2 The second integral family

For the second two-loop integral family which groups the topologies T_4, T_5, T_9 and T_{10} of figure 1, we identify 42 MIs obeying an ϵ-linear system of DEQs:

$$\begin{aligned}
&F_1 = \epsilon^2\mathcal{T}_1, && F_2 = \epsilon^2\mathcal{T}_2, && F_3 = \epsilon^2\mathcal{T}_3,\\
&F_4 = \epsilon^2\mathcal{T}_4, && F_5 = \epsilon^2\mathcal{T}_5, && F_6 = \epsilon^2\mathcal{T}_6,\\
&F_7 = \epsilon^2\mathcal{T}_7, && F_8 = \epsilon^2\mathcal{T}_8, && F_9 = \epsilon^2\mathcal{T}_9,\\
&F_{10} = \epsilon^2\mathcal{T}_{10}, && F_{11} = \epsilon^2\mathcal{T}_{11}, && F_{12} = \epsilon^3\mathcal{T}_{12},\\
&F_{13} = \epsilon^2\mathcal{T}_{13}, && F_{14} = \epsilon^2\mathcal{T}_{14}, && F_{15} = \epsilon^3\mathcal{T}_{15},\\
&F_{16} = \epsilon^2\mathcal{T}_{16}, && F_{17} = \epsilon^2\mathcal{T}_{17}, && F_{18} = \epsilon^3\mathcal{T}_{18},\\
&F_{19} = \epsilon^3\mathcal{T}_{19}, && F_{20} = \epsilon^2\mathcal{T}_{20}, && F_{21} = \epsilon^3\mathcal{T}_{21},\\
&F_{22} = \epsilon^2\mathcal{T}_{22}, && F_{23} = \epsilon^3\mathcal{T}_{23}, && F_{24} = \epsilon^2\mathcal{T}_{24},\\
&F_{25} = \epsilon^3\mathcal{T}_{25}, && F_{26} = (1-2\epsilon)\epsilon^3\mathcal{T}_{26}, && F_{27} = \epsilon^3\mathcal{T}_{27},\\
&F_{28} = \epsilon^2\mathcal{T}_{28}, && F_{29} = \epsilon^3\mathcal{T}_{29}, && F_{30} = \epsilon^2\mathcal{T}_{30},\\
&F_{31} = (1-2\epsilon)\epsilon^3\mathcal{T}_{31}, && F_{32} = \epsilon^3\mathcal{T}_{32}, && F_{33} = \epsilon^4\mathcal{T}_{33},\\
&F_{34} = \epsilon^3\mathcal{T}_{34}, && F_{35} = \epsilon^3\mathcal{T}_{35}, && F_{36} = \epsilon^4\mathcal{T}_{36},\\
&F_{37} = \epsilon^4\mathcal{T}_{37}, && F_{38} = \epsilon^3\mathcal{T}_{38}, && F_{39} = \epsilon^4\mathcal{T}_{39},\\
&F_{40} = \epsilon^4\mathcal{T}_{40}, && F_{41} = \epsilon^4\mathcal{T}_{41},\\
&F_{42} = \epsilon^4\,(\mathcal{T}_{26} + \mathcal{T}_{42}), && && (30)
\end{aligned}$$

where the $\mathcal{T}_i$ are depicted in figure 5. Through the Magnus exponential, we identify the corresponding canonical basis:

$$\begin{aligned}
I_1 &= F_1,\\
I_2 &= -t\,F_2,\\
I_3 &= \lambda_tF_3,\\
I_4 &= -t\,F_4,\\
I_5 &= \frac{1}{2}(\lambda_t - t)\,F_4 - \lambda_t\,F_5,\\
I_6 &= -s\,F_6,\\
I_7 &= 2m^2\,F_6 + (m^2-s)\,F_7,\\
I_8 &= m^2\,F_8,\\
I_9 &= m^2F_9,\\
I_{10} &= -s\,F_{10},\\
I_{11} &= -t\,\lambda_t\,F_{11},\\
I_{12} &= -t\,F_{12},\\
I_{13} &= -t\,m^2\,F_{13},\\
I_{14} &= -m^2(\lambda_t - t)\left(\frac{3}{2}F_{12} + F_{13}\right) - m^2\,\lambda_t\,F_{14},\\
I_{15} &= \lambda_t\,F_{15},\\
I_{16} &= m^2\,\lambda_t\,F_{16},\\
I_{17} &= m^2(t-\lambda_t)\left(\frac{3}{2}F_{15} + F_{16}\right) - m^2\,t\,F_{17},\\
I_{18} &= \lambda_t\,F_{18},
\end{aligned}$$

$$
\begin{aligned}
I_{19} &= (m^2 - s)\,F_{19}\,,\\
I_{20} &= m^2\,(m^2 - s)\,F_{20}\,,\\
I_{21} &= (m^2 - s)\,F_{21}\,,\\
I_{22} &= -\frac{3}{2}s\,F_9 + (s^2 - m^4)\,F_{22}\,,\\
I_{23} &= \lambda_t\,F_{23}\,,\\
I_{24} &= \frac{1}{4}\left(4m^2 - t + \lambda_t\right)(F_4 + 2F_5) + m^2(4m^2 - t)\,F_{24}\,,\\
I_{25} &= \lambda_t\,F_{25}\,,\\
I_{26} &= -t\,F_{26}\,,\\
I_{27} &= s\,t\,F_{27}\,,\\
I_{28} &= -m^4\,t\,F_{27} - m^2(m^2 - s)\,t\,F_{28}\,,\\
I_{29} &= -s\,\lambda_t\,F_{29}\,,\\
I_{30} &= m^4\lambda_t\,F_{29} + m^2\,(m^2 - s)\,\lambda_t\,F_{30}\,,\\
I_{31} &= -(m^2 - s)F_{31} - (m^2 - s)\,(4m^2 - t + \lambda_t)\,F_{32}\,,\\
I_{32} &= (m^2 - s)\lambda_t F_{32}\,,\\
I_{33} &= (m^2 - s - t)\,F_{33}\,,\\
I_{34} &= (m^2 - s)\,\lambda_t\,F_{34}\,,\\
I_{35} &= 2\frac{m^4(m^2 - s)}{2m^2 - t - \lambda_t}\,F_{34} + m^2\,(m^2 - s)F_{35}\,,\\
I_{36} &= \lambda_t\,F_{36}\,,\\
I_{37} &= -t\,(4m^2 - t)\,F_{37}\,,\\
I_{38} &= -(m^2 - s)\,t\,F_{38}\,,\\
I_{39} &= -(m^2 - s)\,tF_{39}\,,\\
I_{40} &= -(m^2 - s)\,t\,\lambda_t\,F_{40}\,,\\
I_{41} &= t\lambda_t\,(F_{40} - F_{41})\,,\\
I_{42} &= (m^2 - t + \lambda_t)\times\\
&\quad\left(\frac{2}{3}F_3 + \frac{1}{4}F_4 + \frac{1}{2}F_5 - \frac{1}{2}\,t\,F_{11} + \frac{5}{2}F_{12} + \frac{5}{3}m^2F_{13}\right.\\
&\quad\left. + \frac{5}{3}m^2F_{14} + 2F_{36} - \frac{1}{2}(m^2 + s)F_{40} + t\,F_{41}\right) +\\
&\quad + m^2\left(\frac{1}{3}F_3 - \frac{1}{2}tF_{11} + \frac{1}{2}F_{12} + \frac{1}{3}m^2F_{13} + \frac{1}{3}m^2F_{14} +\right.\\
&\quad\left. + \frac{1}{2}F_{18} - \frac{1}{2}F_{40}\right) +\\
&\quad - t\,(m^2 - s)\,F_{11} - 2\frac{m^4}{2m^2 - t - \lambda_t}F_{15} + t\,F_{26} +\\
&\quad + \frac{m^2(m^2 - s)(t + \lambda_t)}{2m^2 - t - \lambda_t}\left(\frac{2}{3}F_{29} - F_{34}\right) +\\
&\quad - \frac{2}{3}\frac{m^2\,s\,(t - \lambda_t)}{2m^2 - t - \lambda_t}F_{29} + 2t\,F_{33} + \frac{4}{3}tm^4\frac{m^2 - s}{\lambda_t + t}F_{30} - t\,F_{42}\,,
\end{aligned}
\tag{31}
$$

which satisfies a system of DEQs of the form in eq.(10). We observe that $I_{1,2,6,7,8,10,15,16,17,27,28}$ correspond, respectively, to $I_{1,3,5,6,4,2,13,14,15,23,24}$ of the first integral family, previously discussed.

To determine the solution of the DEQ for the MIs of both two-loop families, we choose proper boundary values for each master integral. The boundary fixing can be achieved either by knowing the integral at some special kinematic point or by demanding the absence of unphysical thresholds that appear in the alphabet of the generic solution, defined in eq. (12). For more details, we forward the reader to Ref.[34].

All results have been numerically checked with the help of the computer codes `GiNaC` and `SecDec`.

7 Towards the non-planar integrals

The complete computation of the NNLO virtual QED corrections to μe scattering requires the evaluation of one last missing four-point topology, which corresponds to the non-planar diagram T_6 of figure 1. We are confident that the previously adopted strategy, based on differential equations, Magnus exponential and regularity conditions, can be efficiently applied to compute the MIs of a simpler vertex integral belonging to same family.

8 Adaptive Integrand Decomposition

The decomposition of multiloop scattering amplitudes in terms of independent functions, together with the subsequent determination of the latter, is a viable alternative to the direct integration which, for non-trivial processes, may require the calculation of a prohibitively large number of complicated Feynman integr*als*.

Decomposing multi-loop amplitudes in terms of independent integrals can become problematic when the number of the scales of the diagrams increases, due to the exchange or to the production of massive particles, or when a large number of external particles are scattered, or when the morphology of the contributing diagrams becomes involved. The integr*and* decomposition algorithm has the advantage of treating scattering amplitudes involving massive particles at the same *price* of amplitudes for massless scattering. The output of the reduction procedure is the partial fractioning of the original integrand, namely the determination of the remainders of the successive division between the numerator and (the partitions of the product of) the denominators. Upon integration, the partial fraction formula correspond to rewrite the original amplitudes as a combination of independent integrals. However, the result of the integrand decomposition represents an intermediate step towards the complete amplitude reduction. In fact, additional relations among those integrals, like integration-by-parts identities, can minimise the number of independent master integrals (MIs) which can appear in the final formulas.

The integrand decomposition algorithm [63–69] played a key role for the automation of one-loop corrections to high-multiplicity scattering processes. The extension of this approach at two-loop and beyond [70–73] has been under intense investigation. The recent developments on the integrand side have been accompanied by important developments for novel derivation of the integral relations needed to identify MIs [74–77], as well as by progress in the ability of computing the latter analytically [41, 42, 78] as well as numerically [51, 79]. This vivid research has been largely due to the deeper understanding of the properties of the integrands of Feynman graph, and of the refined algebraic and differential calculus which control them.

In this section, we recall the main features of the Adaptive Integrand Decomposition (AID) [60–62], implemented in the package AIDA, and show its first applications to the μe-scattering.

Witin AID, the space-time dimension, $d = 4 - 2\epsilon$, are decomposed diagram-by-diagram, as $d = d_\parallel + d_\perp$, into a space spanned by the external momenta flowing in the leg - with dimensions $d_\parallel$ - referred to as parallel (or longitudinal) space, and a space spanned by the external momenta flowing in the leg - with dimensions $d_\perp$ - referred to as orthogonal (or transverse) space. The latter is formed by the union of the four-dimensional complement of the longitudinal space, and of the extra dimensional -2ϵ-space.

In the structure of the Feynman integrals,

$$I^{(\ell)}_{i_1\cdots i_n}[\mathcal{N}] \equiv \int \left(\prod_{i=1}^{\ell} \frac{d^d k_i}{\pi^{d/2}}\right) \mathcal{I}_{i_1\cdots i_n} = \int \left(\prod_{i=1}^{\ell} \frac{d^d k_i}{\pi^{d/2}}\right) \frac{\mathcal{N}_{i_1\cdots i_n}(k_i)}{\prod_j D_j(k_j)}, \tag{32}$$

d-dimensional loop momenta are defined as

$$k_i^\alpha = k_{\parallel i}^\alpha + \lambda_i^\alpha, \tag{33}$$

with

$$k_{\parallel i}^\alpha = \sum_{j=1}^{d_\parallel} x_{ji}\, e_j^\alpha,$$
$$\lambda_i^\alpha = \sum_{j=d_\parallel+1}^{4} x_{ji}\, e_j^\alpha + \mu_i^\alpha,$$
$$\lambda_{ij} \equiv \lambda_i \cdot \lambda_j = \sum_{l=d_\parallel+1}^{4} x_{li}\, x_{lj} + \mu_{ij}. \tag{34}$$

In Eq. (33), $l_{\parallel i}$ is a vector of the $d_\parallel$-dimensional space spanned by the external momenta, and λ_i belongs the $d_\perp$-dimensional orthogonal subspace. Moreover $e^\alpha_{i\ i=1,\ldots,4}$ is a 4-dimensional basis, while μ_i^α lie in the -2ϵ-space. Within this parametrisation, denominators appear to depend on less variables than the numerators, yielding a simplification of the decomposition procedure.

Let us indicate with $\mathbf{z}$ the full set of $\ell(\ell+9)/2$ variables

$$\mathbf{z} = \{\mathbf{x}_{\parallel i}, \mathbf{x}_{\perp i}, \lambda_{ij}\}, \quad i, j = 1, \ldots \ell, \tag{35}$$

where $\mathbf{x}_{\parallel i}$ ($\mathbf{x}_{\perp i}$) are the components of the loop momenta parallel (orthogonal) to the external kinematics, the denominators are reduced to polynomials in the subset of variables

$$\boldsymbol{\tau} = \{\mathbf{x}_\parallel, \lambda_{ij}\}, \quad \boldsymbol{\tau} \subset \mathbf{z}, \tag{36}$$

so that the general r-point integrand has the form

$$\mathcal{I}_{i_1\ldots i_r}(\boldsymbol{\tau}, \mathbf{x}_\perp) \equiv \frac{\mathcal{N}_{i_1\ldots i_r}(\boldsymbol{\tau}, \mathbf{x}_\perp)}{D_{i_1}(\boldsymbol{\tau}) \cdots D_{i_r}(\boldsymbol{\tau})}. \tag{37}$$

Since numerator and denominators depend on different variables, the adaptive integrand decomposition can proceed along the following algorithm:

1. **Divide:** we divide the numerator $\mathcal{N}_{i_1\ldots i_r}(\boldsymbol{\tau}, \mathbf{x}_\perp)$ modulo the Gröbner basis $\mathcal{G}_{i_1\cdots i_r}(\boldsymbol{\tau})$ of the ideal $\mathcal{J}_{i_1\cdots i_r}(\boldsymbol{\tau})$ generated by the set of denominators. The polynomial division is performed by adopting the lexicographic ordering $\lambda_{ij} \ll \mathbf{x}_\parallel$,

$$\mathcal{N}_{i_1\ldots i_r}(\boldsymbol{\tau}, \mathbf{x}_\perp) = \sum_{k=1}^{r} \mathcal{N}_{i_1\ldots i_{k-1} i_{k+1}\ldots i_r}(\boldsymbol{\tau}, \mathbf{x}_\perp) D_{i_k}(\boldsymbol{\tau}) + \Delta_{i_1\ldots i_r}(\mathbf{x}_\parallel, \mathbf{x}_\perp). \tag{38}$$

The Gröbner basis does not need to be explicitly computed, since, with the choice of variables and the ordering described here, the division is equivalent to applying the set of linear relations described above.

2. **Integrate:** Since denominators do not depend on transverse variables, $\mathbf{x}_\perp$, we can integrate the residue $\Delta_{i_1\ldots i_r}$ over transverse directions. This integration is carried out by expressing $\Delta_{i_1\ldots i_r}$ in terms of Gegenbauer polynomials, i.e.,

$$\Delta^{\rm int}_{i_1\ldots i_r}(\boldsymbol{\tau}) = \int d^{(4-d_\parallel)\ell} \boldsymbol{\Theta}_\perp \Delta_{i_1\ldots i_r}(\boldsymbol{\tau}, \boldsymbol{\Theta}_\perp). \tag{39}$$

Where $\Delta^{\rm int}_{i_1\ldots i_r}$ is a polynomial in $\boldsymbol{\tau}$ whose coefficients depend on the space-time dimension d.

3. **Divide:** the structure of the integrated residue suggests a second division. This can be seen from the dependence $\Delta^{\rm int}_{i_1\ldots i_r}$ has on the variables $\boldsymbol{\tau}$. In fact, after applying the division, similarly as in the first step of this algorithm, we get

$$\Delta^{\rm int}_{i_1\ldots i_r}(\boldsymbol{\tau}) = \sum_{k=1}^{r} \mathcal{N}^{\rm int}_{i_1\ldots i_{k-1} i_{k+1}\ldots i_r}(\boldsymbol{\tau}) D_{i_k}(\boldsymbol{\tau}) + \Delta'_{i_1\ldots i_r}(\mathbf{x}_\parallel), \tag{40}$$

where the new residue $\Delta'_{i_1\ldots i_r}(\mathbf{x}_\parallel)$ can only depend on $\mathbf{x}_\parallel$.

At the end of the decomposition, the integrand is written as

$$\mathcal{I}_{i_1\ldots i_r}(q_j) = \sum_{k=0}^{r} \sum_{\{i_1\cdots i_k\}} \frac{\Delta'_{j_1\cdots j_k}(k_j)}{D_{j_1}(q_j) \cdots D_{j_k}(k_j)}. \tag{41}$$

where the residue functions $\Delta'(k_j)$ are polynomials in the irreducible scalar products, whose coefficients may depend on the external kinematics as well as on d (after the first division, the polynomial $\Delta(k_j)$ have no d-dependence).

This algorithm has been succesfully applied to a few cases as shown in in [60], and also to the leading color contribution to the two-loop all-plus five-gluon amplitude [61, 62, 80].

The Mathematica package AIDA takes as input the integrands of Feynman diagrams generated by FEYNARTS [81] and FEYNCALC [82, 83] and applies the AID algorithm to them. The output of AIDA is the expression of the original amplitude, written as sum of Feynamn diagrams, in terms of integrals, which can be further reduced in terms of MIs by automatic software packages like Reduze.

For the μe-scattering, we succesfully applied Aida to the one-loop diagrams shown in fig.6 and to the two-loop planar and non-planar diagrams in fig.7. We considered also the automatic generation of the one-loop renormalisation countertem-diagrams, which can be processed together with the two-loop diagrams, to provide a result where the one-loop UV (sub)divergencies have been removed.

9 Conclusions

The scattering of high-energy muons on atomic electrons has been recently proposed as an ideal framework to determine the leading hadronic contribution to the anomalous magnetic moment of the muon. The ambitious experimental goal of measuring the differential cross section of the $\mu e \to \mu e$ process with an accuracy of 10ppm requires, on the theoretical side, the knowledge of the QED corrections at NNLO. In this proceedings, we reported our investigations on the feasibility of the evaluation of the corrections at NNLO. In particular, we began by considering the two-loop planar box-diagrams contributing to this process. We employed the method of differential equations and of the Magnus exponential series to identify a canonical set of master integrals. Boundary conditions were derived from the regularity requirements at pseudothresholds, or from the knowledge of the integrals at special kinematic points, evaluated by means of auxiliary, simpler systems of differential equations.

The considered master integrals were expressed as a Taylor series around four space-time dimensions, whose coefficients are written as a combination of generalised polylogarithms. We worked in the massless electron approximation, while keeping full dependence on the muon mass. Besides μe scattering, our results are relevant also for crossing-related processes such as muon-pair production at e^+e^--colliders, as well as for the QCD corrections to *top*-pair production at hadron colliders.

The evaluation of the missing contributions due to non-planar box graphs will be the subject of a dedicated, future work – we are confident that the techniques employed here can be systematically applied for that case as well.

At the same time, we considered the generation of the two-loop scattering amplitudes within the novel software Aida, implementing the adaptive integrand decomposition algorithm, and the decomposition of its output in terms of a minimal number of master integrals, by interfacing Aida to existing softwares carrying out the integration-by-parts algebra.

Our preliminary studies show that the evaluation of the virtual NNLO corrections are feasable. We shall report on that in future communications.

References

[1] G. Backenstoss, B.D. Hyams, G. Knop, P.C. Marin, U. Stierlin, Phys. Rev. Lett. **6**, 415 (1961)

[2] G. Backenstoss, B.D. Hyams, G. Knop, P.C. Marin, U. Stierlin, Phys. Rev. **129**, 2759 (1963)

[3] T. Kirk, S. Neddermeyer, Phys. Rev. **171**, 1412 (1968)

[4] P.L. Jain, N.J. Wixon, Phys. Rev. Lett. **23**, 715 (1969)

[5] R.F. Deery, S.H. Neddermeyer, Phys. Rev. **121**, 1803 (1961)

[6] I.B. McDiarmid, M.D. Wilson, Can. J. Phys. **40**, 698 (1962)

[7] N. Chaudhuri, M.S. Sinha, Nuovo Cimento **35**, 13 (1965)

[8] P.D. Kearney, W.E. Hazen, Phys. Rev. **138**, B173 (1965)

[9] K.P. Schuler, AIP Conf. Proc. **187**, 1401 (1989)

[10] D. Adams et al. (Spin Muon), Nucl. Instrum. Meth. **A443**, 1 (2000)

[11] G. Abbiendi et al., Eur. Phys. J. **C77**, 139 (2017), 1609.08987

[12] C.M. Carloni Calame, M. Passera, L. Trentadue, G. Venanzoni, Phys. Lett. **B746**, 325 (2015), 1504.02228

[13] A.I. Nikishov, Sov. Phys. JETP **12**, 529 (1961), [Zh. Eksp. Teor. Fiz.39,757(1960)]

[14] K.E. Eriksson, Nuovo Cimento **19**, 1029 (1961)

[15] K.E. Eriksson, B. Larsson, G.A. Rinander, Nuovo Cimento **30**, 1434 (1963)

[16] P. Van Nieuwenhuizen, Nucl. Phys. **B28**, 429 (1971)

[17] T.V. Kukhto, N.M. Shumeiko, S.I. Timoshin, J. Phys. **G13**, 725 (1987)

[18] D.Yu. Bardin, L. Kalinovskaya (1997), hep-ph/9712310

[19] N. Kaiser, J. Phys. **G37**, 115005 (2010)

[20] E. Derman, W.J. Marciano, Annals Phys. **121**, 147 (1979)

[21] G. D'Ambrosio, Lett. Nuovo Cim. **38**, 593 (1983)

[22] J.C. Montero, V. Pleitez, M.C. Rodriguez, Phys. Rev. **D58**, 097505 (1998), hep-ph/9803450

[23] T. Gehrmann, E. Remiddi, Nucl. Phys. **B601**, 287 (2001), hep-ph/0101124

[24] R. Bonciani, P. Mastrolia, E. Remiddi, Nucl. Phys. **B690**, 138 (2004), hep-ph/0311145

[25] R. Bonciani, A. Ferroglia, JHEP **11**, 065 (2008), 0809.4687

[26] H.M. Asatrian, C. Greub, B.D. Pecjak, Phys. Rev. **D78**, 114028 (2008), 0810.0987

[27] M. Beneke, T. Huber, X.Q. Li, Nucl. Phys. **B811**, 77 (2009), 0810.1230

[28] G. Bell, Nucl. Phys. **B812**, 264 (2009), 0810.5695

[29] T. Huber, JHEP **03**, 024 (2009), 0901.2133

[30] R. Bonciani, A. Ferroglia, T. Gehrmann, D. Maitre, C. Studerus, JHEP **07**, 129 (2008), 0806.2301

[31] R. Bonciani, A. Ferroglia, T. Gehrmann, C. Studerus, JHEP **08**, 067 (2009), 0906.3671

[32] R. Bonciani, A. Ferroglia, T. Gehrmann, A. von Manteuffel, C. Studerus, JHEP **01**, 102 (2011), 1011.6661

[33] R. Bonciani, A. Ferroglia, T. Gehrmann, A. von Manteuffel, C. Studerus, JHEP **12**, 038 (2013), 1309.4450
[34] P. Mastrolia, M. Passera, A. Primo, U. Schubert, JHEP **11**, 198 (2017), 1709.07435
[35] F.V. Tkachov, Phys. Lett. **100B**, 65 (1981)
[36] K. Chetyrkin, F. Tkachov, Nucl.Phys. **B192**, 159 (1981)
[37] S. Laporta, Int.J.Mod.Phys. **A15**, 5087 (2000), hep-ph/0102033
[38] A. Kotikov, Phys.Lett. **B254**, 158 (1991)
[39] E. Remiddi, Nuovo Cim. **A110**, 1435 (1997), hep-th/9711188
[40] T. Gehrmann, E. Remiddi, Nucl. Phys. **B580**, 485 (2000), hep-ph/9912329
[41] J.M. Henn, Phys.Rev.Lett. **110**, 251601 (2013), 1304.1806
[42] M. Argeri, S. Di Vita, P. Mastrolia, E. Mirabella, J. Schlenk et al., JHEP **1403**, 082 (2014), 1401.2979
[43] S. Di Vita, P. Mastrolia, U. Schubert, V. Yundin, JHEP **09**, 148 (2014), 1408.3107
[44] R. Bonciani, S. Di Vita, P. Mastrolia, U. Schubert, JHEP **09**, 091 (2016), 1604.08581
[45] S. Di Vita, P. Mastrolia, A. Primo, U. Schubert, JHEP **04**, 008 (2017), 1702.07331
[46] A. Goncharov, Proceedings of the International Congree of Mathematicians **1,2**, 374 (1995)
[47] E. Remiddi, J. Vermaseren, Int.J.Mod.Phys. **A15**, 725 (2000), hep-ph/9905237
[48] T. Gehrmann, E. Remiddi, Comput.Phys.Commun. **141**, 296 (2001), hep-ph/0107173
[49] J. Vollinga, S. Weinzierl, Comput.Phys.Commun. **167**, 177 (2005), hep-ph/0410259
[50] C.W. Bauer, A. Frink, R. Kreckel (2000), cs/0004015
[51] S. Borowka, G. Heinrich, S.P. Jones, M. Kerner, J. Schlenk, T. Zirke, Comput. Phys. Commun. **196**, 470 (2015), 1502.06595
[52] A. von Manteuffel, C. Studerus (2012), 1201.4330
[53] V.A. Smirnov, Phys. Lett. **B460**, 397 (1999), hep-ph/9905323
[54] J. Tausk, Phys.Lett. **B469**, 225 (1999), hep-ph/9909506
[55] M. Czakon, Phys. Lett. **B664**, 307 (2008), 0803.1400
[56] M. Czakon, A. Mitov, JHEP **01**, 080 (2013), 1210.6832
[57] M. Czakon, A. Mitov, JHEP **12**, 054 (2012), 1207.0236
[58] P. Bärnreuther, M. Czakon, A. Mitov, Phys. Rev. Lett. **109**, 132001 (2012), 1204.5201
[59] M. Czakon, P. Fiedler, A. Mitov, Phys. Rev. Lett. **110**, 252004 (2013), 1303.6254
[60] P. Mastrolia, T. Peraro, A. Primo, JHEP **08**, 164 (2016), 1605.03157
[61] P. Mastrolia, T. Peraro, A. Primo, W.J. Torres Bobadilla, PoS **LL2016**, 007 (2016), 1607.05156
[62] W.J. Torres Bobadilla (2018), 1801.03010
[63] G. Ossola, C.G. Papadopoulos, R. Pittau, Nucl.Phys. **B763**, 147 (2007), hep-ph/0609007
[64] G. Ossola, C.G. Papadopoulos, R. Pittau, JHEP **0707**, 085 (2007), 0704.1271
[65] R.K. Ellis, W.T. Giele, Z. Kunszt, JHEP **03**, 003 (2008), 0708.2398
[66] R. Ellis, W.T. Giele, Z. Kunszt, K. Melnikov, Nucl.Phys. **B822**, 270 (2009), 0806.3467
[67] G. Ossola, C.G. Papadopoulos, R. Pittau, JHEP **0805**, 004 (2008), 0802.1876
[68] P. Mastrolia, G. Ossola, C.G. Papadopoulos, R. Pittau, JHEP **06**, 030 (2008), 0803.3964
[69] P. Mastrolia, E. Mirabella, T. Peraro, JHEP **06**, 095 (2012), [Erratum: JHEP11,128(2012)], 1203.0291
[70] P. Mastrolia, G. Ossola, JHEP **1111**, 014 (2011), 1107.6041
[71] S. Badger, H. Frellesvig, Y. Zhang, JHEP **1204**, 055 (2012), 1202.2019
[72] Y. Zhang, JHEP **1209**, 042 (2012), 1205.5707
[73] P. Mastrolia, E. Mirabella, G. Ossola, T. Peraro, Phys.Lett. **B718**, 173 (2012), 1205.7087
[74] H. Ita (2015), 1510.05626
[75] K.J. Larsen, Y. Zhang, Phys. Rev. **D93**, 041701 (2016), 1511.01071
[76] A. von Manteuffel, R.M. Schabinger, Phys. Lett. **B744**, 101 (2015), 1406.4513
[77] P. Kant, Comput. Phys. Commun. **185**, 1473 (2014), 1309.7287
[78] C.G. Papadopoulos (2014), 1401.6057
[79] A.V. Smirnov, Comput. Phys. Commun. **204**, 189 (2016), 1511.03614
[80] W.J. Torres Bobadilla, Ph.D. thesis, Padua U. (2017), http://paduaresearch.cab.unipd.it/10135/
[81] T. Hahn, Comput.Phys.Commun. **140**, 418 (2001), hep-ph/0012260
[82] R. Mertig, M. Bohm, A. Denner, Comput. Phys. Commun. **64**, 345 (1991)
[83] V. Shtabovenko, R. Mertig, F. Orellana (2016), 1601.01167

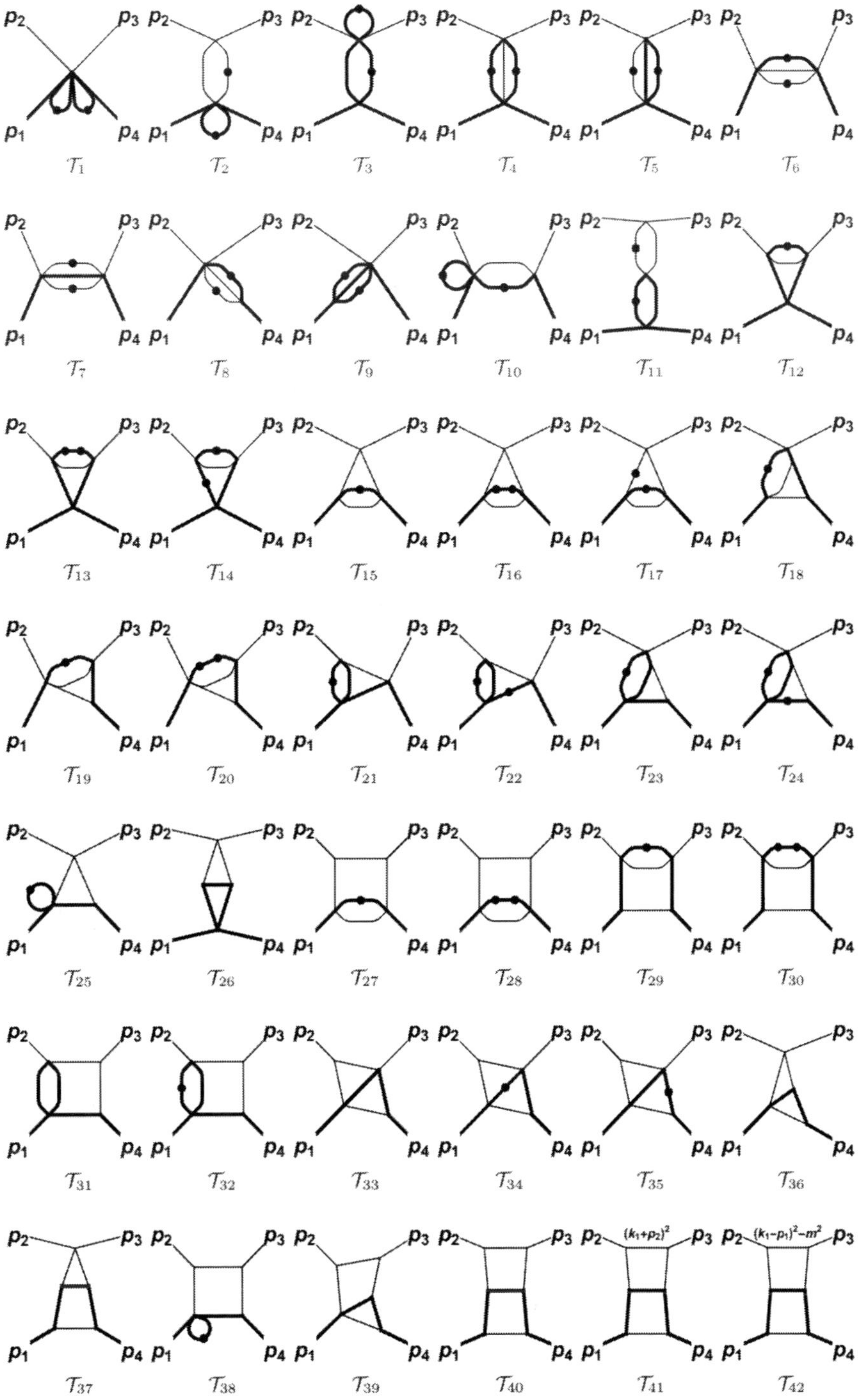

Figure 5: Two-loop MIs $\mathcal{T}_{1,\ldots,42}$ for the second integral family.

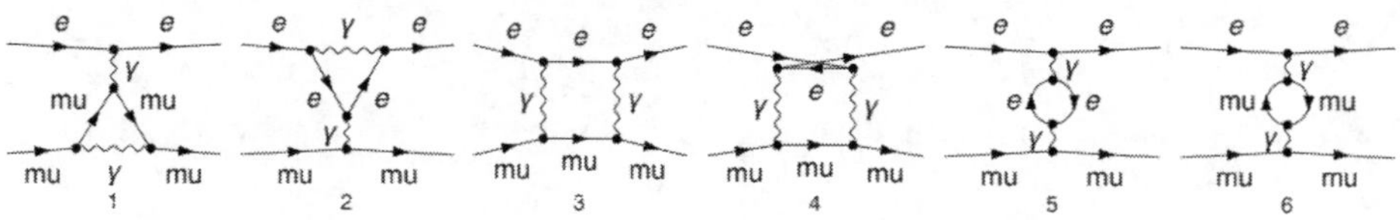

Figure 6: Feynman diagrams contributing to the one-loop μe-scattering amplitude.

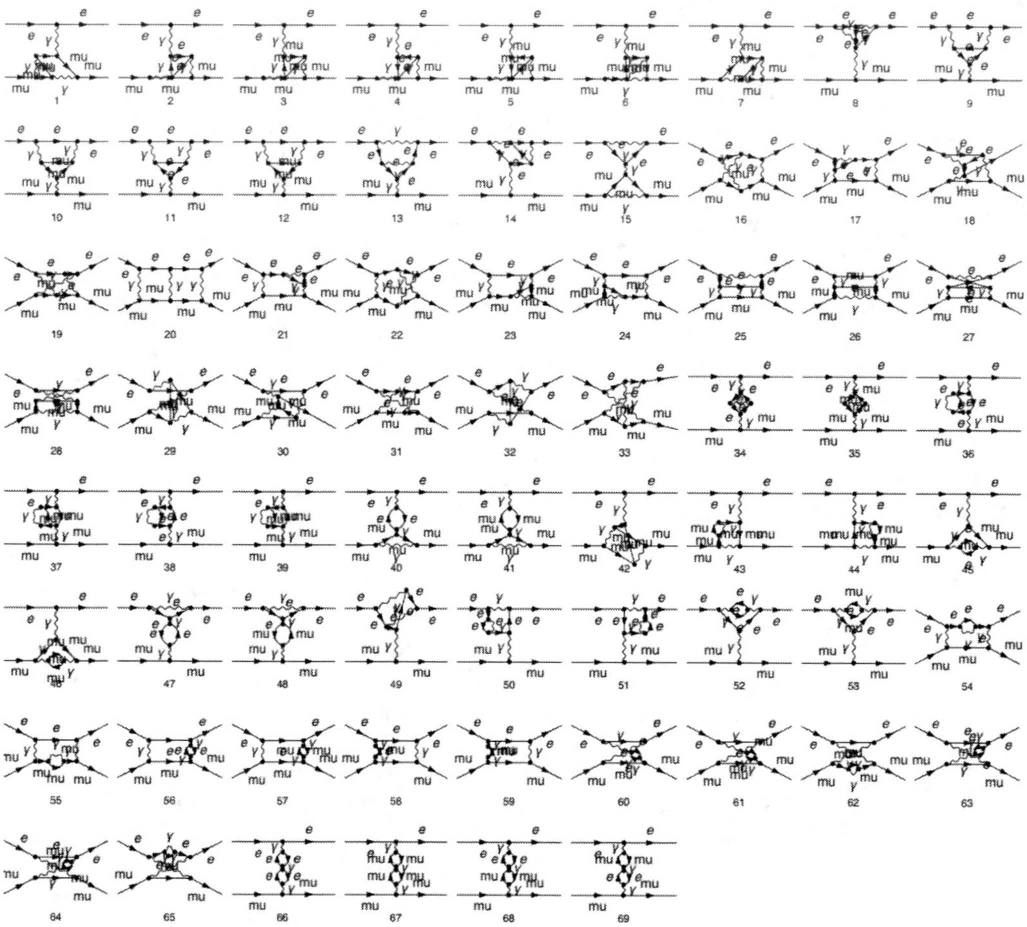

Figure 7: Feynman diagrams contributing to the two-loop μe-scattering amplitude.

 https://doi.org/10.1051/epjconf/201817901015

Leptoquarks in Flavour Physics

Dario Müller[1,2,a]

[1] Paul Scherrer Institut, CH–5232 Villigen PSI, Switzerland
[2] Physik Institut, Universität Zürich, Winterthurstrasse 190,CH-8057 Zürich, Switzerland

Abstract. While the LHC has not directly observed any new particle so far, experimental results from LHCb, BELLE and BABAR point towards the violation of lepton flavour universality in $b \to s\ell^+\ell^-$ and $b \to c\ell\nu$. In this context, also the discrepancy in the anomalous magnetic moment of the muon can be interpreted as a sign of lepton flavour universality violation. Here we discuss how these hints for new physics can also be explained by introducing leptoquarks as an extension of the Standard Model. Indeed, leptoquarks are good candidates to explain the anomaly in the anomalous magnetic moment of the muon because of an m_t/m_μ enhanced contribution giving correlated effects in Z boson decays which is particularly interesting in the light of future precision experiments.

1 Introduction

So far the LHC has not directly observed any particles beyond the ones in the Standard Model of particle physics (SM). While in the SM the Higgs boson couplings to fermions are the only source of lepton flavour universality violation (LFUV), in the past years several hints for additional LFUV have been accumulated. The ratios of semi-leptonic B decays

$$R\left(K^{(*)}\right) = \mathrm{Br}\left[B \to K^{(*)}\mu^+\mu^-\right] \Big/ \mathrm{Br}\left[B \to K^{(*)}e^+e^-\right],$$
$$R\left(D^{(*)}\right) = \mathrm{Br}\left[B \to D^{(*)}\tau\nu\right] \Big/ \mathrm{Br}\left[B \to D^{(*)}\ell\nu\right],$$
$$R\left(J/\Psi\right) = \mathrm{Br}\left[B_c^+ \to J/\Psi\tau^+\nu\right] \Big/ \mathrm{Br}\left[B_c \to J/\Psi\mu^+\nu\right],$$

show sizeable deviations from the SM predictions.
Considering $b \to s\ell^+\ell^-$ transitions, $R(K)$ [1] and $R(K^*)$ [2] show a combined significance for LFUV at the 4σ level [3–7]. If we also take into account all other $b \to s\mu^+\mu^-$ observables, in several scenarios the global fit prefers new physics (NP) above the 5σ level [8].
In $R(D^{(*)})$ BaBar [9], BELLE [10, 11] and LHCb [12, 13] found a significance for LFUV of about 4σ [14]. In this context it is interesting to mention that LHCb recently measured a deviation of about 2σ in $R(J/\Psi)$ [15]. This is consistent with the anomaly in $R(D)$ and $R(D^*)$.
The tension in the anomalous magnetic moment of the muon (a_μ) between measurement [16] and SM prediction is at the 3σ level [17, 18]. Also this discrepancy can be interpreted as a sign of LFUV even though the anomalous magnetic moments are already in the SM flavour non-universal: If we assume that NP coupled with the same strength to electrons and to muons, a_e would be more sensitive for NP. Since no deviation from the SM prediction in a_e has been observed so far, the tension in a_μ can be interpreted as another sign for LFUV induced by NP.
Motivated by these anomalies, we study leptoquarks (LQ) which provide a possible solution of them. Firstly we concentrate on each anomaly itself and discuss which LQ representations suit best for an explanation and consider correlated effects in other observables. Finally, we will investigate if we can explain the described anomalies simultaneously.

2 a_μ

In order to explain the descrepancy in a_μ, a rather large effect of the order of the SM weak interaction contribution is required. Among the set of new particles that can give the desired effect, LQ are interesting candidates [19–23]. Even though the LQ must be rather heavy due to LHC constraints (above 1 TeV), one can still get relevant effects in a_μ. For some representations of LQ, the amplitude can be enhanced by a factor $m_t/m_\mu \simeq O(10^3)$ or $m_b/m_\mu \simeq O(10)$ compared to the SM.
To achieve an enhancement of factor m_t/m_μ, the LQ must couple to left- and right-handed top or anti-top quarks simultaneously. However, not all LQ representations have this feature. Among the 10 scalar and vector LQ representations generating lepton-quark interaction terms that are invariant under the SM gauge group [24], only two scalars can possess this enhancement: An SU(2) singlet Φ_1 and an SU(2) doublet Φ_2 with quantum numbers

$$Q(\Phi_1) : \left(3, 1, -\frac{2}{3}\right), \quad Q(\Phi_2) : \left(\bar{3}, 2, -\frac{7}{3}\right). \quad (1)$$

[a] e-mail: dario.mueller@psi.ch

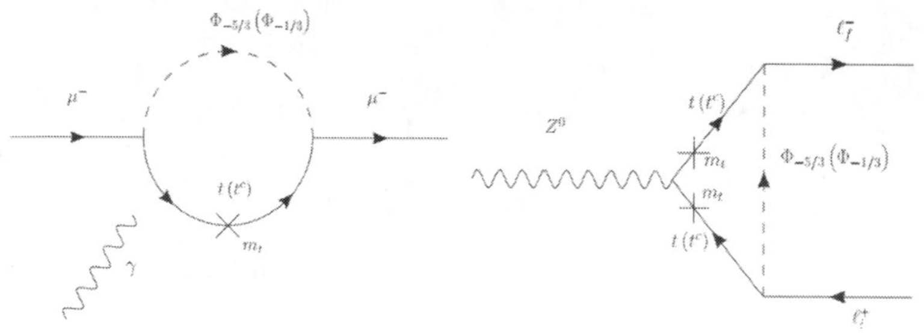

Figure 1. Feynman diagrams that contribute to a_μ and $Z \to \ell_i^+\ell_f^-$. The crosses on the fermion lines stand for chirality flips, involving a top mass.

These representations couple to the fermions of the SM as follows

$$\mathcal{L}_{int} = \left(\lambda_{fi}^{1R}\overline{u_f^c}\ell_i + \lambda_{fi}^{1L}\overline{Q_f^c}i\tau_2 L_i\right)\Phi_1^\dagger, \quad (2)$$

$$\mathcal{L}_{int} = \left(\lambda_{fi}^{2RL}\overline{u_f}L_i + \lambda_{fi}^{2LR}\overline{Q_f}i\tau_2\ell_i\right)\Phi_2^\dagger. \quad (3)$$

Q and L are SU(2) doublets, u and ℓ are SU(2) singlets and c denotes charge conjugation. f and i are flavour indices. The Feynman diagrams involving Φ_1 and Φ_2 are shown left in Fig. 1.
For simplicity, we only consider couplings of top quarks to muons and set $\lambda_{32}^{1L,1R}$ and $\lambda_{32}^{2RL,2LR}$ equal to $\lambda_\mu^{L,R}$. Besides the contribution to a_μ, we also obtain correlated effects in $Z \to \mu^+\mu^-$ (see the right Feynman diagram in Fig. 1), in $b \to s\nu\bar{\nu}$ (for Φ_1) and in $b \to s\mu^+\mu^-$ (for Φ_2). Fig. 2 shows the allowed regions in λ_μ^L-λ_μ^R parameter space for a LQ with mass of 1 TeV, to respect constraints from direct searches. For $Z \to \mu^+\mu^-$ the expected bounds from GigaZ [25] and TLEP [26] are shown as well. Additionally, for Φ_1 the expected BELLE II limit for $B \to K^{(*)}\nu\bar{\nu}$ is taken into account. Φ_2 gives a $C_9 = C_{10}$-like contribution to $b \to s\ell^+\ell^-$ which provides a slight improvement of $1\,\sigma$ in the global fit.
If we also allow for couplings of the top quark to taus or electrons, lepton flavour violating processes as $\tau \to \mu\gamma$ or $Z \to \mu e$ will be possible. A more detailed discussion can be found in Ref. [27].

3 $b \to s\ell^+\ell^-$

Dealing with $b \to s\ell^+\ell^-$ transitions, one usually works with an effective Hamiltonian of the form

$$H_{\text{eff}}^{\ell_f\ell_i} = -\frac{4G_F}{\sqrt{2}}V_{tb}V_{ts}^* \sum_{a=9,10} C_a^{fi}O_a^{fi} + C_a'^{fi}O_a'^{fi}, \quad (4)$$

$$O_{9(10)}^{fi} = \frac{\alpha}{4\pi}[\bar{s}\gamma^\mu P_L b]\,[\bar{\ell}_f\gamma_\mu(\gamma_5)\ell_i], \quad (5)$$

$$O_{9(10)}'^{fi} = \frac{\alpha}{4\pi}[\bar{s}\gamma^\mu P_R b]\,[\bar{\ell}_f\gamma_\mu(\gamma_5)\ell_i]. \quad (6)$$

Then, for several scenarios with NP in C_9 only, $C_9 = -C_{10}$ or $C_9 = -C_9'$, the global fit prefers NP above the $5\,\sigma$ level [8].
We consider the case where NP contributes as $C_9 = -C_{10}$. While in the muon channel the preferred value deviates from the SM prediction with a tension of $5\,\sigma$, in the case with electrons the SM contribution is sufficient to give a good fit to the data. To reach the central value, an $O(10\%)$ NP effect in the muon channel is required. $b \to s\ell^+\ell^-$ is suppressed by a loop and a CKM factor because it is a flavour changing neutral current. Therefore, the effect of NP does not have to be very large to account for the data. There are several suggestions to explain the anomaly in $b \to s\mu^+\mu^-$ with LQ, see e.g. [28–41].
There are three LQ representations that give a $C_9 = -C_{10}$-like contribution to $b \to s\ell^+\ell^-$

$$\Phi_3 : \left(3, 3, -\frac{2}{3}\right), \; \mathcal{L}_{int} = \lambda_{fi}^3\overline{Q_f^c}i\tau_2\left(\tau\cdot\Phi_3\right)^\dagger L_i, \quad (7)$$

$$V_1^\mu : \left(\bar{3}, 1, -\frac{4}{3}\right), \; \mathcal{L}_{int} = \left(\kappa_{fi}^{1R}\overline{d_f}\gamma_\mu\ell_i + \kappa_{fi}^{1L}\overline{Q_f}\gamma_\mu L_i\right)V_1^{\mu\dagger}, \quad (8)$$

$$V_3^\mu : \left(3, 3, \frac{4}{3}\right), \; \mathcal{L}_{int} = \kappa_{fi}^3\overline{Q_f}\gamma_\mu\left(\tau\cdot V_3^\mu\right)L_i. \quad (9)$$

If we only allow for couplings to muons, these LQ will give tree-level effects in $b \to s\mu^+\mu^-$ but contribute at loop-level in other flavour observables. This is why LQ can explain the anomalies in $b \to s\mu^+\mu^-$ transitions but are not in conflict with other observables.
Including couplings to electrons, we get additional effects in $b \to se^+e^-$ as well as in the lepton flavour violating processes $\mu \to e\gamma$ and $b \to s\mu e$. We can write

$$\text{Br}\left[\mu \to e\gamma\right] \propto \left|\chi C_9^{ee} + \frac{C_9^{\mu\mu}}{\chi}\right|^2, \quad (10)$$

$$\text{Br}\left[B \to K\mu e\right] \propto \left|\frac{C_9^{ee}}{\gamma}\right|^2 + \left|\gamma C_9^{\mu\mu}\right|^2, \quad (11)$$

where $\chi = y_{32}/y_{21}$ and $\gamma = y_{21}/y_{22}$, with $y = \lambda$ for scalar LQ and $y = \kappa$ for vector LQ. Note that one can avoid an effect in $\mu \to e\gamma$ completely if $C_9^{\mu\mu} = -\chi^2 C_9^{ee}$. For real values of χ this means that sign$[C_9^{ee}]$ = $-$sign$[C_9^{\mu\mu}]$. In Fig. 3 you find a detailed analysis for the scalar triplet Φ_3. For V_1^μ and V_3^μ the situation is quite similar, albeit the blue bands are more narrow. For more information see Ref. [42].

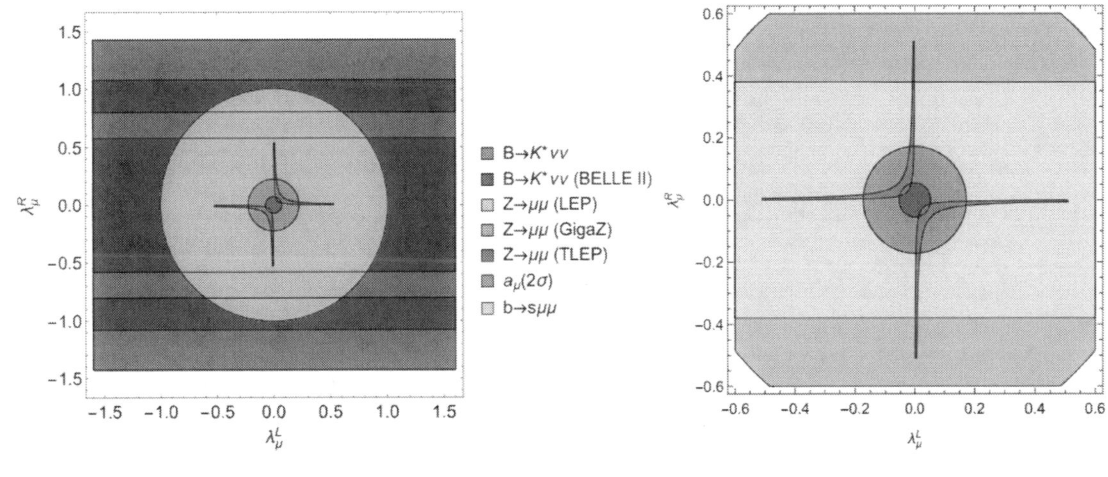

Figure 2. Allowed regions in the λ^L_μ-λ^R_μ parameter space from current limits of LEP and expected future constraints of GigaZ and TLEP with M_{LQ} = 1 TeV. Left: SU(2) singlet Φ_1. Right: SU(2) doublet Φ_2. Note that if the expected sensitivity of TLEP was reached and no excess in the $Z\mu\bar{\mu}$-coupling was found, an explanation of the anomaly in a_μ would strongly be disfavoured.

4 $b \to c\tau\nu$

The transition $b \to c\tau\nu$ is a charged current process and is therefore generated at tree-level in the SM. Since an explanation of $R(D)$ and $R(D^*)$ requires an $O(10\%)$ effect, also the NP must contribute at tree-level. Charged Higgses [44–48], W' bosons [31] and LQ [22, 23, 32, 33, 35, 49, 49–54] are NP candidates. However, B_c lifetime [55, 56] and q^2 distributions [57] in combination with direct searches [58] strongly disfavour charged Higgses and W' bosons.

Hence, we are left with LQ. If we want to explain the excess in $R(D)$ and $R(D^*)$ by modifying the couplings to tau neutrinos and/or taus, we also get effects in $b \to s\tau^+\tau^-$ and $b \to s\nu\bar{\nu}$. It can be shown in a model independent way that these effects are of the order of 10^3 compared to the SM [33]. This is due to the fact that the processes $b \to s\tau^+\tau^-$ and $b \to s\nu\bar{\nu}$ occur at loop-level in the SM, while LQ contribute at tree-level with considerably large couplings in order to explain $R(D^{(*)})$. An $O(10^3)$ effect is in strong conflict with experimental results of $B \to K^{(*)}\nu\bar{\nu}$.

Hence, a model that explains $R(D)$ and $R(D^*)$ must avoid a major effect in $b \to s\nu\bar{\nu}$. One possibility is the vector LQ singlet V_1^μ (see Eq. (8)). Another model was proposed in Ref. [59], where instead of V_1^μ two scalar LQ (the SU(2) singlet Φ_1 in Eq. (2) and the SU(2) triplet Φ_3 in Eq. (7)) are required. If Fig. 4 we can see for which parameters we can explain $R(D^{(*)})$ without violating the limits of $b \to s\nu\bar{\nu}$.

Besides explaining $R(D^{(*)})$, it also enhances $B_s \to \tau^+\tau^-$ up to a factor 10^3 compared to the SM and therefore lies within experimental reach. The current limit is $\mathrm{Br}[B_s \to \tau^+\tau^-] < 6.8 \times 10^{-3}$ [60]. Fig. 5 shows the correlation of $R(D^{(*)})$ and $\mathrm{Br}[B_s \to \tau^+\tau^-]$.

5 Simultaneous Explanation

We may ask if one can explain the three types of anomalies in a_μ, $b \to s\mu^+\mu^-$ and $R(D^{(*)})$ simultaneously.

If we take the model that we proposed in Sec. 4 and insert a discrete symmetry for the couplings $\lambda^{1L}_{jk} = e^{i\pi j}\lambda^3_{jk} \equiv \lambda^L_{jk}$ the effect in $b \to s\nu\bar{\nu}$ will be cancelled exactly. If we include $b \to s\mu^+\mu^-$, $B \to D^{(*)}\mu\nu/B \to D^{(*)}e\nu$ will be affected. However, these observables remain compatible with experiments. Additionally the LFV process $\tau \to \mu\gamma$ and $b \to s\tau\mu$ appear. Both processes give constraints on our parameter space. On the left-hand side of Fig. 6 the allowed regions for M_{LQ} = 1 TeV are shown, under the assumption that $b \to s\mu^+\mu^-$ matches the central value of the global fit. As we can see easily, the explanation of $R(D^{(*)})$ requires $\lambda^L_{33}/\lambda^L_{32} > 1$. On the right-hand side of Fig. 6 we need $\lambda^L_{33}/\lambda^L_{32} < 0.6$ to explain a_μ without getting in conflict with experimental constraints of $\tau \to \mu\gamma$. With this contradiction, we cannot explain all three anomalies simultaneously. However, if we abandon anyone of them, it is possible to explain the other two anomalies. If we abandon the assumption $\lambda^{1L}_{jk} = e^{i\pi j}\lambda^3_{jk}$ one can avoid the discussed constraints and explain all three anomalies.

Another good candidate is the vector LQ singlet V_1^μ because it does not affect the $b \to s\nu\bar{\nu}$ transitions but gives a $C_9 = -C_{10}$ effect in $b \to s\ell^+\ell^-$. It can also be used to explain $R(D^{(*)})$. In a_μ we can obtain an m_b/m_μ contribution which is not as large as for the two scalar LQ Φ_1 and Φ_2 (see Sec.2), but still may be significant.

However, one faces the problem that massive vector bosons without a Higgs mechanism are not renormalizable. For $\ell' \to \ell\gamma$ in Sec. 3 this does not cause a problem because this process is finite in unitary gauge. Recently, models containing renormalizable massive vector LQ were proposed [61–64]. In Ref. [62] a Pati-Salam model in combination with vector-like fermions is phe-

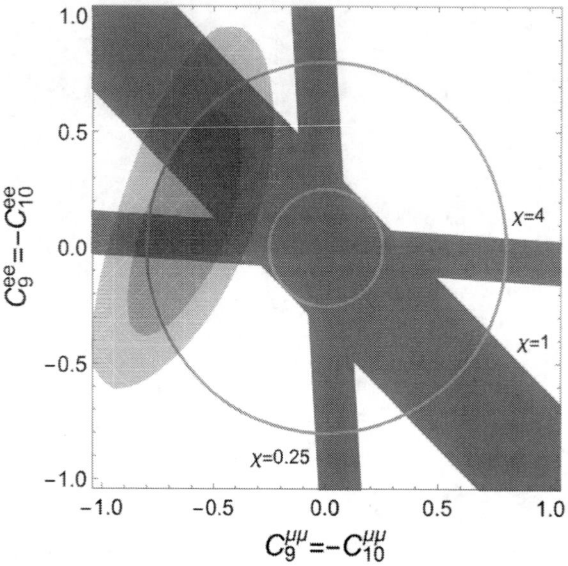

Figure 3. The darker (lighter) red region shows the global fit at 1 σ (2 σ) for $b \to se^+e^-$ and $b \to s\mu^+\mu^-$ transitions. In blue, the allowed region constrained by the current limit $\text{Br}[\mu \to e\gamma] < 4.2 \times 10^{-13}$ [43] with $\chi = \{1/4, 1, 4\}$ for the scalar triplet Φ_3 with $M_{LQ} = 1\text{TeV}$. The outer (inner) green contour shows the case where $\text{Br}[\text{B} \to \text{K}\mu e] = 2 \times 10^{-8}(0.2 \times 10^{-8})$. For real values of χ only the region with $C_9^{\mu\mu} < 0$ and $C_9^{ee} > 0$ seems to be compatible with experimental limits. Indeed this is also where the central value of the global fit lies. However, if in the future the global fit prefers $C_9^{\mu\mu} < 0$ and $C_9^{ee} < 0$, LQ with real couplings will not be able to explain $b \to s\ell^+\ell^-$ transitions without violating the constraints from $\mu \to e\gamma$.

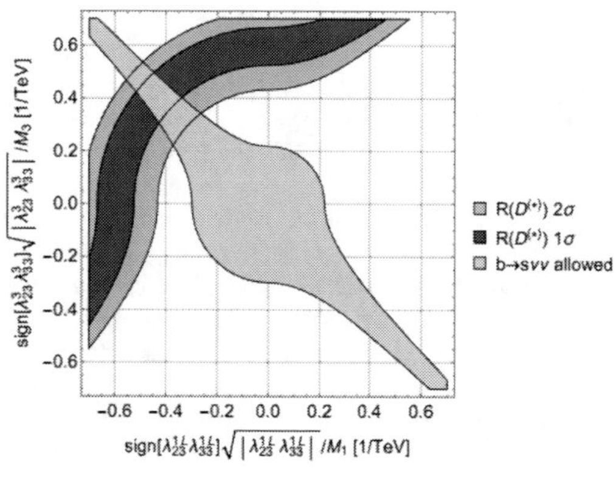

Figure 4. In lighter (darker) blue you see the 1 σ (2 σ) preferred region of the weighted sum of $R(D)$ and $R(D^*)$, in yellow the allowed region of $b \to s\nu\bar{\nu}$. $M_{1(3)}$ is the mass of the LQ Φ_1 (Φ_3).

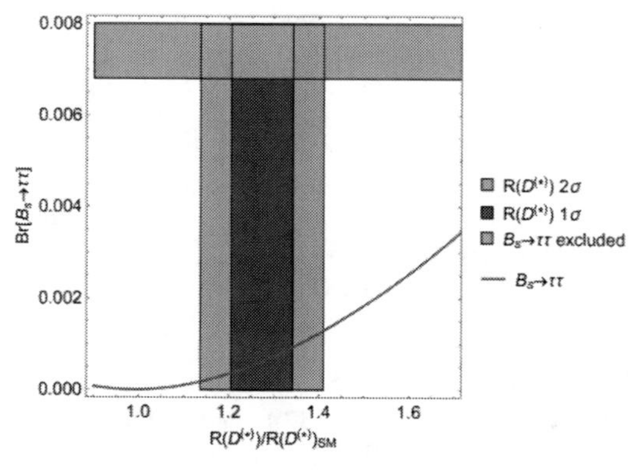

Figure 5. In blue we have the weighted sum of $R(D)$ and $R(D^*)$ and the red line depicts the correlated prediction of $\text{Br}[\text{B}_s \to \tau^+\tau^-]$. The grey region shows the current experimental bounds.

nomenologically consistent and leads to a massive vector LQ. This vector LQ has the properties of V_1^μ in Eq. (8).

6 Conclusion

In these proceedings we reviewed which LQ are good candidates to explain flavour anomalies: For the deviation in a_μ the scalar LQ Φ_1 and Φ_2 give an m_t/m_μ enhanced contribution. This is a useful feature to explain the rather large discrepancy. Considering $b \to s\ell^+\ell^-$ transitions, the three LQ representations Φ_3, V_1^μ and V_3^μ are required by the fit, because they give $C_9 = -C_{10}$. $R(D^{(*)})$ can be explained either by V_1^μ or, as we proposed, by a combination of a scalar LQ singlet Φ_1 and scalar LQ triplet Φ_3 without getting in conflict with $b \to s\nu\bar{\nu}$ observables. We also showed that this model is able to explain any two of the three anomalies (a_μ, $b \to s\mu^+\mu^-$ and $R(D^{(*)})$ simultaneously.

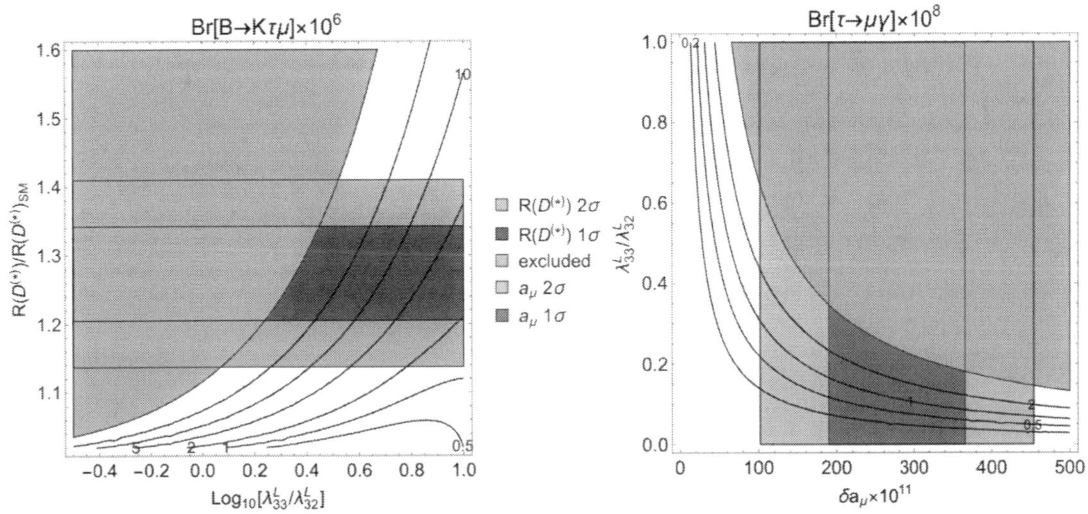

Figure 6. Left: In blue we have the weighted sum of $R(D)$ and $R(D^*)$. The contour lines show $\mathrm{Br}[B \to K\tau\mu]\times 10^6$. Right: The anomaly in a_μ combined with the contour lines depicting $\mathrm{Br}[\tau \to \mu\gamma] \times 10^8$. In both plots the grey region is excluded by experiments.

Acknowledgments

D.M. thanks the organizing committee, and especially Giancarlo D'Ambrosio, for the invitation and for giving the opportunity to present his work.
D.M. is supported by an Ambizione Grant of the Swiss National Science Foundation (PZ00P2_154834).

References

[1] R. Aaij *et al.* [LHCb Collaboration], Phys. Rev. Lett. **113** (2014) 151601 doi:10.1103/PhysRevLett.113.151601 [arXiv:1406.6482 [hep-ex]].

[2] R. Aaij *et al.* [LHCb Collaboration], JHEP **1708** (2017) 055 doi:10.1007/JHEP08(2017)055 [arXiv:1705.05802 [hep-ex]].

[3] W. Altmannshofer, P. Stangl and D. M. Straub, Phys. Rev. D **96** (2017) no.5, 055008 doi:10.1103/PhysRevD.96.055008 [arXiv:1704.05435 [hep-ph]].

[4] G. D'Amico, M. Nardecchia, P. Panci, F. Sannino, A. Strumia, R. Torre and A. Urbano, JHEP **1709** (2017) 010 doi:10.1007/JHEP09(2017)010 [arXiv:1704.05438 [hep-ph]].

[5] G. Hiller and I. Nisandzic, Phys. Rev. D **96** (2017) no.3, 035003 doi:10.1103/PhysRevD.96.035003 [arXiv:1704.05444 [hep-ph]].

[6] L. S. Geng, B. Grinstein, S. Jäger, J. Martin Camalich, X. L. Ren and R. X. Shi, Phys. Rev. D **96** (2017) no.9, 093006 doi:10.1103/PhysRevD.96.093006 [arXiv:1704.05446 [hep-ph]].

[7] T. Hurth, F. Mahmoudi, D. Martinez Santos and S. Neshatpour, Phys. Rev. D **96** (2017) no.9, 095034 doi:10.1103/PhysRevD.96.095034 [arXiv:1705.06274 [hep-ph]].

[8] B. Capdevila, A. Crivellin, S. Descotes-Genon, J. Matias and J. Virto, arXiv:1704.05340 [hep-ph].

[9] J. P. Lees *et al.* [BaBar Collaboration], Phys. Rev. Lett. **109** (2012) 101802 doi:10.1103/PhysRevLett.109.101802 [arXiv:1205.5442 [hep-ex]].

[10] M. Huschle *et al.* [Belle Collaboration], Phys. Rev. D **92** (2015) no.7, 072014 doi:10.1103/PhysRevD.92.072014 [arXiv:1507.03233 [hep-ex]].

[11] S. Hirose *et al.* [Belle Collaboration], arXiv:1709.00129 [hep-ex].

[12] R. Aaij *et al.* [LHCb Collaboration], Phys. Rev. Lett. **115** (2015) no.11, 111803 Erratum: [Phys. Rev. Lett. **115** (2015) no.15, 159901] doi:10.1103/PhysRevLett.115.159901, 10.1103/PhysRevLett.115.111803 [arXiv:1506.08614 [hep-ex]].

[13] F. Betti [LHCb Collaboration], arXiv:1705.10651 [hep-ex].

[14] Y. Amhis *et al.*, arXiv:1612.07233 [hep-ex].

[15] R. Aaij *et al.* [LHCb Collaboration], arXiv:1711.05623 [hep-ex].

[16] G. W. Bennett *et al.* [Muon g-2 Collaboration], Phys. Rev. D **73** (2006) 072003 doi:10.1103/PhysRevD.73.072003 [hep-ex/0602035].

[17] A. Nyffeler, Phys. Rev. D **94** (2016) no.5, 053006 doi:10.1103/PhysRevD.94.053006 [arXiv:1602.03398 [hep-ph]].

[18] F. Jegerlehner and A. Nyffeler, Phys. Rept. **477** (2009) 1 doi:10.1016/j.physrep.2009.04.003 [arXiv:0902.3360 [hep-ph]].

[19] A. Djouadi, T. Kohler, M. Spira and J. Tutas, Z. Phys. C **46** (1990) 679. doi:10.1007/BF01560270

[20] S. Davidson, D. C. Bailey and B. A. Campbell, Z. Phys. C **61** (1994) 613 doi:10.1007/BF01552629 [hep-ph/9309310].

[21] G. Couture and H. Konig, Phys. Rev. D **53** (1996) 555 doi:10.1103/PhysRevD.53.555 [hep-ph/9507263].

[22] M. Bauer and M. Neubert, Phys. Rev. Lett. **116** (2016) no.14, 141802 doi:10.1103/PhysRevLett.116.141802 [arXiv:1511.01900 [hep-ph]].

[23] D. Das, C. Hati, G. Kumar and N. Mahajan, Phys. Rev. D **94** (2016) 055034 doi:10.1103/PhysRevD.94.055034 [arXiv:1605.06313 [hep-ph]].

[24] W. Buchmuller, R. Ruckl and D. Wyler, Phys. Lett. B **191** (1987) 442 Erratum: [Phys. Lett. B **448** (1999) 320]. doi:10.1016/S0370-2693(99)00014-3, 10.1016/0370-2693(87)90637-X

[25] H. Baer *et al.*, arXiv:1306.6352 [hep-ph].

[26] M. Bicer *et al.* [TLEP Design Study Working Group], JHEP **1401** (2014) 164 doi:10.1007/JHEP01(2014)164 [arXiv:1308.6176 [hep-ex]].

[27] E. Coluccio Leskow, G. D'Ambrosio, A. Crivellin and D. Müller, Phys. Rev. D **95** (2017) no.5, 055018 doi:10.1103/PhysRevD.95.055018 [arXiv:1612.06858 [hep-ph]].

[28] B. Gripaios, M. Nardecchia and S. A. Renner, JHEP **1505** (2015) 006 doi:10.1007/JHEP05(2015)006 [arXiv:1412.1791 [hep-ph]].

[29] S. Fajfer and N. Košnik, Phys. Lett. B **755** (2016) 270 doi:10.1016/j.physletb.2016.02.018 [arXiv:1511.06024 [hep-ph]].

[30] D. Bečirević, S. Fajfer and N. Košnik, Phys. Rev. D **92** (2015) no.1, 014016 doi:10.1103/PhysRevD.92.014016 [arXiv:1503.09024 [hep-ph]].

[31] A. Greljo, G. Isidori and D. Marzocca, JHEP **1507** (2015) 142 doi:10.1007/JHEP07(2015)142 [arXiv:1506.01705 [hep-ph]].

[32] L. Calibbi, A. Crivellin and T. Ota, Phys. Rev. Lett. **115** (2015) 181801 doi:10.1103/PhysRevLett.115.181801 [arXiv:1506.02661 [hep-ph]].

[33] R. Alonso, B. Grinstein and J. Martin Camalich, JHEP **1510** (2015) 184 doi:10.1007/JHEP10(2015)184 [arXiv:1505.05164 [hep-ph]].

[34] R. Barbieri, G. Isidori, A. Pattori and F. Senia, Eur. Phys. J. C **76** (2016) no.2, 67 doi:10.1140/epjc/s10052-016-3905-3 [arXiv:1512.01560 [hep-ph]].

[35] D. Bečirević, S. Fajfer, N. Košnik and O. Sumensari, Phys. Rev. D **94** (2016) no.11, 115021 doi:10.1103/PhysRevD.94.115021 [arXiv:1608.08501 [hep-ph]].

[36] D. Bečirević, N. Košnik, O. Sumensari and R. Zukanovich Funchal, JHEP **1611** (2016) 035 doi:10.1007/JHEP11(2016)035 [arXiv:1608.07583 [hep-ph]].

[37] D. Bečirević and O. Sumensari, JHEP **1708** (2017) 104 doi:10.1007/JHEP08(2017)104 [arXiv:1704.05835 [hep-ph]].

[38] D. Aloni, A. Dery, C. Frugiuele and Y. Nir, JHEP **1711** (2017) 109 doi:10.1007/JHEP11(2017)109 [arXiv:1708.06161 [hep-ph]].

[39] B. Chauhan, B. Kindra and A. Narang, arXiv:1706.04598 [hep-ph].

[40] O. Sumensari, arXiv:1705.07591 [hep-ph].

[41] C. H. Chen, T. Nomura and H. Okada, Phys. Lett. B **774** (2017) 456 doi:10.1016/j.physletb.2017.10.005 [arXiv:1703.03251 [hep-ph]].

[42] A. Crivellin, D. Mueller, A. Signer and Y. Ulrich, arXiv:1706.08511 [hep-ph].

[43] A. M. Baldini *et al.* [MEG Collaboration], Eur. Phys. J. C **76** (2016) no.8, 434 doi:10.1140/epjc/s10052-016-4271-x [arXiv:1605.05081 [hep-ex]].

[44] A. Crivellin, C. Greub and A. Kokulu, Phys. Rev. D **86** (2012) 054014 doi:10.1103/PhysRevD.86.054014 [arXiv:1206.2634 [hep-ph]].

[45] M. Tanaka and R. Watanabe, Phys. Rev. D **87** (2013) no.3, 034028 doi:10.1103/PhysRevD.87.034028 [arXiv:1212.1878 [hep-ph]].

[46] A. Celis, M. Jung, X. Q. Li and A. Pich, JHEP **1301** (2013) 054 doi:10.1007/JHEP01(2013)054 [arXiv:1210.8443 [hep-ph]].

[47] A. Crivellin, A. Kokulu and C. Greub, Phys. Rev. D **87** (2013) no.9, 094031 doi:10.1103/PhysRevD.87.094031 [arXiv:1303.5877 [hep-ph]].

[48] A. Crivellin, J. Heeck and P. Stoffer, Phys. Rev. Lett. **116** (2016) no.8, 081801 doi:10.1103/PhysRevLett.116.081801 [arXiv:1507.07567 [hep-ph]].

[49] R. Barbieri, C. W. Murphy and F. Senia, Eur. Phys. J. C **77** (2017) no.1, 8 doi:10.1140/epjc/s10052-016-4578-7 [arXiv:1611.04930 [hep-ph]].

[50] B. Chauhan and B. Kindra, arXiv:1709.09989 [hep-ph].

[51] I. Doršner, S. Fajfer, D. A. Faroughy and N. Košnik, JHEP **1710** (2017) 188 doi:10.1007/JHEP10(2017)188 [arXiv:1706.07779 [hep-ph]].

[52] O. Popov and G. A. White, Nucl. Phys. B **923** (2017) 324 doi:10.1016/j.nuclphysb.2017.08.007 [arXiv:1611.04566 [hep-ph]].

[53] G. Hiller, D. Loose and K. Schönwald, JHEP **1612** (2016) 027 doi:10.1007/JHEP12(2016)027 [arXiv:1609.08895 [hep-ph]].

[54] S. Sahoo, R. Mohanta and A. K. Giri, Phys. Rev. D **95** (2017) no.3, 035027 doi:10.1103/PhysRevD.95.035027 [arXiv:1609.04367 [hep-ph]].

[55] A. G. Akeroyd and C. H. Chen, Phys. Rev. D **96** (2017) no.7, 075011 doi:10.1103/PhysRevD.96.075011 [arXiv:1708.04072 [hep-ph]].

[56] M. Freytsis, Z. Ligeti and J. T. Ruderman, Phys. Rev. D **92** (2015) no.5, 054018 doi:10.1103/PhysRevD.92.054018 [arXiv:1506.08896 [hep-ph]].

[57] A. Celis, M. Jung, X. Q. Li and A. Pich, Phys. Lett. B **771** (2017) 168 doi:10.1016/j.physletb.2017.05.037 [arXiv:1612.07757 [hep-ph]].

[58] D. A. Faroughy, A. Greljo and J. F. Kamenik, Phys. Lett. B **764** (2017) 126 doi:10.1016/j.physletb.2016.11.011 [arXiv:1609.07138 [hep-ph]].

[59] A. Crivellin, D. Müller and T. Ota, JHEP **1709** (2017) 040 doi:10.1007/JHEP09(2017)040 [arXiv:1703.09226 [hep-ph]].

[60] R. Aaij *et al.* [LHCb Collaboration], Phys. Rev. Lett. **118** (2017) no.25, 251802 doi:10.1103/PhysRevLett.118.251802 [arXiv:1703.02508 [hep-ex]].

[61] L. Di Luzio, A. Greljo and M. Nardecchia, Phys. Rev. D **96** (2017) no.11, 115011 doi:10.1103/PhysRevD.96.115011 [arXiv:1708.08450 [hep-ph]].

[62] L. Calibbi, A. Crivellin and T. Li, arXiv:1709.00692 [hep-ph].

[63] M. Bordone, C. Cornella, J. Fuentes-Martin and G. Isidori, arXiv:1712.01368 [hep-ph].

[64] R. Barbieri and A. Tesi, arXiv:1712.06844 [hep-ph].

Hadronic Leading Order Contribution to the Muon $g-2$

Daisuke Nomura[1,2,a] (for the KNT collaboration [1])

[1] KEK Theory Center, Tsukuba, Ibaraki 305-0801, Japan
[2] Yukawa Institute for Theoretical Physics, Kyoto University, Kyoto 606-8502, Japan

Abstract. We calculate the Standard Model (SM) prediction for the muon anomalous magnetic moment. By using the latest experimental data for $e^+e^- \to$ hadrons as input to dispersive integrals, we obtain the values of the leading order (LO) and the next-to-leading-order (NLO) hadronic vacuum polarisation contributions as $a_\mu^{\rm had,\,LO\,VP} = (693.27 \pm 2.46) \times 10^{-10}$ and $a_\mu^{\rm had,\,NLO\,VP} = (-9.82 \pm 0.04) \times 10^{-10}$, respectively. When combined with other contributions to the SM prediction, we obtain $a_\mu(\rm SM) = (11659182.05 \pm 3.56) \times 10^{-10}$, which is deviated from the experimental value by $\Delta a_\mu \equiv a_\mu(\exp) - a_\mu(\rm SM) = (27.05 \pm 7.26) \times 10^{-10}$. This means that there is a 3.7 σ discrepancy between the experimental value and the SM prediction. We also discuss another closely related quantity, the running QED coupling at the Z-pole, $\alpha(M_Z^2)$. By using the same $e^+e^- \to$ hadrons data as input, our result for the 5-flavour quark contribution to the running QED coupling at the Z pole is $\Delta\alpha_{\rm had}^{(5)}(M_Z^2) = (276.11 \pm 1.11) \times 10^{-4}$, from which we obtain $\alpha^{-1}(M_Z^2) = 128.946 \pm 0.015$.

1 Introduction

The anomalous magnetic moment of the muon, a_μ, also known as the muon $g-2$, is an extremely important quantity in particle physics since it can be used to probe/constrain new physics beyond the Standard Model (SM). Experimentally, it has been measured to an extremely high precision [2], and the current world average reads [3]

$$a_\mu(\exp) = (11659209.1 \pm 6.3) \times 10^{-10} . \quad (1)$$

This value should be compared to the SM prediction for a_μ. According to Ref. [1], the most recent value of the SM prediction is

$$a_\mu(\rm SM) = (11659182.05 \pm 3.56) \times 10^{-10} . \quad (2)$$

The difference Δa_μ between Eqs. (1) and (2) is

$$\Delta a_\mu \equiv a_\mu(\exp) - a_\mu(\rm SM) = (27.05 \pm 7.26) \times 10^{-10} , \quad (3)$$

which means a 3.7 σ discrepancy. This deviation may be due to a contribution from physics beyond the SM, which makes a_μ extremely important. There are two experiments which aim to improve the experimental uncertainty by a factor of 4 [4, 5], which further enhances the importance of this quantity.

2 Standard Model contribution to a_μ

The SM contribution to the muon $g-2$ can be conveniently separated into three pieces: the QED, electroweak (EW), and hadronic contributions,

$$a_\mu(\rm SM) = a_\mu(\rm QED) + a_\mu(\rm EW) + a_\mu(\rm had) . \quad (4)$$

The QED contribution is a perturbatively calculable quantity, and has been calculated up to and including 5-loop diagrams [6]:

$$a_\mu(\rm QED) = (11658471.8971 \pm 0.007) \times 10^{-10} . \quad (5)$$

We should note that the 4-loop contributions are already confirmed by other groups [7, 8]. In addition, the 5-loop contribution is very small ($\sim 0.5 \times 10^{-10}$), and hence the QED contribution does not pose any major problem to the muon $g-2$ in the near future.

The EW contribution is also perturbatively calculable, and the current value reads [9]:

$$a_\mu(\rm EW) = (15.36 \pm 0.10) \times 10^{-10} . \quad (6)$$

The most problematic contribution is the hadronic contributions. The hadronic contributions can be written as the sum of a few terms:

$$a_\mu(\rm had) = a_\mu^{\rm had,\,LO\,VP} + a_\mu^{\rm had,\,NLO\,VP} + a_\mu^{\rm had,\,NNLO\,VP} + a_\mu^{\rm had,\,LbL} + a_\mu^{\rm had,\,NLO\,LbL} , \quad (7)$$

where the LO, NLO and NNLO hadronic vacuum polarisation (VP) contributions in the first line can be calculated by using dispersive integrals, whilst to compute the light-by-light (LbL) contributions in the second line we have to rely on hadronic models to some extent.

According to our recent evaluation [1], the values of $a_\mu^{\rm had,\,LO\,VP}$ and $a_\mu^{\rm had,\,NLO\,VP}$ are

$$a_\mu^{\rm had,\,LO\,VP} = (693.27 \pm 2.46) \times 10^{-10} , \quad (8)$$

[a] e-mail: dnomura@post.kek.jp

Table 1. Breakdown of the SM prediction for the muon $g-2$, together with the experimental value and the deviation between the experimental value and the SM prediction. The numbers are given in units of 10^{-10}.

QED contributions	11 658 471.8971 (0.007)	Ref. [6]
EW contributions	15.36 (0.10)	Ref. [9]
hadronic contributions		
LO hadronic VP contributions	693.27 (2.46)	Ref. [1]
NLO hadronic VP contributions	−9.82 (0.04)	Ref. [1]
NNLO hadronic VP contributions	1.24 (0.01)	Ref. [10]
hadronic l-by-l contributions	9.8 (2.6)	Ref. [11]
hadronic l-by-l NLO contributions	0.3 (0.2)	Ref. [12]
Standard Model prediction, a_μ(SM)	11 659 182.05 (3.56)	Ref. [1]
experimental value, a_μ(exp)	11 659 209.1 (6.3)	Refs. [2, 3]
difference, Δa_μ ($\equiv a_\mu(\text{exp}) - a_\mu(\text{SM})$)	27.05 (7.26), 3.7 σ	Ref. [1]

and

$$a_\mu^{\text{had, NLO VP}} = (-9.82 \pm 0.04) \times 10^{-10}\,, \tag{9}$$

respectively. The value of the NNLO hadronic VP contribution quoted in Ref. [10] is

$$a_\mu^{\text{had, NNLO VP}} = (1.24 \pm 0.01) \times 10^{-10}\,. \tag{10}$$

For the LbL contributions, we use the value of $a_\mu^{\text{had, LbL}}$ in Ref. [11]:

$$a_\mu^{\text{had, LbL}} = (9.8 \pm 2.6) \times 10^{-10}\,, \tag{11}$$

and the value of $a_\mu^{\text{had, NLO LbL}}$ in Ref. [12]:

$$a_\mu^{\text{had, NLO LbL}} = (0.3 \pm 0.2) \times 10^{-10}\,. \tag{12}$$

By adding Eqs. (5), (6), (8), (9), (10), (11) and (12) we obtain the SM prediction in Eq. (2), resulting in the 3.7 σ deviation between the experimental value and the SM prediction, Eq. (3), as summarised in Table 1.

By comparing the various contributions to the SM prediction, Eqs. (5), (6), (8), (9), (10), (11) and (12), we note that the uncertainty from the LO hadronic VP contribution is one of the largest. It is therefore extremely important to evaluate this term as accurately as possible. In the next section we discuss the evaluation of the LO hadronic VP contribution in Ref. [1].

3 Leading Order Hadronic Vacuum Polarisation Contribution to a_μ

In this section we discuss the LO hadronic VP contribution to a_μ. We first give an overview of this contribution, and then discuss recent update given in Ref. [1], often comparing the result with the one given in Ref. [13], which is the previous analysis in which two of the KNT collaboration are involved. Below we refer to Ref. [13] as HLMNT.

3.1 Overview of LO hadronic VP contribution

The LO hadronic VP contribution can be calculated by using the dispersion integral,

$$a_\mu^{\text{had, LO VP}} = \frac{\alpha^2}{3\pi^2}\int_{s_{th}}^{\infty}\frac{ds}{s}R(s)K(s)\,, \tag{13}$$

where $s_{th} = m_\pi^2$, $R(s)$ is the hadronic R-ratio,

$$R(s) = \frac{\sigma^0_{\text{had},\gamma}(s)}{\sigma_{\text{pt}}(s)} \equiv \frac{\sigma^0_{\text{had},\gamma}(s)}{4\pi\alpha^2/(3s)}\,, \tag{14}$$

and $K(s)$ is a kernel function. The superscript '0' means that the cross section should not include VP radiative corrections, and the subscript 'γ' implies that the cross section should be inclusive with respect to final state radiations (FSRs). We have to exclude VP radiative corrections since we must avoid possible double-counting with higher order hadronic diagrams.

To understand the s-dependence of the kernel function $K(s)$, it is useful to define another function $\hat{K}(s)$ by

$$\hat{K}(s) \equiv \frac{3s}{m_\mu^2}K(s)\,. \tag{15}$$

The function $\hat{K}(s)$ is a monotonically increasing function with $\hat{K}(m_\pi^2) = 0.40$, $\hat{K}(4m_\pi^2) = 0.63$ and $\hat{K}(s) \to 1$ for $s \to \infty$. The factor $1/s$ in the integrand of Eq. (13), together with the s-dependence of $K(s)$ as $K(s) = \hat{K}(s)m_\mu^2/(3s) = O(1) \times m_\mu^2/(3s)$, makes the low energy hadronic data important.

In Table 2, we show contributions to $a_\mu^{\text{had, LO VP}}$ from important channels. As clearly seen from the table, the contribution to the mean value as well as the uncertainty from the $\pi^+\pi^-$ channel completely dominates over the others. This can be seen also from the pie charts, Fig. 1, where the importance of the $\pi^+\pi^-$ channel is represented as the large area of the green region (the region $0.6 < \sqrt{s} < 0.9$ GeV) in the upper panel. It follows that the good input data in this channel is crucial for this analysis.

Table 2. Contributions from important channels to the dispersion integrals Eqs. (13) and (29). The numbers from the energy region $\sqrt{s} \leq 1.937$ GeV include contributions from data as well as near-threshold contributions to 2π, 3π and $\pi^0\gamma$ channels. The numbers in the table are taken from Ref. [1].

channel	$a_\mu^{\text{had, LO VP}} \times 10^{10}$	$\Delta\alpha_{\text{had}}^{(5)}(M_Z^2) \times 10^4$
Contributions from $\sqrt{s} \leq 1.937$ GeV		
$\pi^+\pi^-$	503.84 ± 1.97	34.27 ± 0.12
$\pi^+\pi^-\pi^0$	47.80 ± 0.89	4.77 ± 0.08
K^+K^-	23.03 ± 0.22	3.37 ± 0.03
$\pi^+\pi^-2\pi^0$	19.39 ± 0.78	5.00 ± 0.20
$2\pi^+2\pi^-$	14.87 ± 0.20	4.02 ± 0.05
$K_S^0K_L^0$	13.04 ± 0.19	1.77 ± 0.03
$\pi^0\gamma$	4.58 ± 0.10	0.36 ± 0.01
$KK\pi$	2.71 ± 0.12	0.89 ± 0.04
$KK\pi\pi$	1.93 ± 0.08	0.75 ± 0.03
⋮	⋮	⋮
Contributions from $1.937 \leq \sqrt{s} \leq 11.199$ GeV		
Inclusive channel	43.67 ± 0.67	82.82 ± 1.05
Narrow Resonance Contributions		
J/ψ	6.26 ± 0.19	7.07 ± 0.22
ψ'	1.58 ± 0.04	2.51 ± 0.06
$\Upsilon(1S-4S)$	0.09 ± 0.00	1.06 ± 0.02
Contributions from $\sqrt{s} \geq 11.199$ GeV		
pQCD	2.07 ± 0.00	124.79 ± 0.10
Total	693.27 ± 2.46	276.11 ± 1.11

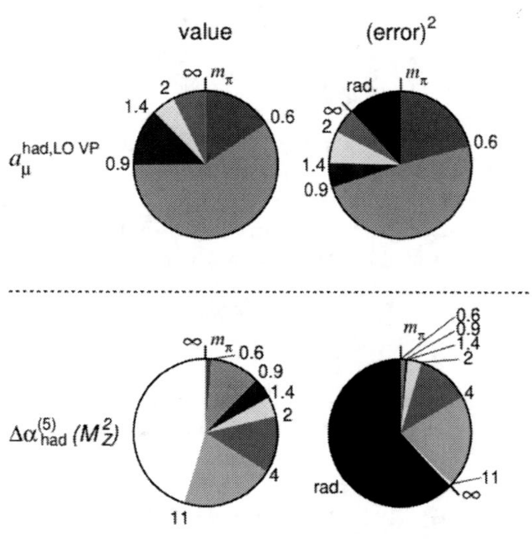

Figure 1. Upper panel: pie charts for the mean value and the error-squared of the LO hadronic VP contribution to the muon $g-2$, $a_\mu^{\text{had, LO VP}}$. Lower panel: pie charts for the mean value and the error-squared of the 5-flavour quark contribution $\Delta\alpha_{\text{had}}^{(5)}(M_Z^2)$ to the QED coupling at the Z pole, $\alpha(M_Z^2)$.

3.2 Update in Input Data

Since the HLMNT paper [13], in the most important $\pi^+\pi^-$ channel, two major data sets have appeared: one from the KLOE experiment [14], and the other from the BESIII collaboration [15]. To correctly compute the dispersive integral, we have to take into account correlations between data sets. In the case of the KLOE data, there are non-negligible correlations with their previous data sets [16, 17]. These correlations are correctly summarised in the covariance matrix in Ref. [18], which is used in the KNT analysis [1]. By using these new data sets, together with our improved data-combination algorithm, which we will discuss in the next subsection, we obtain an improvement of more than 30% in this channel compared to the HLMNT analysis [13], as shown in Table 3.

In the $\pi^+\pi^-\pi^0$ channel, a new data set appeared from the SND collaboration [19], which covers the energy range $1.05 \leq \sqrt{s} \leq 2.00$ GeV.

In the K^+K^- channel, new precise data has been published from BaBar [20] as well as from CMD-3 [21] and SND [22]. Thanks to these new data, we have about a factor of 2 improvement in the uncertainty compared to the HLMNT analysis.

In the $\pi^+\pi^-\pi^0\pi^0$ channel, a new data set appeared from BaBar [23]. In the $\pi^+\pi^-\pi^+\pi^-$ channel, new data sets have been published from BaBar [24] and CMD-3 [25]. In both channels, these new precise data sets allow us to significantly reduce uncertainties from these channels.

These are just examples of new input data. For a full list of new data, see our paper [1].

3.3 Improvements in Data Combination Algorithm

Compared to the HLMNT analysis, we have improved the data combination algorithm.

To evaluate the dispersive integral Eq. (13), it is not a good idea to integrate over raw experimental data. If we would use this method, then the obtained uncertainty of the dispersive integral would be dominated by the contributions of those data points which have large uncertainties. Instead, we should make full use of good accurate data. To do so, we gather near-by data points into a bin (which we call a 'cluster') and integrate over the clusters to evaluate the dispersive integral. In this way, we can reduce influences from inaccurate data points.

In the HLMNT paper, when combining data, the authors assumed a constant cross section across the width of each cluster. In the KNT analysis, we linearly interpolate the cross sections between adjacent clusters, which allows us a more stable determination of the combined cross section. The concrete procedure is the following: We first determine cluster centres E_m and the cross section value R_m at $E = E_m$ by

$$R_m = \left[\sum_{i=1}^{N^{(m)}} \frac{R_i^{(m)}}{(\mathrm{d}\tilde{R}_i^{(m)})^2}\right]\left[\sum_{i=1}^{N^{(m)}} \frac{1}{(\mathrm{d}\tilde{R}_i^{(m)})^2}\right]^{-1}, \quad (16)$$

$$E_m = \left[\sum_{i=1}^{N^{(m)}} \frac{E_i^{(m)}}{(\mathrm{d}\tilde{R}_i^{(m)})^2}\right]\left[\sum_{i=1}^{N^{(m)}} \frac{1}{(\mathrm{d}\tilde{R}_i^{(m)})^2}\right]^{-1}, \quad (17)$$

Table 3. Difference between the KNT18 analysis [1] and the HLMNT11 analysis [13]. Similarly to Table 2, the numbers from the energy region $\sqrt{s} \leq 2$ GeV include contributions from data as well as near-threshold contributions to 2π, 3π and $\pi^0\gamma$ channels. Note that although the default transition point between the sum of exclusive channels and the inclusive measurement is 1.937 GeV in Ref. [1], in this table we take the transition point at 2 GeV for comparison. This table is extracted from Table 3 of Ref. [1]. For a full table, see Ref. [1].

channel	KNT18 [1]	HLMNT11 [13]	Diff
Contributions from $\sqrt{s} \leq 2$ GeV			
$\pi^+\pi^-$	503.86 ± 1.97	505.77 ± 3.09	-2.78
$\pi^+\pi^-\pi^0$	47.83 ± 0.89	47.52 ± 0.99	0.31
K^+K^-	23.05 ± 0.22	22.15 ± 0.46	0.90
$\pi^+\pi^-2\pi^0$	19.80 ± 0.79	20.37 ± 1.26	-0.57
$2\pi^+2\pi^-$	15.17 ± 0.21	14.65 ± 0.47	0.52
$K_S^0K_L^0$	13.05 ± 0.19	13.33 ± 0.16	-0.28
$\pi^0\gamma$	4.58 ± 0.10	4.66 ± 0.14	-0.08
$KK\pi$	2.80 ± 0.12	2.77 ± 0.15	0.03
$KK\pi\pi$	2.42 ± 0.09	3.31 ± 0.58	-0.89
⋮	⋮	⋮	
Contributions from $2 \leq \sqrt{s} \leq 11.199$ GeV			
Inclusive channel	41.27 ± 0.62	41.40 ± 0.87	-0.13
Narrow Resonance Contributions			
J/ψ	6.26 ± 0.19	6.24 ± 0.16	0.02
ψ'	1.58 ± 0.04	1.56 ± 0.05	0.02
$\Upsilon(1S-4S)$	0.09 ± 0.00	0.10 ± 0.00	-0.01
Contributions from $\sqrt{s} \geq 11.199$ GeV			
pQCD	2.07 ± 0.00	2.06 ± 0.00	0.01
Total	693.27 ± 2.46	694.91 ± 4.27	-1.64

where $E_i^{(m)}$ and $R_i^{(m)}$ are the energy and the cross section value of the i-th data point which gives a contribution to the m-th cluster. $d\tilde{R}_i^{(m)}$ is the uncertainty of $R_i^{(m)}$ which is defined as

$$d\tilde{R}_i^{(m)} = \sqrt{(dR_{i;\mathrm{stat}}^{(m)})^2 + (dR_{i;\mathrm{sys}}^{(m)})^2}\,, \tag{18}$$

where $dR_{i;\mathrm{stat}}^{(m)}$ and $dR_{i;\mathrm{sys}}^{(m)}$ are the statistical and systematic uncertainties of $R_i^{(m)}$, respectively. $N^{(m)}$ is the total number of the data points which contribute to the m-th cluster. We now assume a linear interpolation between adjacent clusters. In this case, the interpolant cross section value $\mathcal{R}_m^i$ at $\sqrt{s} = E_i^{(m)}$ is given by,

$$\mathcal{R}_m^i = \begin{cases} R_m + \frac{E_i^{(m)}-E_m}{E_{m+1}-E_m}(R_{m+1} - R_m)\,, & \text{if } E_i^{(m)} > E_m \\ R_{m-1} + \frac{E_i^{(m)}-E_{m-1}}{E_m-E_{m-1}}(R_m - R_{m-1})\,, & \text{if } E_i^{(m)} < E_m \end{cases}. \tag{19}$$

In addition, we have improved the χ^2 function itself. In the HLMNT paper, to construct the χ^2 function, the authors take into account the normalisation uncertainties of experimental data by using the 'penalty trick.' Explicitly, the χ^2 function used in the HLMNT paper is

$$\chi^2_{\mathrm{HLMNT}}(R_m, f_k) = \sum_{k=1}^{N_{\mathrm{exp}}}\left(\frac{1-f_k}{df_k}\right)^2$$

$$+\left\{\sum_{m=1}^{N_{\mathrm{clu}}}\sum_{i=1}^{N_{(k,m)}}\left(\frac{R_i^{(k,m)} - f_k R_m}{dR_i^{(k,m)}}\right)^2\right\}_{\mathrm{w/out\ cov.\ mat.}}$$

$$+\sum_{m,n=1}^{N_{\mathrm{clu}}}\sum_{i=1}^{N_{(k,m)}}\sum_{j=1}^{N_{(k,n)}}\left(R_i^{(k,m)} - f_k R_m\right)C^{-1}(m_i, n_j)\left(R_j^{(k,n)} - f_k R_n\right)\,, \tag{20}$$

where f_k is the overall normalisation factor of the k-th experiment. $R_i^{(k,m)}$ and $dR_i^{(k,m)}$ are the R-values and their errors from k-th experiment which contributes to the m-th cluster. The authors of the HLMNT paper determined the values of f_k and R_m by minimising this χ^2 function.

Originally, this method is proposed to avoid a potential bias [26–28]. However, in Ref. [29], it is pointed out that this method could also suffer from a potential bias. An alternative way to construct the χ^2 function, which is proposed in Ref. [29] and used in the KNT analysis, is to determine the combined cross sections by iteration. We first construct the χ^2 function χ_1^2 for the first iteration as

$$\chi_1^2 = \sum_{i=1}^{N_{\mathrm{tot}}}\sum_{j=1}^{N_{\mathrm{tot}}}(R_i^{(m)} - \mathcal{R}_m^i)\mathbf{C}_1^{-1}(i^{(m)}, j^{(n)})(R_j^{(n)} - \mathcal{R}_n^j)\,, \tag{21}$$

where $N_{\rm tot}$ is the total number of data points. $\mathcal{R}^i_m$ is the interpolant cross section as a function of R_m (see Eq. (19)), and $\mathbf{C}_1^{-1}$ is the inverse matrix of the covariance matrix $\mathbf{C}_1$ for the first iteration, which is given as the sum of the statistical covariance matrix $C^{\rm stat}(i^{(m)}, j^{(n)})$ and the systematic covariance matrix $C^{\rm sys}(i^{(m)}, j^{(n)})$,

$$\mathbf{C}_1(i^{(m)}, j^{(n)}) = C^{\rm stat}(i^{(m)}, j^{(n)}) + \frac{C^{\rm sys}(i^{(m)}, j^{(n)})}{R_i^{(m)} R_j^{(n)}} \mathcal{R}^{i,0}_m \mathcal{R}^{j,0}_n \,, \tag{22}$$

where $\mathcal{R}^{i,0}_m$ is a linearly interpolated cross section value determined by using Eq. (16) as input to Eq. (19). We call the R_m values which minimise the function χ_1^2 as $R_m = R^1_m$, which are used for the input to the next turn of the iterations. From R^1_m, we calculate the interpolant values $\mathcal{R}^i_m$ by substituting $R_m = R^1_m$ in Eq. (19), and call the output values of $\mathcal{R}^i_m$ as $\mathcal{R}^{i,1}_m$. We now construct the covariance matrix $\mathbf{C}_2$ by

$$\mathbf{C}_2(i^{(m)}, j^{(n)}) = C^{\rm stat}(i^{(m)}, j^{(n)}) + \frac{C^{\rm sys}(i^{(m)}, j^{(n)})}{R_i^{(m)} R_j^{(n)}} \mathcal{R}^{i,1}_m \mathcal{R}^{j,1}_n \,, \tag{23}$$

and construct the χ^2 function for the second iteration χ_2^2 as

$$\chi_2^2 = \sum_{i=1}^{N_{\rm tot}} \sum_{j=1}^{N_{\rm tot}} (R_i^{(m)} - \mathcal{R}^i_m) \mathbf{C}_2^{-1}(i^{(m)}, j^{(n)}) (R_j^{(n)} - \mathcal{R}^j_n) \,, \tag{24}$$

where $\mathcal{R}^i_m$ is the interpolant cross section as a function of R_m. We find the set of R_m which minimises χ_2^2, which we call R^2_m. Similarly, we construct the χ^2 function for the I-th iteration as

$$\chi_I^2 = \sum_{i=1}^{N_{\rm tot}} \sum_{j=1}^{N_{\rm tot}} (R_i^{(m)} - \mathcal{R}^i_m) \mathbf{C}_I^{-1}(i^{(m)}, j^{(n)}) (R_j^{(n)} - \mathcal{R}^j_n) \,, \tag{25}$$

where $\mathcal{R}^i_m$ is the interpolant cross section as a function of R_m, and

$$\mathbf{C}_I(i^{(m)}, j^{(n)}) = C^{\rm stat}(i^{(m)}, j^{(n)}) + \frac{C^{\rm sys}(i^{(m)}, j^{(n)})}{R_i^{(m)} R_j^{(n)}} \mathcal{R}^{i,I-1}_m \mathcal{R}^{j,I-1}_n \,. \tag{26}$$

We refer to the set of R_m which minimises χ_I^2 as R^I_m. In this way, we repeat this procedure until we reach convergence. In our case, convergence is achieved after only a few steps.

We should note here that since all the χ^2 functions Eqs. (21), (24) and (25) are quadratic functions of R_m, and hence do not introduce a bias [29]. We have also checked that the result of the HLMNT analysis does not change very much even if we adopt the new iterative χ^2 minimisation algorithm.

3.4 Result for LO Hadronic VP Contribution to a_μ

By using the new data input and the new data combination algorithm, we obtain Eq. (8) for $a_\mu^{\rm had,\,LO\,VP}$. A similar result has been obtained in Ref. [30], although there are slight differences in some individual channels. Ref. [31] quotes slightly smaller value for $a_\mu^{\rm had,\,LO\,VP}$.

Once the $a_\mu^{\rm had,\,LO\,VP}$ is calculated, we can instantly compute $a_\mu^{\rm had,\,NLO\,VP}$ since we have only to evaluate similar dispersion relations. Our result for $a_\mu^{\rm had,\,NLO\,VP}$ is Eq. (8). When combined with other SM contributions to a_μ such as the QED contribution, we obtain Eq. (2). The difference Δa_μ between the experimental determination Eq. (1) and the SM prediction (2) is

$$\Delta a_\mu = a_\mu({\rm exp}) - a_\mu({\rm SM}) = (27.05 \pm 7.26) \times 10^{-10} \,, \tag{27}$$

which could be due to a contribution from new physics beyond the SM.

4 Hadronic Contributions to QED Coupling at the Z-pole

Another closely related quantity is the 5-flavour quark contributions $\Delta\alpha^{(5)}_{\rm had}(M_Z^2)$ to the running QED coupling at the Z-pole, $\alpha(M_Z^2)$. $\alpha(M_Z^2)$ is related to $\Delta\alpha^{(5)}_{\rm had}(M_Z^2)$ by the relation,

$$\alpha(M_Z^2) = \frac{\alpha}{1 - \Delta\alpha_{\rm lep}(M_Z^2) - \Delta\alpha^{(5)}_{\rm had}(M_Z^2) - \Delta\alpha_{\rm top}(M_Z^2)} \,, \tag{28}$$

where $\Delta\alpha_{\rm lep}(M_Z^2)$ and $\Delta\alpha_{\rm top}(M_Z^2)$ are the leptonic and top-quark contributions, respectively. While the leptonic and top-quark contributions can be evaluated perturbatively, to evaluate $\Delta\alpha^{(5)}_{\rm had}(M_Z^2)$, we have to rely on the dispersion relation,

$$\Delta\alpha^{(5)}_{\rm had}(M_Z^2) = -\frac{\alpha M_Z^2}{3\pi} {\rm P} \int_{s_{th}}^{\infty} {\rm d}s \frac{R(s)}{s(s - M_Z^2)} \,. \tag{29}$$

Fortunately, this integral can be evaluated by using the same R-ratio data as used in the evaluation of $a_\mu^{\rm had,\,LO\,VP}$.

By a similar analysis as in the case of $a_\mu^{\rm had,\,LO\,VP}$, we obtain

$$\Delta\alpha^{(5)}_{\rm had}(M_Z^2) = (276.11 \pm 1.11) \times 10^{-4} \,. \tag{30}$$

The contributions from major channels are given in Table 2.

The value of the leptonic contributions is calculated up to and including 4-loop order in Ref. [32], which quotes the value,

$$\Delta\alpha_{\rm lep}(M_Z^2) = (314.979 \pm 0.002) \times 10^{-4} \,. \tag{31}$$

The top-quark contribution is computed in Refs. [33, 34] and the current value is

$$\Delta\alpha_{\rm top}(M_Z^2) = (-0.7180 \pm 0.0054) \times 10^{-4} \,, \tag{32}$$

where we have used the values $\alpha_s(M_Z) = 0.1182(12)$ [3] and $m_t = 173.1 \pm 0.6$ GeV [3] as input parameters. By substituting Eqs. (30), (31) and (32) into Eq. (28), we obtain

$$\alpha^{-1}(M_Z^2) = 128.946 \pm 0.015 \,. \tag{33}$$

5 Conclusions

We have evaluated the LO and NLO hadronic VP contributions to the muon $g-2$. By using the latest experimental data for $e^+e^- \to$ hadrons as input, we obtain the values $a_\mu^{\rm had,\,LO\,VP} = (693.27 \pm 2.46) \times 10^{-10}$ and $a_\mu^{\rm had,\,NLO\,VP} = (-9.82 \pm 0.04) \times 10^{-10}$, respectively. When combined with other contributions to the SM prediction, we obtain $a_\mu({\rm SM}) = (11659182.05 \pm 3.56) \times 10^{-10}$, which is deviated from the experimental value by $\Delta a_\mu = a_\mu({\rm exp}) - a_\mu({\rm SM}) = (27.05 \pm 7.26) \times 10^{-10}$. This means a 3.7 σ discrepancy between the experimental value and the SM prediction. This could be due to a contribution from new physics beyond the SM.

We have discussed another closely related quantity, the running QED coupling at the Z-pole, $\alpha(M_Z^2)$, as well. By using the same $e^+e^- \to$ hadrons data as input, our result for the 5-flavour quark contribution to the running QED coupling at the Z pole is $\Delta\alpha_{\rm had}^{(5)}(M_Z^2) = (276.11 \pm 1.11) \times 10^{-4}$, from which we obtain $\alpha^{-1}(M_Z^2) = 128.946 \pm 0.015$.

Acknowledgements

D. N. thanks the organisers for inviting him to the inspiring workshop in the beautiful surroundings of Capri Island. The work of D. N. is supported by JSPS KAKENHI grant numbers JP16K05323 and JP17H01133.

References

[1] A. Keshavarzi, D. Nomura and T. Teubner, arXiv:1802.02995 [hep-ph].

[2] G. W. Bennett *et al.* [Muon ($g-2$) Collaboration], Phys. Rev. Lett. **92**, 161802 (2004); Phys. Rev. **D73**, 072003 (2006).

[3] C. Patrignani *et al.* [Particle Data Group], Chin. Phys. **C40**, 100001 (2016) and 2017 update.

[4] J. Grange *et al.* [Muon g-2 Collaboration], arXiv:1501.06858 [physics.ins-det].

[5] T. Mibe [J-PARC g-2 Collaboration], Chin. Phys. **C34**, 745 (2010).

[6] T. Aoyama, T. Kinoshita and M. Nio, Phys. Rev. **D97**, 036001 (2018).

[7] S. Laporta, Phys. Lett. **B772**, 232 (2017); these proceedings.

[8] A. Kurz *et al.*, Nucl. Phys. **B879**, 1 (2014); Phys. Rev. **D92**, 073019 (2015); Phys. Rev. **D93**, 053017 (2016).

[9] C. Gnendiger, D. Stöckinger and H. Stöckinger-Kim, Phys. Rev. **D88**, 053005 (2013).

[10] A. Kurz, T. Liu, P. Marquard and M. Steinhauser, Phys. Lett. **B734**, 144 (2014).

[11] A. Nyffeler, Phys. Rev. **D94**, 053006 (2016).

[12] G. Colangelo, M. Hoferichter, A. Nyffeler, M. Passera and P. Stoffer, Phys. Lett. **B735**, 90 (2014).

[13] K. Hagiwara, R. Liao, A. D. Martin, D. Nomura and T. Teubner, J. Phys. **G 38**, 085003 (2011).

[14] D. Babusci *et al.* [KLOE Collaboration], Phys. Lett. **B720**, 336 (2013).

[15] M. Ablikim *et al.* [BESIII Collaboration], Phys. Lett. **B753**, 629 (2016).

[16] F. Ambrosino *et al.* [KLOE Collaboration], Phys. Lett. **B670**, 285 (2009).

[17] F. Ambrosino *et al.* [KLOE Collaboration], Phys. Lett. **B700**, 102 (2011).

[18] A. Anastasi *et al.* [KLOE-2 Collaboration], arXiv:1711.03085 [hep-ex].

[19] V. M. Aul'chenko *et al.* [SND Collaboration], J. Exp. Theor. Phys. **121**, 27 (2015).

[20] J. P. Lees *et al.* [BaBar Collaboration], Phys. Rev. **D88**, 032013 (2013).

[21] E. A. Kozyrev *et al.* [CMD-3 Collaboration], Phys. Lett. **B779**, 64 (2018).

[22] M. N. Achasov *et al.* [SND Collaboration], Phys. Rev. **D94**, 112006 (2016).

[23] J. P. Lees *et al.* [BaBar Collaboration], Phys. Rev. **D96**, 092009 (2017).

[24] J. P. Lees *et al.* [BaBar Collaboration], Phys. Rev. **D85**, 112009 (2012).

[25] R. R. Akhmetshin *et al.* [CMD-3 Collaboration], Phys. Lett. **B768**, 345 (2017).

[26] G. D'Agostini, Nucl. Instrum. Meth. **A346**, 306 (1994).

[27] T. Takeuchi, Prog. Theor. Phys. Suppl. **123**, 247 (1996).

[28] V. Blobel, eConf **C 030908**, MOET002 (2003).

[29] R. D. Ball *et al.* [NNPDF Collaboration], JHEP **1005**, 075 (2010).

[30] M. Davier, A. Hoecker, B. Malaescu and Z. Zhang, Eur. Phys. J. **C77**, 827 (2017).

[31] F. Jegerlehner, arXiv:1711.06089 [hep-ph].

[32] C. Sturm, Nucl. Phys. **B874**, 698 (2013).

[33] K. G. Chetyrkin, J. H. Kuhn and M. Steinhauser, Phys. Lett. **B371**, 93 (1996); Nucl. Phys. **B482**, 213 (1996); Nucl. Phys. **B505**, 40 (1997).

[34] J. H. Kuhn and M. Steinhauser, Phys. Lett. **B437**, 425 (1998).

Hadronic light-by-light scattering contribution to the muon $g-2$ on the lattice

Nils Asmussen[1], *Antoine* Gérardin[2], *Jeremy* Green[3], *Oleksii* Gryniuk[1], *Georg* von Hippel[1], *Harvey B.* Meyer[1,2], *Andreas* Nyffeler[1,*], *Vladimir* Pascalutsa[1], and *Hartmut* Wittig[1,2]

[1]PRISMA Cluster of Excellence and Institut für Kernphysik, Johannes Gutenberg-Universität Mainz, D-55099 Mainz, Germany
[2]Helmholtz Institut Mainz, D-55099 Mainz, Germany
[3]John von Neumann Institute for Computing (NIC), DESY, Platanenallee 6, D-15738 Zeuthen, Germany

Abstract. We briefly review several activities at Mainz related to hadronic light-by-light scattering (HLbL) using lattice QCD. First we present a position-space approach to the HLbL contribution in the muon $g-2$, where we focus on exploratory studies of the pion-pole contribution in a simple model and the lepton loop in QED in the continuum and in infinite volume. The second part describes a lattice calculation of the double-virtual pion transition form factor $\mathcal{F}_{\pi^0\gamma^*\gamma^*}(q_1^2, q_2^2)$ in the spacelike region with photon virtualities up to 1.5 GeV^2 which paves the way for a lattice calculation of the pion-pole contribution to HLbL. The third topic involves HLbL forward scattering amplitudes calculated in lattice QCD which can be described, using dispersion relations (HLbL sum rules), by $\gamma^*\gamma^* \to$ hadrons fusion cross sections and then compared with phenomenological models.

1 Introduction

The anomalous magnetic moment of the muon has served for many years as a precision test of the Standard Model [1–4] and it has also played an important role in many presentations at this meeting. There is a discrepancy of 3 − 4 standard deviations between experiment and theory for some time now. This could be a signal of New Physics [1, 5], but the uncertainties in the theory prediction from hadronic vacuum polarization (HVP) and hadronic light-by-light scattering (HLbL) make it difficult to draw firm conclusions. In view of upcoming four-fold more precise new experiments at Fermilab and J-PARC [6], these hadronic contributions need to be better controlled.

The improvement for HVP looks straightforward, with more precise experimental data from various experiments on hadronic cross-sections as input for a dispersion relation [7]. But also lattice QCD is getting more and more precise [8], and, hopefully, also a new method using muon-electron scattering to measure the running of α and the HVP in the spacelike region [9], will be feasible at some point with the required precision.

On the other hand, the HLbL contribution to the muon $g-2$, see Fig. 1, has only been calculated using models so far [1, 4, 10] and the frequently used estimates from Refs. [1, 11, 12] (revised slightly in Ref. [2]) suffer from uncontrollable uncertainties. In view of this, dispersion relations have been proposed a few years ago [13–15] (see also the very recent new proposal in Ref. [16]) to determine the presumably numerically dominant contributions from a single neutral pion-pole (light pseudoscalar-pole) and from the two-pion intermediate state (pion-loop) based on input from experimental data for the dispersion relations. Still some modelling will be needed to estimate the contributions from multi-pion intermediate states, like the axial-vector contribution.

Finally, lattice QCD was proposed some time ago as a model-independent, first principle approach to the HLbL contribution in the muon $g-2$ and some promising progress has been achieved recently by the RBC-UKQCD collaboration [17].

Independently, also the lattice group at Mainz has studied HLbL in recent years. We used complementary approaches to tackle the full HLbL contribution in the muon $g-2$ with a new position-space approach [18–20], studied the pion transition form factor (TFF) with two virtual photons on the lattice to evaluate the pion-pole contribution to HLbL [21] and we analyzed HLbL forward scattering amplitudes that can be compared using dispersion relations (HLbL sum rules) [22] to phenomenological models for photon fusion processes [18, 23, 24]. These three topics will be discussed in the following three sections. For all the details, we refer to the quoted papers.

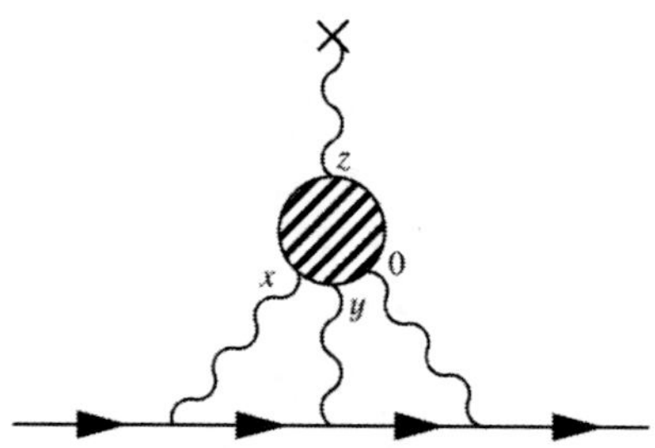

Figure 1. HLbL scattering contribution to the muon $g-2$.

*Speaker, e-mail: nyffeler@uni-mainz.de

2 Position-space approach to HLbL in the muon $g-2$ on the lattice

The HLbL scattering contribution to the anomalous magnetic moment of the muon in Fig. 1 can be split into a perturbative QED kernel $\bar{\mathcal{L}}$ that describes the muon and photon propagators and a non-perturbative QCD four-point function $i\widehat{\Pi}$ (denoted by a blob in the Feynman diagram) that will be evaluated on the lattice.

The projection on the muon $g-2$ yields the master formula (in Euclidean space notation) [18, 19]

$$a_\mu^{\rm HLbL} = \frac{me^6}{3}\int d^4y \int d^4x \, \underbrace{\bar{\mathcal{L}}_{[\rho,\sigma];\mu\nu\lambda}(x,y)}_{\rm QED} \, \underbrace{i\widehat{\Pi}_{\rho;\mu\nu\lambda\sigma}(x,y)}_{\rm QCD}, \quad (1)$$

with the spatial moment of the four-point function

$$i\widehat{\Pi}_{\rho;\mu\nu\lambda\sigma}(x,y) = -\int d^4z \, z_\rho \, \langle j_\mu(x) j_\nu(y) j_\sigma(z) j_\lambda(0)\rangle. \quad (2)$$

We evaluate the QED kernel in the continuum and in infinite volume and thereby avoid $1/L^2$ finite-volume effects from the massless photons. Since Lorentz covariance is manifest in our approach, the eight-dimensional integral in Eq. (1) can be reduced, after contracting all indices, to a three-dimensional integral over the Lorentz invariants x^2, y^2 and $x \cdot y$.

The QED kernel $\bar{\mathcal{L}}_{[\rho,\sigma];\mu\nu\lambda}(x,y)$ can be decomposed into several tensors

$$\bar{\mathcal{L}}_{[\rho,\sigma];\mu\nu\lambda}(x,y) = \sum_{A={\rm I,II,III}} \mathcal{G}^A_{\delta[\rho\sigma]\mu\alpha\nu\beta\lambda} T^{(A)}_{\alpha\beta\delta}(x,y). \quad (3)$$

The $\mathcal{G}^A_{\delta[\rho\sigma]\mu\alpha\nu\beta\lambda}$ are traces of gamma matrices and just yield sums of products of Kronecker deltas. The tensors $T^{(A)}_{\alpha\beta\delta}$ are decomposed into a scalar S, vector V and tensor T part

$$T^{(\rm I)}_{\alpha\beta\delta}(x,y) = \partial^{(x)}_\alpha(\partial^{(x)}_\beta + \partial^{(y)}_\beta)V_\delta(x,y), \quad (4)$$

$$T^{(\rm II)}_{\alpha\beta\delta}(x,y) = m\partial^{(x)}_\alpha\Big(T_{\beta\delta}(x,y) + \frac{1}{4}\delta_{\beta\delta}S(x,y)\Big), \quad (5)$$

$$T^{(\rm III)}_{\alpha\beta\delta}(x,y) = m(\partial^{(x)}_\beta + \partial^{(y)}_\beta)\Big(T_{\alpha\delta}(x,y) + \frac{1}{4}\delta_{\alpha\delta}S(x,y)\Big). \quad (6)$$

These can be parametrized by six weight functions

$$S(x,y) = \bar{\mathfrak{g}}^{(0)}, \quad (7)$$

$$V_\delta(x,y) = x_\delta \bar{\mathfrak{g}}^{(1)} + y_\delta \bar{\mathfrak{g}}^{(2)}, \quad (8)$$

$$T_{\alpha\beta}(x,y) = (x_\alpha x_\beta - \frac{x^2}{4}\delta_{\alpha\beta})\,\bar{\mathfrak{l}}^{(1)} + (y_\alpha y_\beta - \frac{y^2}{4}\delta_{\alpha\beta})\,\bar{\mathfrak{l}}^{(2)} + (x_\alpha y_\beta + y_\alpha x_\beta - \frac{x\cdot y}{2}\delta_{\alpha\beta})\,\bar{\mathfrak{l}}^{(3)}, \quad (9)$$

that depend on the three variables x^2, $x \cdot y = |x||y|\cos\beta$ and y^2. The semi-analytical expressions for the weight functions have been precomputed to about 5 digits precision and stored on a three-dimensional grid. For illustration, the expression for the weight functions $\bar{\mathfrak{g}}^{(2)}$ has been given in Ref. [19]. In Fig. 2 we show the two weight functions $\bar{\mathfrak{g}}^{(1)}$ and $\bar{\mathfrak{g}}^{(2)}$ as a function of $|x| < 12$ fm, for a fixed value of $|y| = 0.506$ fm and three values of $\cos\beta$. Plots for all six weight functions can be found in Ref. [20].

To test our semi-analytical expressions and the software for the QED kernel, we have computed the π^0-pole contribution to HLbL in a simple vector-meson dominance (VMD) model as well as the lepton-loop contribution to $a_\mu^{\rm LbL}$ in QED, where the results are well known. To this aim, we first derived analytical expressions for these contributions to the four-point function $i\widehat{\Pi}_{\rho;\mu\nu\lambda\sigma}(x,y)$ in position-space, see Ref. [20] for details.

In figure 3 we plot the integrand $f(|y|)$ of the final integration over $|y|$ of the HLbL contribution $a_\mu^{\rm HLbL} = \int_0^\infty d|y|\, f(|y|)$ (LbL in QED) for these two examples, after contracting all Lorentz indices and the integrations over $|x|$ and $\cos\beta$ have been performed in Eq. (1).

From the plot of the integrand for the pion-pole contribution in Fig. 3 one observes that this contribution to $a_\mu^{\rm HLbL}$ is remarkably long-range with a long negative tail at large $|y|$. One expects an exponential decay $\sim e^{-\tilde{c}m_\pi|y|}$ of the correlation function. But this seems to be countered by some non-negligible power-like behavior $|y|^n$. For pion masses $m_\pi = 300-900$ MeV we reproduce the known results, which can be easily obtained from the three-dimensional integral representation in momentum space given in Ref. [1], at the percent level. On the other hand, for the physical pion mass, one will need rather large lattices of the order of $5-10$ fm to capture the negative tail at large $|y|$ in a QCD lattice simulation. Hopefully we can correct for finite-size effects on this contribution, by computing the relevant neutral pion transition form factor on the same lattice ensembles, see Ref. [21] and Section 3.

For the lepton-loop in QED the behavior of the integrand for small $|y|$ is compatible with $f(|y|) \propto |y|\log^2(|y|)$. This is quite steep and means that we probe the QED kernel at small distances. In addition, the height of the positive peak grows with smaller masses m_l of the lepton in the loop. Furthermore there is again a long negative tail at large $|y|$ which demands the use of a large size for the grid where the weight functions have been calculated, in particular for a lepton in the loop that is lighter than the muon. For $m_l = m_\mu, 2m_\mu$ we reproduce the analytically known results for $a_\mu^{\rm LbL}$ in QED [25] at the percent level, see Table 1. On the other hand, for the lightest lepton mass $m_l = m_\mu/2$, some further refinements of our numerical evaluation are needed. Once this is achieved, we plan to make the QED kernel publicly available.

Table 1. Results ($\times 10^{11}$), precision and deviation for the lepton-loop contribution to LbL in QED with our approach compared to the known results [25]. The first uncertainty originates from the three-dimensional numerical integration, the second from the extrapolation of the integrand to small $|y|$.

m_l/m_μ	$a_\mu^{\rm LbL}$ [25]	$a_\mu^{\rm LbL}$	Prec.	Dev.
1/2	1229.07	1257.5(6.2)(2.4)	0.5%	2.3%
1	464.97	470.6(2.3)(2.1)	0.7%	1.2%
2	150.31	150.4(0.7)(1.7)	1.2%	0.06%

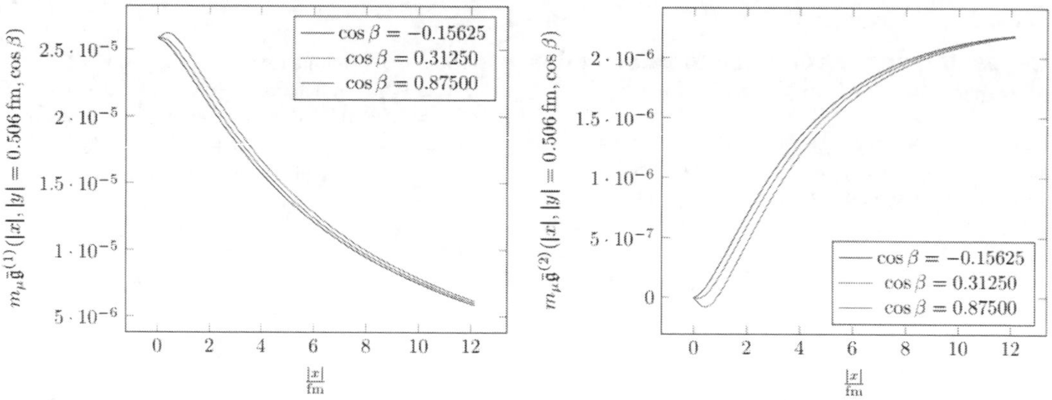

Figure 2. The $|x|$ dependence of the weight functions $\bar{\mathfrak{g}}^{(1)}$ (left) and $\bar{\mathfrak{g}}^{(2)}$ (right) for $|y| = 0.506$ fm and three values of $\cos\beta$. Apart from the scale, the weight functions $\bar{\mathfrak{g}}^{(0)}$, $\bar{\mathfrak{l}}^{(1)}$ and $\bar{\mathfrak{l}}^{(2)}$ have the same shape as $\bar{\mathfrak{g}}^{(1)}$, whereas $\bar{\mathfrak{l}}^{(2)}$ looks similar to $\bar{\mathfrak{g}}^{(2)}$.

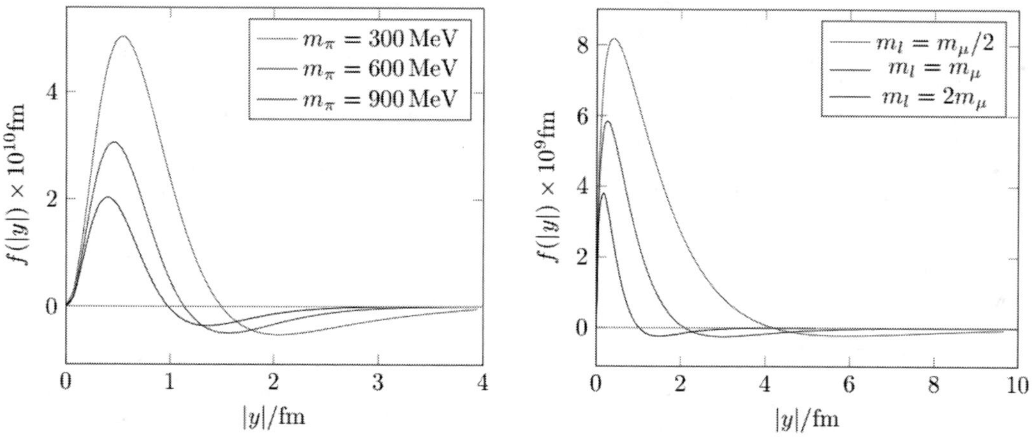

Figure 3. (Left) Integrand of the pion-pole contribution a_μ^{HLbL} for a simple VMD transition form factor for three different values of the pion mass. (Right) Integrand of the lepton-loop contribution a_μ^{LbL} in QED for three different lepton masses m_l in the loop.

3 Lattice calculation of the pion transition form factor $\mathcal{F}_{\pi^0\gamma^*\gamma^*}(q_1^2, q_2^2)$

The pion transition form factor can be defined from the following correlation function in Euclidean space, using the methods proposed and used before in Ref. [26]

$$\begin{aligned} M_{\mu\nu}^E(p, q_1) &= -\int \mathrm{d}\tau\, e^{\omega_1\tau} \int \mathrm{d}^3 z\, e^{-i\vec{q}_1\vec{z}} \\ &\quad \times \langle 0 | T\left\{J_\mu(\vec{z}, \tau) J_\nu(\vec{0}, 0)\right\} |\pi(p)\rangle \\ &= \epsilon_{\mu\nu\alpha\beta}\, q_{1\alpha}\, q_{2\beta}\, \mathcal{F}_{\pi^0\gamma^*\gamma^*}(q_1^2, q_2^2), \qquad (10) \end{aligned}$$

provided the photon virtualities satisfy $q_{1,2}^2 < \min(M_\rho^2, 4m_\pi^2)$ to avoid poles in the analytical continuation from the original definition of the form factor in Minkowski space to Euclidean space. The free real parameter ω_1 denotes the zeroth (energy) component of the four-momentum $q_1 = (\omega_1, \vec{q}_1)$.

The main object to compute on the lattice is the three-point function

$$\begin{aligned} & C_{\mu\nu}^{(3)}(\tau, t_\pi; \vec{p}, \vec{q}_1, \vec{q}_2) \qquad (11) \\ &= a^6 \sum_{\vec{x},\vec{z}} \langle T\left\{J_\nu(\vec{0}, t_f) J_\mu(\vec{z}, t_i) P(\vec{x}, t_0)\right\}\rangle\, e^{i\vec{p}\vec{x}}\, e^{-i\vec{q}_1\vec{z}}. \end{aligned}$$

Here $\tau = t_i - t_f$ is the time separation between the two vector currents and $t_\pi = \min(t_f - t_0, t_i - t_0)$. The matrix element in Eq. (10) with an on-shell pion is obtained by considering the limit of large t_π. With the definitions

$$A_{\mu\nu}(\tau) = \lim_{t_\pi \to +\infty} C_{\mu\nu}^{(3)}(\tau, t_\pi)\, e^{E_\pi t_\pi}, \qquad (12)$$

$$\widetilde{A}_{\mu\nu}(\tau) = \begin{cases} A_{\mu\nu}(\tau) & \tau > 0 \\ A_{\mu\nu}(\tau)\, e^{-E_\pi\tau} & \tau < 0 \end{cases}, \qquad (13)$$

one obtains

$$M_{\mu\nu}^{\mathrm{E}} = \frac{2E_\pi}{Z_\pi} \int_{-\infty}^{\infty} \mathrm{d}\tau\, e^{\omega_1\tau}\, \widetilde{A}_{\mu\nu}(\tau). \qquad (14)$$

The overlap factor Z_π and the pion energy E_π can be obtained from the asymptotic behavior of the pseudoscalar two-point function.

For the calculation of $\mathcal{F}_{\pi^0\gamma^*\gamma^*}(q_1^2, q_2^2)$ on the lattice in Ref. [21] we used eight CLS (Coordinated Lattice Simulations) lattice ensembles [27] with $n_f = 2$ dynamical quarks with three different lattice spacings $a = (0.048, 0.065, 0.075)$ fm and pion masses in the range $194 - 437$ MeV.

We choose the pion rest frame $\vec{p} = 0$, with photons back-to-back spatially $\vec{q}_2 = -\vec{q}_1$, where the kinematical range accessible on the lattice is given by $q_1^2 = \omega_1^2 - \vec{q}_1^{\,2}$, $q_2^2 = (m_\pi - \omega_1)^2 - \vec{q}_1^{\,2}$. With multiple values of $|\vec{q}_1|^2 = \left(\frac{2\pi}{L}\right)^2 |\vec{n}|^2$, $|\vec{n}|^2 = 1, 2, 3, \ldots$ we obtain mostly spacelike photon virtualities up to $|q_{1,2}^2| \approx 1.5$ GeV^2, as can be seen in Fig. 4 (left). In practice, discrete values of ω_1 have been used to sample the momenta. Note that on the lattice it is actually easier to access the double-virtual TFF $\mathcal{F}_{\pi^0\gamma^*\gamma^*}(q_1^2, q_2^2)$, in particular near the diagonal $q_1^2 = q_2^2$, than the single-virtual form factor $\mathcal{F}_{\pi^0\gamma^*\gamma^*}(q_{1,2}^2, 0)$ along the two axis, in contrast to the situation in experiments [28].

From the theoretical side, the form factor is constrained by the chiral anomaly such that $\mathcal{F}_{\pi^0\gamma^*\gamma^*}(0,0) = 1/(4\pi^2 F_\pi)$ [29] (in the chiral limit). For the single-virtual form factor one expects the Brodsky-Lepage (BL) behavior $\mathcal{F}_{\pi^0\gamma^*\gamma^*}(-Q^2, 0) \xrightarrow[Q^2\to\infty]{} 2F_\pi/Q^2$ [30]. The precise value of the prefactor is, however, under debate. The double-virtual form factor, where both momenta become simultaneously large, has been computed using the OPE at short distances. In the chiral limit the result reads $\mathcal{F}_{\pi^0\gamma^*\gamma^*}(-Q^2, -Q^2) \xrightarrow[Q^2\to\infty]{} 2F_\pi/(3Q^2)$ [31].

In order to get a result for the double-virtual TFF $\mathcal{F}_{\pi^0\gamma^*\gamma^*}(q_1^2, q_2^2)$ in the continuum and for the physical pion mass, we fit our lattice data, obtained for the eight ensembles with different lattice spacings a and pion masses m_π, with three simple models: vector-meson dominance (VMD), lowest-meson dominance (LMD) and LMD+V, see Refs. [32, 33] for details about these models. The often used VMD model fulfills the BL behavior for the single-virtual case (and describes quite well the available experimental data below $2 - 3$ GeV^2), but falls off too fast for the double virtual case $\mathcal{F}^{\mathrm{VMD}}_{\pi^0\gamma^*\gamma^*}(-Q^2, -Q^2) \sim 1/Q^4$. The LMD model is constructed in such a way that it fulfills the OPE constraint, but fails to reproduce the Brodsky-Lepage behavior. Finally, the LMD+V models is a generalization of the LMD model and contains two vector resonances ρ and ρ'. It can be made to fulfill both the BL and the OPE constraints, for the prize of a large number of free parameters.

In order to reduce the number of fit parameters for all the models, a global fit is performed where all lattice ensembles are fitted simultaneously assuming a linear dependence of each model parameter on the lattice spacing a and on the squared pion mass m_π^2, see Ref. [21] for all the details and results of the fits.

The fits for the VMD and the LMD model are also used to perform the integration in Eq. (14) up to infinite τ, see Fig. 4 (right). For $|\tau| < 1.3$ fm the lattice data are used. The dependence on these models for large τ is small, but the behavior for small τ is very different. In fact, both the lattice data and the LMD model show a cusp at $\tau = 0$, which is related to the OPE in the double-virtual case, whereas the VMD model is smooth at $\tau = 0$, since the VMD model falls off too fast at large Q^2.

The VMD model leads to a poor description of our data, with $\chi^2/\mathrm{d.o.f.} = 2.9$ (uncorrelated fit), especially in the double-virtual case and at large Euclidean momenta, see Fig. 5. The normalization $\mathcal{F}^{\mathrm{VMD}}_{\pi^0\gamma^*\gamma^*}(0,0)$ is off from the value expected from the chiral anomaly and the fitted vector meson mass does not agree with the ρ-mass.

On the other hand, both the LMD model and the LMD+V model lead to a quite good fit. The LMD model has a $\chi^2/\mathrm{d.o.f.} = 1.3$ (uncorrelated fit) and leads to a determination of the chiral anomaly at the 7% level, with a value in agreement with expectations. The fitted vector meson mass is close to the rho-meson mass. Furthermore, the fit result for another parameter that is related to the OPE is compatible with the theoretical expections. Despite the fact that the LMD model fails to reproduce the BL behavior for the single-virtual form factor, this does not seem to affect the global fit, since there are only few lattice data points at rather low momenta in the single virtual case, see Figs. 4 and 5, i.e. one is not yet sensitive to the asymptotic behavior.

Finally, the LMD+V model has a $\chi^2/\mathrm{d.o.f.} = 1.4$ (uncorrelated fit), however, only after the model parameters that are related to the two vector meson masses, the BL behavior and the OPE constraint, have been fixed, otherwise no stable fit was obtained. The chiral anomaly is reproduced with 9% accuracy and two other fitted model parameters are close to results obtained in phenomenological analyses of the TFF [33].

The form factors in the three models, extrapolated to the physical point, are shown in Fig. 6. In the single-virtual case, the LMD+V model is in quite good agreement with the experimental data from Ref. [28]. The LMD model starts to deviate already at $Q^2 = 1$ GeV^2. In the double-virtual case, the LMD and LMD+V models are quite similar and already close to their asymptotic behavior at the largest point $Q^2 \sim 1.5$ GeV^2 where we have lattice data.

Using the LMD+V model at the physical point from our fit, the pion-pole contribution to HLbL in the muon $g - 2$ can be obtained from the three-dimensional integral representation from Ref. [1], with the result [21],

$$a_{\mu;\mathrm{LMD+V}}^{\mathrm{HLbL};\pi^0} = (65.0 \pm 8.3) \times 10^{-11}\,. \qquad (15)$$

Note that the given error is only statistical. No attempt has been made to estimate the systematical errors from using different fit models. Since the relevant momentum range in the pion-pole contribution is below 1 GeV [34], i.e. where most of our lattice data points are obtained, the extrapolation to large momenta with the LMD+V or the LMD model does not have a large effect on the final result. The result in Eq. (15) is fully consistent with most model calculations which yield results in the range $a_\mu^{\mathrm{HLbL};\pi^0} = (50 - 80) \times 10^{-11}$, but with rather arbitrary, model-dependent error estimates, see Refs. [1, 10, 34] and references therein.

https://doi.org/10.1051/epjconf/201817901017

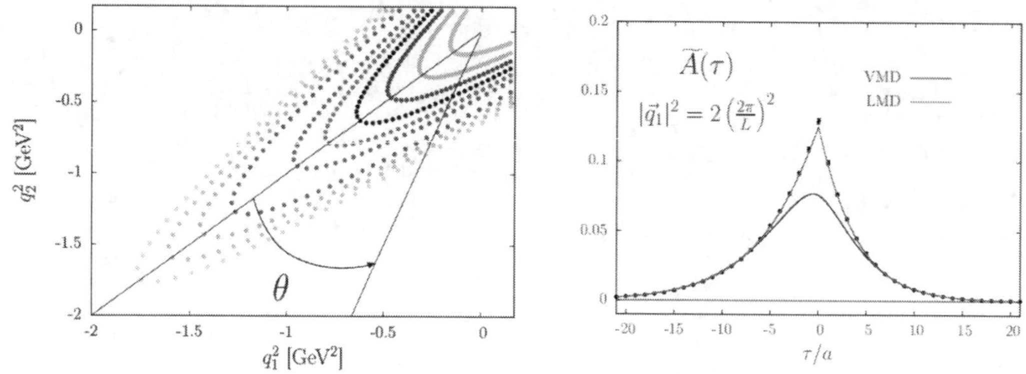

Figure 4. (Left) Kinematic reach in the photon virtualities (q_1^2, q_2^2) with the pion at rest, for a lattice resolution of $64^3 \times 128$ with $a = 0.048$ fm and $m_\pi = 268$ MeV. (Right) The function $\widetilde{A}(\tau)$ (black points) and the VMD (blue line) and LMD (red line) fits used to describe the tail of the function at large τ for another lattice ensemble with $a = 0.065$ fm and $m_\pi = 270$ MeV.

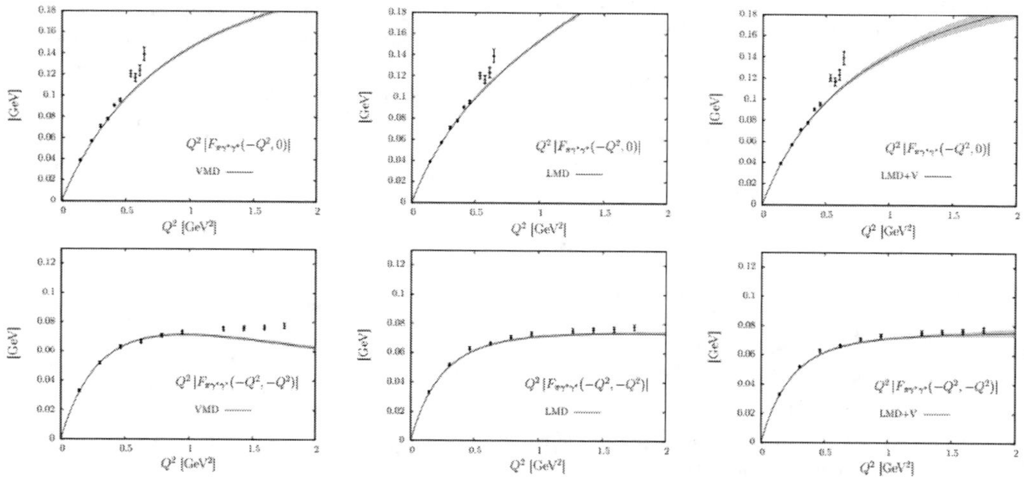

Figure 5. Comparison of the VMD, LMD and LMD+V fits for a lattice ensemble with $a = 0.048$ fm and $m_\pi = 268$ MeV. The red line corresponds to the results from our global fit. Note that the points at different Q^2 are correlated.

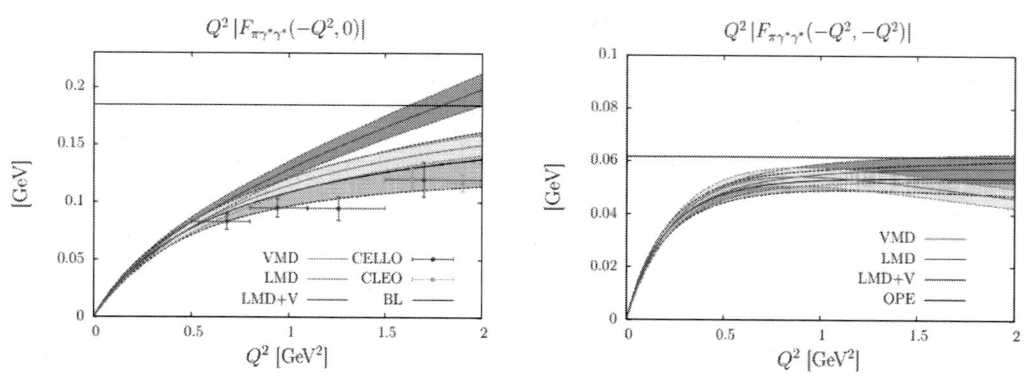

Figure 6. Extrapolation of the lattice data to the continuum and the physical pion mass for the VMD, LMD and LMD+V models. (Left) Single-virtual form factor, compared with experimental results [28] and the expectation for the asymptotic behavior according to Brodsky-Lepage [30]. (Right) Double-virtual form factor at $Q_1^2 = Q_2^2$ and the expectation from the OPE [31].

4 HLbL forward scattering amplitudes in lattice QCD

Using parity and time-reversal invariance of QCD, there are eight independent light-by-light forward scattering amplitudes describing the process $\gamma^*(\lambda_1, q_1)\gamma^*(\lambda_2, q_2) \to \gamma^*(\lambda'_1, q_1)\gamma^*(\lambda'_2, q_2)$, where q_i and $\lambda_i^{(\prime)} = 0, \pm$ are the momenta and helicities of the virtual photons ($i = 1, 2$). Six amplitudes are even and two are odd functions of the crossing-symmetric variable $\nu = q_1 \cdot q_2$. The forward scattering amplitudes $\mathcal{M}_{\lambda'_1\lambda'_2\lambda_1\lambda_2}$ can be related via the optical theorem to two-photon fusion amplitudes $\mathcal{M}_{\lambda_1\lambda_2}$ for the process $\gamma^*(\lambda_1, q_1)\gamma^*(\lambda_2, q_2) \to X(p_X)$ as follows:

$$\begin{aligned} W_{\lambda'_1\lambda'_2,\lambda_1\lambda_2} &= \operatorname{Im} \mathcal{M}_{\lambda'_1\lambda'_2,\lambda_1\lambda_2} \\ &= \frac{1}{2}\int d\Gamma_X (2\pi)^4 \delta(q_1 + q_2 - p_X) \\ &\quad \times \mathcal{M}_{\lambda_1\lambda_2}(q_1, q_2, p_X)\, \mathcal{M}^*_{\lambda'_1\lambda'_2}(q_1, q_2, p_X)\,. \end{aligned} \tag{16}$$

Unitarity and analyticity then allow one to write dispersion relations in ν at fixed values of the virtualities $Q_i^2 = -q_i^2$. Performing one subtraction to get a faster convergence and to suppress higher resonance states, one obtains the following sum rules (omitting all dependence on the virtualities and the helicities):

$$\mathcal{M}_{\text{even}}(\nu) = \mathcal{M}_{\text{even}}(0) + \frac{2\nu^2}{\pi}\int_{\nu_0}^{\infty} d\nu' \frac{W_{\text{even}}(\nu')}{\nu'(\nu'^2 - \nu^2 - i\epsilon)}, \tag{17}$$

$$\mathcal{M}_{\text{odd}}(\nu) = \nu\mathcal{M}'_{\text{odd}}(0) + \frac{2\nu^3}{\pi}\int_{\nu_0}^{\infty} d\nu' \frac{W_{\text{odd}}(\nu')}{\nu'^2(\nu'^2 - \nu^2 - i\epsilon)}, \tag{18}$$

with $\nu_0 = (Q_1^2 + Q_2^2)/2$. For later use, we introduce the following notation for the subtracted even $\overline{\mathcal{M}}(q_1^2, q_2^2, \nu) = \mathcal{M}(q_1^2, q_2^2, \nu) - \mathcal{M}(q_1^2, q_2^2, 0)$ and odd $\overline{\mathcal{M}}(q_1^2, q_2^2, \nu) = \mathcal{M}(q_1^2, q_2^2, \nu) - \nu\mathcal{M}'(q_1^2, q_2^2, 0)$ amplitudes.

As proposed first in Ref. [23] for the forward scattering amplitude $\mathcal{M}_{TT}$ and extended recently to all eight amplitudes in Ref. [24], the scattering amplitudes on the left-hand sides of Eqs. (17)-(18) can be computed from the correlation function of four vector currents on the lattice. On the other hand, the right-hand sides of Eqs. (17)-(18) are related to the two-photon fusion processes using Eq. (16). The latter are described by single-meson TFFs which can be parametrized by simple models, as sketched below. Fitting the lattice data then determines the model parameters and the TFFs. This will allow one to calculate the contributions from single-meson poles to the HLbL contribution in the muon $g-2$.

The lattice simulations are performed on five CLS lattice ensembles [27] with two degenerate light dynamical quarks at two lattice spacings $a = 0.048$ fm (one ensemble) and $a = 0.065$ fm (four ensembles) and pion masses in the range from 194 – 437 MeV. For all ensembles the fully connected and for two ensembles also the leading quark-disconnected diagrams contributing to the four-point function are taken into account. For each ensemble, the correlation function is computed at up to three values of Q_1^2 below 0.8 GeV^2 and for all values of $Q_2^2 \leq 4$ GeV^2. The result for a subset of four scattering amplitudes is shown in Fig. 7.

The photon fusion reaction on the right-hand sides of Eqs. (17)-(18) can produce any C-parity-even state X. The main contribution is expected from the pseudoscalar (0^{-+}), scalar (0^{++}), axial-vector (1^{++}) and tensor (2^{++}) mesons, where we consider in each channel only the lightest state. We are working with two degenerate dynamical quarks and fit our phenomenological model to only the fully-connected diagrams. To compensate for this, we include only the contributions from isovector mesons, multiplied by a factor of 34/9 [10, 24].

There is one TFF for the pseudoscalars, two each for the scalars and axial-vectors and four for the tensor mesons [22]. For the pseudoscalars, we take the TFF from Ref. [21] evaluated on the same lattice ensembles. The other TFFs are parametrized as follows

$$F_X(Q_1^2, Q_2^2) = \frac{F_X(0,0)}{\left(1 + Q_1^2/\Lambda_X^2\right)^n \left(1 + Q_2^2/\Lambda_X^2\right)^n}, \tag{19}$$

where we assume a monopole ansatz ($n = 1$) for the scalars and a dipole ansatz ($n = 2$) for the axial-vectors and tensor mesons, parametrized by the mass scale Λ_X. We assume one common mass for the scalar TFFs, one mass for the TFFs of the axial-vectors and four different masses for the TFFs of the tensor mesons. These six masses will be considered as free fit parameters.

The normalization of the TFFs is given by the two-photon decay width and is taken from experiment where available, e.g. for the scalars $\Gamma_{\gamma\gamma} = \frac{\pi\alpha^2}{4} m_S \left[F^T_{S\gamma^*\gamma^*}(0,0)\right]^2$ (an appropriately defined effective two-photon width is employed for the axial-vectors [22]). Since not all normalizations have been measured, further input from dispersive sum rules from Ref. [15] is used. Furthermore, for the pseudoscalars, again the lattice data from Ref. [21] are used.

As an example, the pseudoscalar contribution to the cross-section σ_{TT} of two transversely polarized photons is given in the narrow-width approximation by

$$\sigma_{TT} = 8\pi^2\delta(s - m_P^2)\frac{\Gamma_{\gamma\gamma}}{m_P}\frac{2\sqrt{X}}{m_P^2}\left[\frac{F_{\mathcal{P}\gamma^*\gamma^*}(Q_1^2, Q_2^2)}{F_{\mathcal{P}\gamma^*\gamma^*}(0,0)}\right]^2, \tag{20}$$

where $X = \nu^2 - Q_1^2 Q_2^2$ is the virtual-photon flux factor. Similar results can be obtained for the other mesons where we assume a Breit-Wigner shape for the resonances [24].

In Fig. 7 the results for four scattering amplitudes of a combined fit to all eight amplitudes with the phenomenological model described above is shown for one lattice ensemble. For four ensembles, the χ^2/d.o.f. is quite good, between 1.13 – 1.35. The fit for the ensemble with the heaviest pion mass $m_\pi = 437$ MeV is, however, not very satisfactory and that ensemble is left out in the chiral extrapolation to the physical pion mass.

The relative contribution of the different mesons to the individual scattering amplitudes was also studied in Ref. [24]. The pseudoscalar and tensor mesons give the dominant contribution to the amplitudes $\overline{\mathcal{M}}_{TT}$, $\overline{\mathcal{M}}^{\tau}_{TT}$ and $\overline{\mathcal{M}}^{a}_{TT}$ that involve two transverse photons. The pseudoscalar meson does not contribute to $\overline{\mathcal{M}}_{TL}$, $\overline{\mathcal{M}}_{LT}$ where

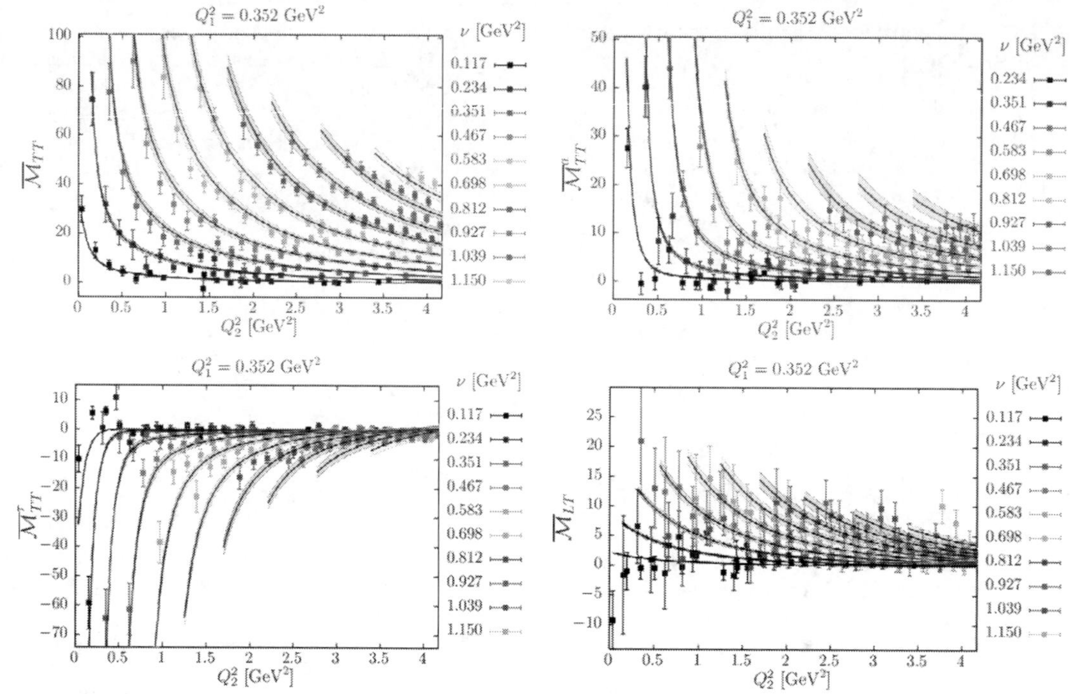

Figure 7. The forward scattering amplitudes $\overline{\mathcal{M}}_{TT}$, $\overline{\mathcal{M}}^a_{TT}$, $\overline{\mathcal{M}}^\tau_{TT}$ and $\overline{\mathcal{M}}_{LT}$ ($\times 10^6$) for a lattice ensemble with $m_\pi = 314$ MeV and lattice spacing $a = 0.065$ fm as function of Q_2^2 for a fixed $Q_1^2 = 0.352$ GeV^2 and different values of ν. The curves with error-bands represent the result of the fit of the phenomenological model described in the text to all eight scattering amplitudes with χ^2/d.o.f. = 1.15.

the main contribution are from axial and tensor mesons. In the amplitudes $\overline{\mathcal{M}}^a_{TL}$, $\overline{\mathcal{M}}^\tau_{TL}$, $\overline{\mathcal{M}}_{LL}$ scalar, axial and tensor mesons contribute significantly. The contribution from $\gamma^*\gamma^* \to \pi^+\pi^-$, evaluated with scalar QED dressed with a monopole vector form factor, is always small compared to the other channels.

The lattice simulations are performed for ensembles away from the physical quark masses. The pion and ρ-meson masses are set to their lattice values obtained from the exponential decay of the pseudoscalar and vector two-point functions. For other resonances, we assume a constant shift in the masses $m_X = m_X^{\rm phys} + (m_\rho^{\rm lat} - m_\rho^{\rm exp})$. In Table 2 we compare the results of the masses in the TFFs, extrapolated to the physical pion mass, to experimental and phenomenological determinations [15, 35]. The agreement is reasonably good for M_S, $M_T^{(2)}$, $M_T^{(0,L)}$, although the scalar mass on the lattice is a bit high. There are quite strong tensions for M_A, $M_T^{(1)}$, $M_T^{(0,T)}$. In particular the latter two tensor masses on the lattice are almost a factor two larger than the phenomenological determinations. See Ref. [24] for a more detailed discussion and potential reasons for the disagreements. Note in particular, that we have not yet performed the continuum limit.

Overall, we get a good description of the lattice data with the lattice determination of the pion TFF [21] and the simple monopole or dipole ansätze from Eq. (19) for the various TFF's with one resonance in each channel.

Table 2. Chiral extrapolation to the physical pion mass for the scalar monopole mass M_S, the axial dipole mass M_A and the four tensor dipole masses corresponding to different helicities, compared to experimental or phenomenological determinations. All masses are given in GeV. The ensembles have a finite lattice spacing of $a = 0.065$ fm and the ensemble with the largest pion mass has been excluded in the chiral extrapolation.

	Lattice	Experiment
M_S	1.04(14)	0.796(54)
M_A	1.32(07)	1.040(80)
$M_T^{(2)}$	1.35(24)	1.222(66)
$M_T^{(1)}$	1.69(16)	0.916(20)
$M_T^{(0,T)}$	1.96(09)	1.051(36)
$M_T^{(0,L)}$	0.67(19)	0.877(66)

5 Conclusions

Lattice QCD can provide a model-independent, first-principle calculation of the HLbL contribution to the muon $g-2$. Recent first preliminary results by RBC-UKQCD [17] and the lattice group at Mainz [18–21, 23, 24] look very promising. Hopefully, in a few years time when the final result from the Fermilab experiment will be published, an estimate with 10% uncertainty (combined statistical and controlled systematics errors) can be reached, which would match the expected experimental precision, if one assumes that the central value for

HLbL is close to current model estimates $a_\mu^{\text{HLbL}} \approx 100 \times 10^{-11}$ [1, 2, 11, 12].

Since HLbL involves a rank-four tensor with many independent momenta (or space-time points in position space), it is a very complicated object. We therefore think it makes sense to have as many tests and observables as possible on the lattice, not just the final number for a_μ^{HLbL} itself. Therefore the calculation of the pion (light pseudoscalar) transition form factor and of HLbL forward scattering amplitudes, combined with HLbL sum rules, will be a valuable tool to compare different lattice calculations, once they become available.

We are grateful for the use of CLS lattice ensembles. We acknowledge the use of computing time on the JUGENE and JUQUEEN machines located at Forschungszentrum Jülich, Germany. The correlation functions were computed on the clusters "Wilson" at the Institute of Nuclear Physics, University of Mainz, "Clover" at the Helmholtz-Institute Mainz and "Mogon" at the University of Mainz, Germany. This work was partially supported by the Deutsche Forschungsgemeinschaft (DFG) through the Collaborative Research Center "The Low-Energy Frontier of the Standard Model" (SFB 1044).

References

[1] F. Jegerlehner and A. Nyffeler, Phys. Rept. **477**, 1 (2009).

[2] F. Jegerlehner, EPJ Web Conf. **118**, 01016 (2016); F. Jegerlehner, arXiv:1705.00263 [hep-ph].

[3] S. Laporta, Phys. Lett. B **772**, 232 (2017); S. Laporta, talk at this meeting.

[4] M. Knecht, talk at this meeting.

[5] D. Stöckinger, talk at this meeting.

[6] T. Gorringe, G. Marshall, talks at this meeting.

[7] D. Nomura, V. Sauli, F. Ignatov, A. Denig, talks at this meeting.

[8] M. Marinkovic, C. Lehner, talks at this meeting.

[9] L. Trentadue, U. Marconi, P. Mastrolia, F. Piccinini, talks at this meeting.

[10] J. Bijnens and J. Relefors, JHEP **1609**, 113 (2016); J. Bijnens, talk at this meeting, arXiv:1712.09787 [hep-ph].

[11] J. Prades, E. de Rafael and A. Vainshtein, Adv. Ser. Direct. High Energy Phys. **20**, 303 (2009).

[12] A. Nyffeler, Phys. Rev. D **79**, 073012 (2009).

[13] G. Colangelo *et al.*, JHEP **1409**, 091 (2014); G. Colangelo *et al.*, Phys. Lett. B **738**, 6 (2014); G. Colangelo *et al.*, JHEP **1509**, 074 (2015); Phys. Rev. Lett. **118**, 232001 (2017); JHEP **1704**, 161 (2017); G. Colangelo, talk at this meeting.

[14] V. Pauk and M. Vanderhaeghen, arXiv:1403.7503 [hep-ph]; Phys. Rev. D **90**, 113012 (2014).

[15] I. Danilkin and M. Vanderhaeghen, Phys. Rev. D **95**, 014019 (2017).

[16] F. Hagelstein and V. Pascalutsa, arXiv:1710.04571 [hep-ph].

[17] M. Hayakawa, *et al.*, PoS LAT **2005**, 353 (2006); T. Blum, M. Hayakawa and T. Izubuchi, PoS LATTICE **2012**, 022 (2012); T. Blum *et al.*, Phys. Rev. Lett. **114**, 012001 (2015); T. Blum *et al.*, Phys. Rev. D **93**, 014503 (2016); T. Blum *et al.*, Phys. Rev. Lett. **118**, 022005 (2017); Phys. Rev. D **96**, 034515 (2017); C. Lehner, talk at this meeting.

[18] J. Green *et al.*, PoS LATTICE **2015**, 109 (2016).

[19] N. Asmussen *et al.*, PoS LATTICE **2016**, 164 (2016).

[20] N. Asmussen *et al.*, arXiv:1711.02466 [hep-lat].

[21] A. Gérardin, H. B. Meyer and A. Nyffeler, Phys. Rev. D **94**, 074507 (2016).

[22] V. Pascalutsa and M. Vanderhaeghen, Phys. Rev. Lett. **105**, 201603 (2010); V. Pascalutsa, V. Pauk and M. Vanderhaeghen, Phys. Rev. D **85**, 116001 (2012).

[23] J. Green *et al.*, Phys. Rev. Lett. **115**, 222003 (2015).

[24] A. Gérardin *et al.*, arXiv:1712.00421 [hep-lat].

[25] S. Laporta and E. Remiddi, Phys. Lett. B **301**, 440 (1993); M. Passera, private communication.

[26] X. d. Ji and C. w. Jung, Phys. Rev. Lett. **86**, 208 (2001); Phys. Rev. D **64** (2001) 034506; J. J. Dudek and R. G. Edwards, Phys. Rev. Lett. **97**, 172001 (2006); X. Feng *et al.*, Phys. Rev. Lett. **109**, 182001 (2012).

[27] P. Fritzsch *et al.*, Nucl. Phys. B **865**, 397 (2012).

[28] H. J. Behrend *et al.* [CELLO Collaboration], Z. Phys. C **49**, 401 (1991); J. Gronberg *et al.* [CLEO Collaboration], Phys. Rev. D **57**, 33 (1998); B. Aubert *et al.* [BaBar Collaboration], Phys. Rev. D **80**, 052002 (2009); S. Uehara *et al.* [Belle Collaboration], Phys. Rev. D **86**, 092007 (2012).

[29] S. L. Adler, Phys. Rev. **177**, 2426 (1969); J. S. Bell and R. Jackiw, Nuovo Cim. A **60**, 47 (1969).

[30] G. P. Lepage and S. J. Brodsky, Phys. Lett. B **87**, 359 (1979); Phys. Rev. D **22**, 2157 (1980); S. J. Brodsky and G. P. Lepage, Phys. Rev. D **24**, 1808 (1981).

[31] V. A. Nesterenko and A. V. Radyushkin, Sov. J. Nucl. Phys. **38**, 284 (1983); V. A. Novikov *et al.*, Nucl. Phys. B **237**, 525 (1984).

[32] B. Moussallam, Phys. Rev. D **51**, 4939 (1995); M. Knecht *et al.*, Phys. Rev. Lett. **83**, 5230 (1999).

[33] M. Knecht and A. Nyffeler, Eur. Phys. J. C **21**, 659 (2001); K. Melnikov and A. Vainshtein, Phys. Rev. D **70**, 113006 (2004).

[34] M. Knecht and A. Nyffeler, Phys. Rev. D **65**, 073034 (2002); A. Nyffeler, Phys. Rev. D **94**, 053006 (2016).

[35] P. Achard *et al.* [L3 Collaboration], Phys. Lett. B **526**, 269 (2002); P. Achard *et al.* [L3 Collaboration], JHEP **0703**, 018 (2007); M. Masuda *et al.* [Belle Collaboration], Phys. Rev. D **93**, 032003 (2016).

Charged lepton flavour violation searches at the Paul Scherrer Institut: Status of the MEGII and Mu3e experiments

Angela Papa[1,a]

[1] Paul Scherrer Institut, 5232 Villigen (Switzerland)
On behalf of the MEGII and Mu3e collaborations

Abstract.
The MEG experiment has recently set a new upper limit on the branching ratio of the $\mu^+ \to e^+\gamma$ decay, $\mathcal{B}(\mu^+ \to e^+\gamma) < 4.2 \times 10^{-13}$ (at 90% confidence level) and un upgrade of the experiment (the MEGII experiment) is ongoing with the aim of improving the single event sensitivity (SES) by one order of magnitude with respect to the previous MEG experiment's SES. The strong scientific motivation associated with the charged Lepton Flavour Violation (cLFV) searches pushes also towards searching for the complementary muon cLFV $\mu^+ \to e^+e^+e^-$ decay with the Mu3e experiment aiming at a SES improved by at least three orders of magnitude with respect to the previous SINDRUM experiment's SES (phase I) up to an ultimate SES of few $\times 10^{-16}$. Both experiments will be hosted at the Paul Scherrer Institut which delivers the most intense continuous low energy muon beam in the world up to few $\times 10^8$ μ/s. The status of both the MEGII and Mu3e experiments is given.

1 Introduction

Lepton flavour violation (LFV) research is currently one of the most exciting branches of particle physics due to its high sensitivity to unveil New Physics (NP) [1]. There are both strong theoretical motivations and experimental evidences which suggest that the Standard Model (SM), the theoretical framework describing the fundamental particles and interactions and recently enriched by the discovery of the Higgs boson, is not the ultimate theory and must be extended [2]. The unknown physics needed to address the left open question of the SM is what is called New Physics.

In the SM lepton flavour violation processes like $\mu^+ \to e^+\gamma$ are not allowed. Actually in the SM the lepton flavour is an accidental symmetry, not related to the gauge theory structure and naturally violated in the SM extensions. Furthermore the observation of neutrino oscillations has clearly demonstrated that neutral lepton flavour is not conserved [3–5]. This implies that charged LFV (cLFV) processes, such as $\mu^+ \to e^+\gamma$ or $\mu^+ \to e^+e^+e^-$ decays, can occur in simple extended SM versions which take into account for neutrino oscillations (i.e. including Dirac neutrinos), although strongly suppressed (i.e. $\mathcal{B}(\mu^+ \to e^+\gamma) \approx 10^{-54}$). On the other hand, Beyond SM (BSM) extensions enhance the predictions for cLFV branching ratios at levels that can be measured by the experiments currently in preparation [6–10]. Due to the fact that at these SES levels there are not contributions coming from SM processes (i.e. SM background free) an evidence of such a decays would be a clear signature of NP (behind neutrino oscillations) and therefore cLFV searches turn out to be ideal probes for NP hints.

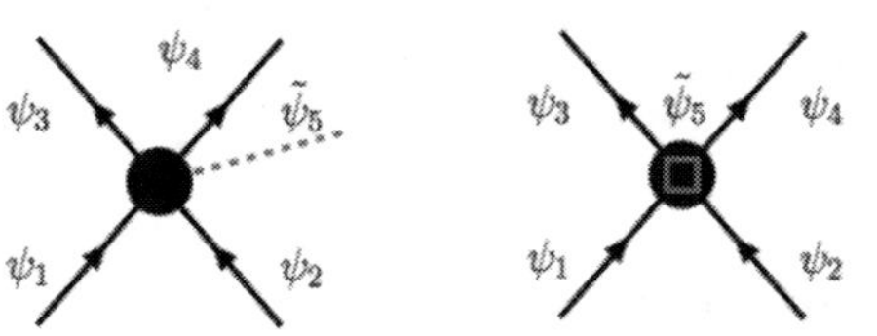

Figure 1. Complementarity between direct (left) and indirect (right) BSM particle searches.

The strong complementarity between direct and indirect NP search with accelerator machines is schematically shown in Figure 1. The first case (Figure 1, left) refers to the direct searches that can be performed at energy frontiers where the new particles (red dashed line) would be *directly* produced in the final state while the second case (Figure 1, right) indicates the indirect searches, typically performed at the precision and intensity frontiers and at which cLFV searches belong too, where the contribution of the new particles (red line) would *indirectly* appear enhancing the probability of processes that otherwise

[a] e-mail: angela.papa@psi.ch

 https://doi.org/10.1051/epjconf/201817901018

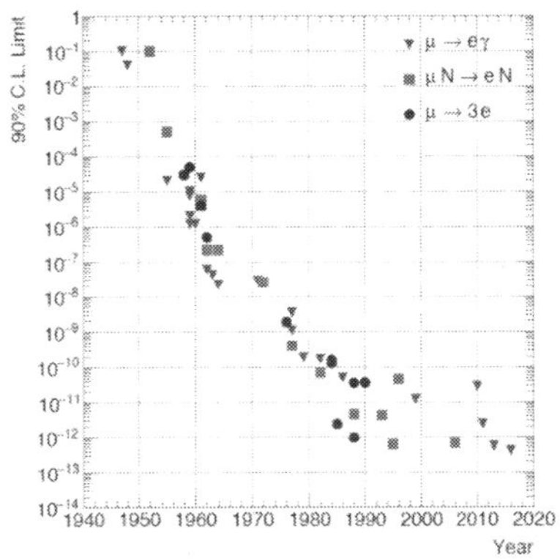

Figure 2. History of cLFV experiments with muons.

would be strongly suppressed or never occur.

The most general approach to describe the NP under the assumption that the NP characteristic energy scale is well above the energies explored so far is to write an effective lagrangian made of by the sum of the SM lagrangian and all the other new terms, suppressed by inverse powers of the new heavy mass scale Λ [11–13]:

$$\mathcal{L}_{eff} = \mathcal{L}_{SM} + \sum_{d>4} \frac{c_n^{(d)}}{\Lambda^{d-4}} O^{(d)} \quad (1)$$

where O are the operators, d is the mass dimension and c_n dimensionless coefficients. As it follows from eq. 1 searching for strongly suppressed or forbidden processes offers the unique possibility to probe otherwise unreachable and unexploited new physics energy scale. Following the approach of the effective lagrangian and assuming NP natural coupling the current upper limits on muon cLFV processes translates in new energy scale limits $\Lambda > O(100)$ TeV, independently of the detailed form of the operator responsible for the cLFV process [14, 15].

Muonic rare channels such as the $\mu^+ \to e^+\gamma$ decay, the $\mu^+ \to e^+e^+e^-$ decay and $\mu^- N \to e^- N$ conversion in nuclei are the most promising and complementary cLFV processes (often referred to as "golden muonic channels" [1, 16–20]): (a) The tremendous muon beam intensities (already available: up to few $\times 10^8$ μ/s (continuous, DC) [21, 22], available soon: $O(10^{11})$ μ/s (pulsed) [23, 24] and understudy: $O(10^{10})$ μ/s (DC) [25, 26]), implying for huge statistical samples, together with ultimate performing detectors allow for astonishing muonic cLFV SES; (b) The combined phenomenological analysis of these three processes allows for discriminating the underlying operators generating a potential signal, given different process sensitivities to the different operators. Figure 2 shows the history of cLFV experiments based on the golden muonic channels.

Two of the three golden muonic channels can be studied at PSI which delivers the world's most intense continuous muon beam uniquely suited to study coincidence-type experiments as $\mu^+ \to e^+\gamma$ and $\mu^+ \to e^+e^+e^-$ decay searches, where there is more than one particle in the final state.

The MEG experiment searches for the $\mu^+ \to e^+\gamma$ decay ([27, 28]) and has recently set the most stringent upper limit on its branching ratio $\mathcal{B}(\mu^+ \to e^+\gamma) < 4.2 \times 10^{-13}$ [29–32]. It is a factor 30 improvement over the previous limit set by the MEGA experiment [33] and also the strongest bound on any forbidden decay particle. The strong physics motivation to further explore the $\mu^+ \to e^+\gamma$ decay has led the collaboration to decide upon an upgrade of the experiment, with the aim to improve the sensitivity by at least one order of magnitude. The MEG upgrade (MEGII) has been approved at PSI and by the Institutions of the international collaboration [34], and is now underway [35].

Following the mentioned complementary approach the Mu3e experiment at PSI will search for the $\mu^+ \to e^+e^+e^-$ decay aiming at a sensitivity of a few $\times 10^{-16}$ [36] (current upper limit $\mathcal{B}(\mu^+ \to e^+e^+e^-) < 1.0 \times 10^{-12}$ [37]) and COMET [23] in Japan and Mu2e [24] in US will search for the $\mu^- N \to e^- N$ conversion aiming at final sensitivity of few $\times 10^{-17}$(current upper limit $\mathcal{B}(\mu$ Au $\to$ e Au$) < 7 \times 10^{-13}$ [38]).

2 The MEGII experiment

A schematic view of the MEGII apparatus is shown in Figure 3.

In MEGII, surface (positive) muons with a momentum of 29 MeV/c are stopped in a thin slanted polyethylene target (thickness 140 μm; angle 15 deg), located at the center of a magnetic spectrometer.

The signature of a $\mu^+ \to e^+\gamma$ decay at rest is a back-to-back, mono-energetic, time coincident γ and e^+. The signal event is identified by five observables: The gamma energy E_γ, the positron energy E_e, the relative gamma-positron time $t_{e\gamma}$, the relative gamma-positron angles $\theta_{e\gamma}$ and $\phi_{e\gamma}$. There are two main background sources, the dominant being the accidental coincidences between a high energy positron from the main muon decay $\mu^+ \to e^+\nu\bar{\nu}$ (Michel decay) and a high energy photon from positron annihilation-in-flight or bremsstrahlung or from the radiative muon decay (RMD) $\mu^+ \to e^+\nu\bar{\nu}\gamma$. The other source comes from the RMD itself when neutrinos take off a small amount of energy.

All the γ kinematics variables (energy E_γ, time t_γ and interaction point X_γ) are measured using a liquid Xenon (LXe) calorimeter. All the e^+ kinematics variables are measured by a spectrometer made of single cylindrical active drift chamber CDCH and a highly segmented pixelated Timing Counter pTC mounted inside a gradient magnetic field, provided by a COnstant Bending RAdius superconducting magnet (COBRA) with a magnetic field of 1.25 T at the center. The performance of the experiment is continuously monitored by a variety of calibration methods. All signals are recorded with a custom designed waveform digitizers (DRS), which includes a flexible tigger system to select $\mu^+ \to e^+\gamma$ candidate events together with calibration data.

 https://doi.org/10.1051/epjconf/201817901018

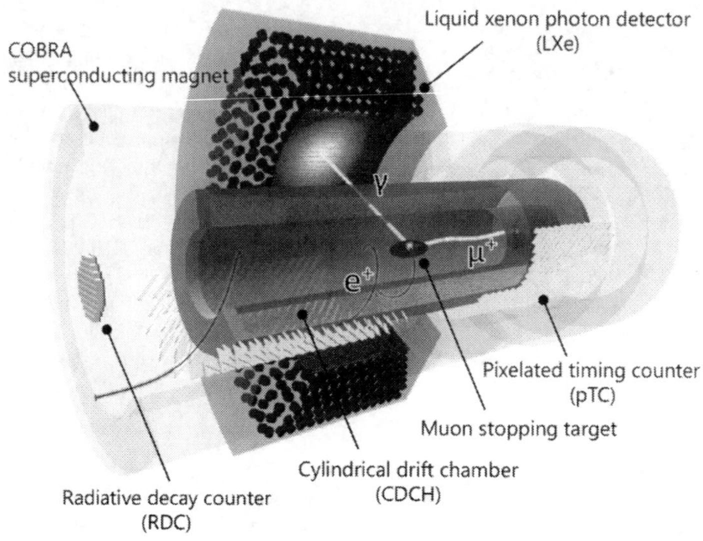

Figure 3. MEGII experimental setup layout (3D view). An example of a $\mu^+ \to e^+\gamma$ decay event is shown.

The LXe calorimeter is a "C-shaped" detector made of $\approx$ 900 liters of liquid xenon where the photosensors, both photomultipliers and silicon-photomultipliers, are directly immersed. The photosensors have been developed in collaboration with Hamamatsu to work in the VUV region where the xenon scintillation light is emitted ($\approx 175 \pm 5$ nm) and at a temperature of 165 K. It is the largest liquid xenon detector in the world based only on scintillation light to preserve the fast Xe scintillating response ($\tau_\alpha = 4, 20$ ns, $\tau_e = 45$ ns). The upgrade of the LXe calorimeter involves a denser allocation of photo-detectors on the front face, replacing the current PMTs (Hamamatsu R9869) 2 inch diameter with smaller SiPM (MPPC Hamamatsu VUV sensitive) 12×12 mm^2 (see Figure 4). The better uniformity would imply better energy and position resolutions with a higher impact on the so called shallow events, which are characterized by a less depth conversion point. The reduced amount of material in front of the Xe itself improves the detector performances as well. The γ entrance face will be extended along the beam direction by 10% at each side, increasing also the acceptance. The comparison between the achieved MEG and expected MEGII LXe calorimeter performances are summarized in Table 1[39–41].

The new spectrometer has been designed to maintain all the benefits of the former one and to overcome previous limitations [42–44]. The main feature is the increased spectrometer granularity which allows to work at higher beam rate and provide better resolutions for all the positron kinematical variables. The positron trajectory is measured up to the point where the positron reaches the new TC tiles, with minimum presence of passive material and an increased number of hits per track, both in the chamber and in the TC. These aspects immediately have an impact on the detector acceptance and efficiency, which is increased by more than a factor 2, on the positron momentum and angular resolutions and on the positron tim-

Figure 4. The inner face of the LXe calorimeter assembled with 2 inch. PMT × 216 (MEG) and 12×12 mm^2 SiPMs × 4096 (MEGII).

Table 1. Comparison of the γ kinematical variable resolutions (in σ) with the MEG (measured) and MEGII (expected) LXe calorimeter.

	MEG	MEGII
u [mm]	5	2.4
v [mm]	5	2.2
w [mm]	6	3.1
E (w < 2 cm) [%]	2.4	1.1
E (w > 2 cm) [%]	1.7	1.0
t [ps]	67	60

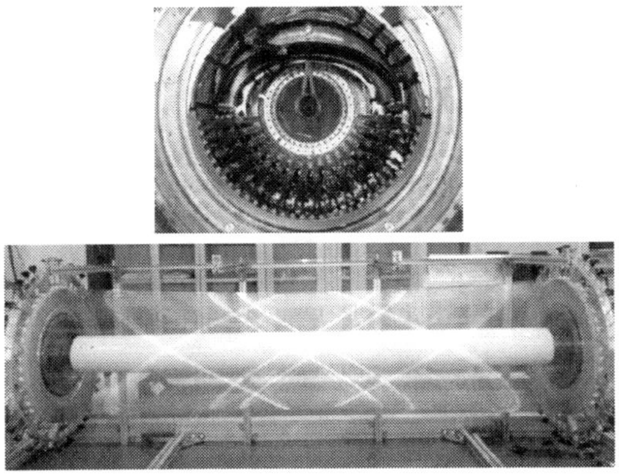

Figure 5. The MEG segmented drift chamber (front view, top). The MEGII unique volume CDCH (lateral view, bottom).

Table 2. Comparison of the e^+ kinematical variable resolutions (in σ) with the MEG (measured) and MEGII (expected) spectrometer.

	MEG	MEGII
E_{core} [keV]	306	130
θ_{core} [mrad]	9.4	5.3
ϕ_{core} [mrad]	8.7	3.7
t [ps]	70	35
Tracking efficiency [%]	65	78
CDCH-pTC matching efficiency [%]	45	90

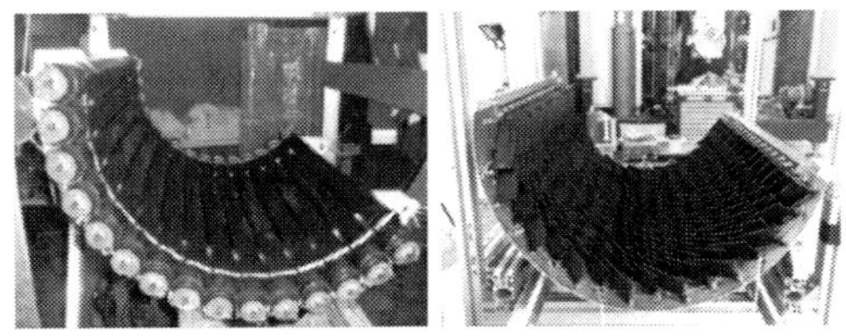

Figure 6. A section of the MEG (left) and MEGII (right) TC.

ing. The new drift chamber will be a 1.91 m long cylinder with inner and outer radii of 17 cm and 29 cm respectively filled with an ultra-low gas mixture of helium and isobutane (C_4H_{10}) in the ratio 90:10. The wires will follow a stereo configuration. Each drift cell (with an approximately squared shape, with a variable side size from 6.6 mm to 9.0 mm moving from the inner towards the outermost layer) is made of a 20 μm diameter gold plated W sense wire surrounded by 40 μm diameter silver plated Al field wires in a ratio of 5:1. Two guard wires layers (50 μm diameter silver-plated Al) are added at proper radii and voltages for gain equalising of the innermost and outermost layers. The total number of wires would be more than 13 000 for an equivalent radiation length per track turn of about 1.6×10^{-3} X_0. Figure 5 shows the previous MEG segmented DCH (top) and the preparation of the MEGII unique volume CDCH (bottom).

The new TC is made by a large number of small ultra-fast scintillator plates coupled with SiPM, with improved timing resolution and acceptance (see Figure 6). The TC will be made by two sections (Upstream and Downstream) each one made of 256 plates of BC422 of different size ($120 \times 40(50) \times 5$ mm^3) readout by SiPM (AdvanSiD ASD-NUM3S-P-50-High-Gain). The high TC segmentation allows to reach a better timing resolution due to multi-hits events (typically one event would be characterised by 7-10 hits), a reduced ambiguity in the positron path length and less scintillation light propagation effects. The use of SiPM overcomes the limited performances of the PMT in magnetic fields, together with a lower transit time spread and higher quantum efficiency [45, 46].

A comparison of the performances of the MEG (measured) and MEGII (expected) spectrometer is given in Table 2.

A large number of calibration and monitoring methods have been introduced and commissioned for the MEG experiment and will be inherited into MEGII with some modifications accordingly with the increased complexity of the upgraded experiment [47–55]. They allow to know in depth the detector behaviour, crucial for building the probability density functions of the observables used to identify a signal event. Furthermore they ensure that the beam characteristics and the detector performances are reached and maintained over the time. Their complexity follow from the fact that some of them require dedicated beams, which means different particles (positrons, pions, protons) at different energies with respect to the normal MEG muon beam, dedicated targets, auxiliary detectors up to a devoted 1 MeV Cockcroft-Walton accelerator. Their variety includes the possibility of producing both quasi mono-chromatic γ and e^+ events at the MEG signal region with the unique opportunity to study the detector with real data using signal-like events. The major MEGII calibration and monitoring methods are summarized in Table 3.

New auxiliary detectors will be added to the framework of the MEGII experiment, among the others (a) the radiative decay counter RDC, enabling to increase the capability of rejecting the accidental background by tagging low energy positrons associated with the high energy photons in the signal region; (b) a sampling beam monitoring detector based on scintillating fibres coupled to SiPM able to measure both beam profile and rates when needed (Sampling SciFi) [54]. Fig. 7 shows the commissioned RDC and Sampling SciFi detectors. The RDC detector has been already integrated into the MEGII apparatus.

Finally, MEG II requires improvements on the existing DAQ system in terms of an increased number of channels and a higher bandwidth of the waveform digitization system: The new DAQ board (WaveDREAM) has been developed, which combines both the waveform digitizing technology using the DRS4 chip as well as the trigger and splitter functionality of our current system (TDAQ system) [56–60]. More than thousands channels have been tested during the 2017 pre-engineering run. It included also the

Table 3. The calibration and monitoring methods of the MEGII experiment.

Process	Energy	Main Purpose	Frequency
Cosmic rays, $\mu^{\pm}$ from atmospheric showers	Wide spectrum $\mathcal{O}$(GeV)	LXe-DCH relative position	annually
		DCH alignment	
		TC energy and time offset calibration	
		LXe purity	on demand
Charge exchange, $\pi^- p \rightarrow \pi^0 n$, $\pi^0 \rightarrow \gamma\gamma$	55, 83, 129 MeV γ	LXe energy scale/resolution	annually
Radiative μ–decay, $\mu \rightarrow e\gamma\nu\nu$	> 40 MeV γ	LXe-TC relative timing	continuously
	> 45 MeV e^+	Normalisation	
Normal μ–decay, $\mu \rightarrow e\nu\nu$	52.83 MeV end-point e^+	DCH energy scale/resolution	continuously
		DCH and target alignment	
		Normalisation	
Mott positrons, $e^+ X \rightarrow e^+ X$	$\approx$ 52.8 MeV e^+	DCH energy scale/resolution	annually
		DCH alignment	
Proton accelerator, $^{7}Li(p, \gamma)^{8}Be$	14.8, 17.6 MeV γ	LXe uniformity/purity	weekly
Proton accelerator, $^{11}B(p, \gamma)^{12}C$	4.4, 11.6, 16.1 MeV γ	TC intercalibration/ LXe-TC timing	weekly
Neutron generator, $^{58}Ni(n, \gamma)^{59}Ni$	9 MeV γ	LXe energy scale	weekly
Radioactive source, $^{241}Am(\alpha, \gamma)^{237}Np$	5.5 MeV α	LXe PMT calibration/purity	weekly
Radioactive source, $^{9}Be(\alpha_{^{241}Am}, n)^{12}C^{*}$	4.4 MeV γ	LXe energy scale	on demand
Radioactive source, $^{57}Co(EC, \gamma)^{57}Fe$	X-rays	LXe-spectrometer alignment	annually
LED	VUV region	LXe PMT calibration	continuously
Laser	401 nm	pTC SiPM calibration	continuously

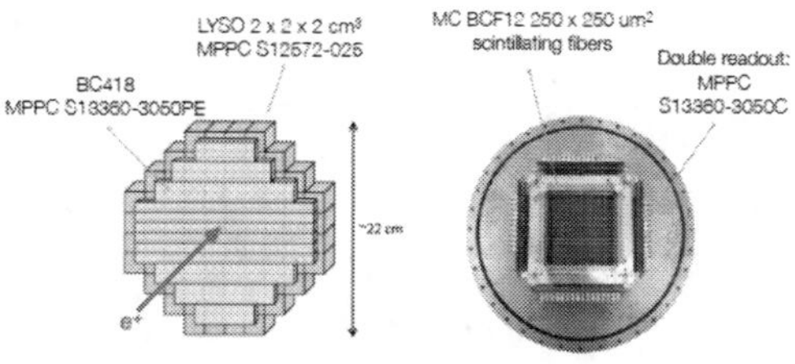

Figure 7. The radiative decay counter (left) and the sampling scintillating fibre detector (right).

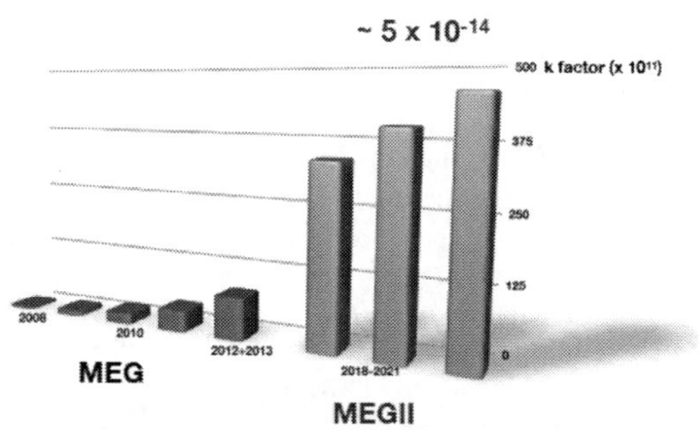

Figure 9. Accumulated (MEG) and projected (MEGII) statistics for a final MEGII SES of $\approx 5 \times 10^{-14}$ in $\mathcal{O}(500)$ DAQ days.

Figure 8. A full assembled TDAQ crate.

trigger electronics and several trigger algorithms, successful tested and delivered. Figure 8 shows one TDAQ crate, which includes 1 trigger board and 16 DAQ boards for a total of 256 channels (16 channels/board).

The achieved (MEG) and projected (MEGII) inverse single event sensitivity is plotted in Figure 9 as a function of the time. The final SES is expected to be of $\approx 5 \times 10^{-14}$ for a total experiment lifetime of $\mathcal{O}(500)$ DAQ days.

3 The Mu3e experiment

The Mu3e experiment aims at reaching an ultimate sensitivity of few $\times 10^{-16}$ (or even better) to the branching fraction of the $\mu^+ \rightarrow e^+e^+e^-$ decay [36], four orders of magnitude better than the current limit $\mathcal{B}(\mu^+ \rightarrow e^+e^+e^-) < 10^{-12}$ set by the SINDRUM experiment [37].

Positive muons are stopped in a thin double cone target and two positrons and one electron in time coincidence, with a total energy equal to the muon mass and with a null vectorial sum of all the decay particle momenta are searched in the final state as the event signature.

https://doi.org/10.1051/epjconf/201817901018

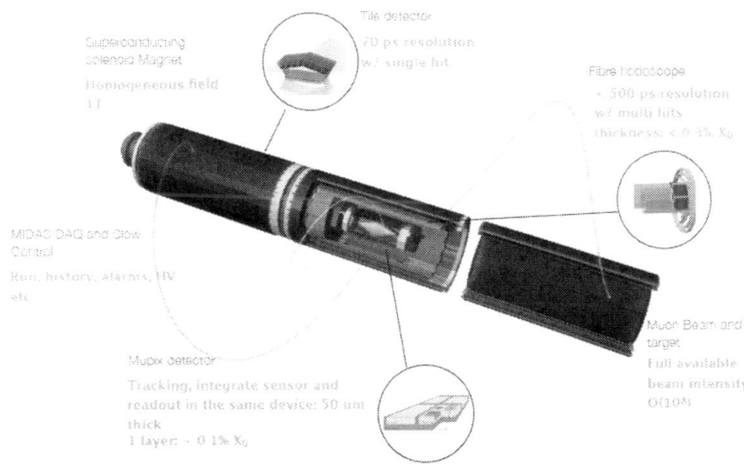

Figure 10. Mu3e experimental setup layout (3D view). An example of a $\mu^+ \to e^+e^+e^-$ decay event is shown.

There are two many kinds of backgrounds: (a) Overlays of different processes producing three tracks resembling a $\mu^+ \to e^+e^+e^-$ decay event (accidental background), the most important represented by two Michel positrons one of those undertaking a Bhabha scattering with electrons in the target material, leading to an electron-positron pair from a common vertex; (b) Radiative decays with internal conversion $\mu^+ \to e^+e^+e^-\nu\bar{\nu}$ with a small energy fraction carried away by the neutrinos: This process is the most serious background for the a $\mu^+ \to e^+e^+e^-$ decay search.

In order be sensitive to the signal and to reject the background excellent momentum resolutions are mandatory to suppress the internal conversion backgrounds and good vertex and timing resolutions are needed to efficiently reduce combinatorial backgrounds. To match those requests the experiment has been design as the following.

The particle momenta will be measured using a novel tracker, made of silicon devices (High-Voltage Monolithic Active Pixel, HV-MAPS [61–63]) for a momentum resolution of about 300 keV/c over most of the detector acceptance and momentum range, a vertex resolution of $\approx$ 200 μm and a relative fast response $O(10)$ ns. The HV-MAPs have the unique characteristics of integrating the active sensor and the readout in the same device. In current designs, signal amplification and shaping is performed in the pixel itself, whereas the digitisation is carried out in the periphery, a small part at the bottom edge of the sensor. The sensor has digital, zero-suppressed data output via a fast serial link of 1.25 Gbits^{-1}. They have pixels with a dimension of 80×80 μm^2, a thickness of 50 μm, a time resolution better than 20 ns, a detection efficiency $\epsilon > 99$ % and a power consumption lower than 350 mW/cm^2. The pixel tracker is made by a central station, made of two inner pixel layers and two outer pixel layers and the re-curl stations, a double copy of the outer pixel layers, displaced upstream and downstream.

Figure 11. MuPix7: The first small-scale tracker chip prototype which includes all the Mu3e functionalities (left). The experimental setup used extensively during the 2017 test beam.

The timing is measured precisely by the combination of two detectors: (a) The scintillating fibre hodoscope, a cylindrical time detector device mounted underneath the central outer pixel layers at radius of 6 cm and with a length of 36 cm. The thinnest available fibres (with a diameter of 250 μm) are arranged to form ribbons of 3-4 fibre layers coupled to SiPM arrays for a timing measurement better than 1 ns and a full detection efficiency; (b) The scintillating tiles, made by plastic scintillators (EJ228)

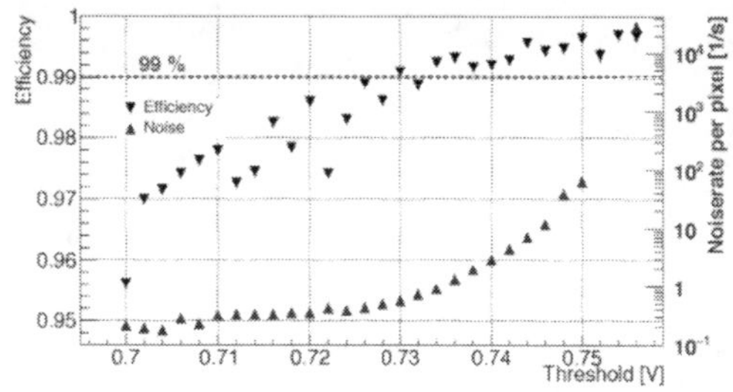

Figure 12. MuPix7: Detection Efficiency.

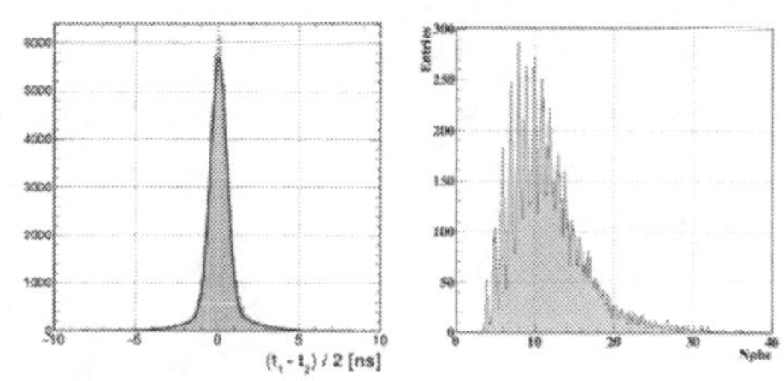

Figure 13. Example of a SciFi prototype made of three layers of fibres readout by silicon photomultipliers (see the text for details). The timing spectrum (left) and the typical charge spectrum (right) are shown for minimum ionizing particles going though the detector.

$6.5 \times 6.5 \times 5$ mm^3 coupled to SiPMs for a precise timing measurement (< 100 ps) at the end of the positron/electron path.

The track and timing information is read out triggerless to a GPU based online filter farm, via three stages of FPGA driven data acquisition boards and an optical link network [64].

All the sub-detectors feature a full detection efficiency (approaching 100%). They are mounted inside a superconductive magnet which provides a solenoidal field with a default magnetic field of 1 T. Figure 10 shows the schematic view of the experimental set-up.

The preparation of the experiment will follow a staged approach with two major phases. The first phase aims to achieve a SES of $O(10^{-15})$ three orders of magnitude improvement in the sensitivity with respect to the previous SINDRUM experiment. The muon beam intensity will increase accordingly, up to exploiting the maximum available beam intensity of few $\times 10^8$ muons/s. The second phase aims to reach the ultimate sensitivity of $O(10^{-16})$. For this phase a continuous muon stopping rate in the GHz region is required and feasibility studies are ongoing at PSI to meet also that need (the High Intensity Muon Beam project) [25, 26].

The Mu3e collaboration is concluding the detector *R&D* phase and is starting the prototyping phase. The major results associated with the *R&D* are shortly summarized here.

Figure 11 shows a pictures of the MuPix7 prototype: It is the first small-scale HV-MAP chip which includes all Mu3e functionalities. It has an active area of 2.9×3.2 mm^2 with a pixel size of 103×80 μm^2, has been operated up to 1MHz yielding a full detection efficiency well above 99 % and a timing resolution smaller than 20 ns (see Figure 12). These results have been obtained during an extensive test beam campaign exposing the detector at different particles, rates and energies [66]. The first large area prototype, MuPix8 with an active area of 20×10 mm^2 has been now produced and ready to be tested. MuPix8 is only a factor two less in active area with respect to the final expected chip size needed for the experiment (with an active area

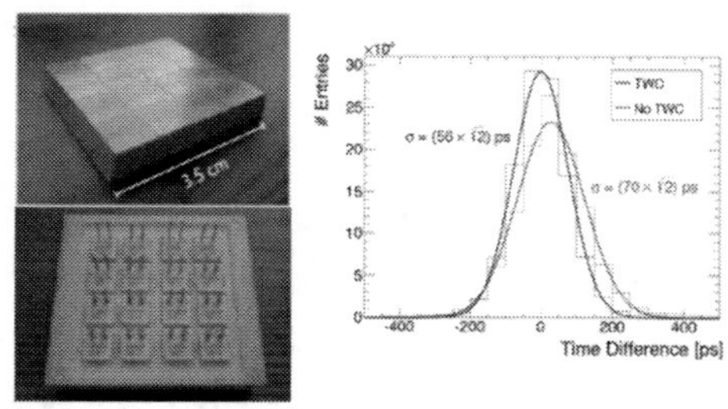

Figure 14. An half submodule of the Tile detector is shown (left) and the relative timing spectrum with the detector fired with minimum ionizing particles (right).

of 20×20 mm^2). MuPix8 represents the most important milestone towards the production of the final Mu3e chip.

Figure 13 shows the timing resolution obtained with a scintillating fibre prototype made of three staggered layers of Saint Gobain BCF12 multi-clad square 250 μm fibres readout at both ends by silicon photomultipliers. The main challenge of this detector is providing a timing measurement better than 1 ns with full detection efficiency and with a minimum amount of material ($< 0.003X_0$). Each fibre is readout individually by a S13360-1350CS. The signals are acquired and digitized with a DAQ acquisition system based on the DRS4 chip at 5 GSample/s [67]. Offline all waveforms associated to the same event and involving more than one fibre are summed up. The timing resolution is quoted as the difference between the left and right signals divided by a factor of 2. A timing resolution of $\sigma_t \approx 572 \pm 10$ ps has been obtained when minimum ionizing particles interact with the detector, with a detection efficiency $\epsilon > 96\%$ [54, 68].

Figure 14 (left) shows an example of an half submodule of the Tile detector. Each submodule is made by 32 elements. Each module is made by 14 submodules for a total of 7 modules per re-curl station. The half submodule is made of 16 BC408 $7.5 \times 8.5 \times 5$ mm^3 elements, each one coupled to a silicon photomultiplier Hamamatsu S10362-33-050C and readout with the STiC2 DAQ system [69, 70]. A timing resolution of $O(56)$ ps with full detection efficiency have been measured, as shown in Figure 14 (right) [71].

In conclusion the performances of all sub-detector protoypes measured during the 2017 campaign beam test successfully fulfilled the Mu3e demands, proven the suitability of the chosen technologies for the usage in the experiment. The collaboration is currently finalizing the detector design and moving to the large scale prototyping phase. This phase will be followed by the preparation for the construction and commissioning of the experiment expected for the 2020.

References

[1] Y. Kuno and Y. Okada, Rev. Mod. Phys. **73** 151 (2001);
[2] J. Ellis, Phil. Trans. R. Soc. A **370**, 818 (2012)
[3] Y. Fukuda *et al.* (Super-Kamiokande Collaboration), Phys. Rev. Lett. **81**, 1562 (1998);
[4] Q. R. Ahmad *et al.* (SNO Collaboration), Phys. Rev. Lett. **87**, 071301 (2001);
[5] K. Eguchi *et al.* (KamLAND Collaboration), Phys. Rev. Lett. **90**, 021802 (2003);
[6] S.T. Petcov, Sov. J. Nucl. Phys. **25**, 340 (1977);
[7] R. Barbieri and L. Hall Nucl. Phys. B **338**, 212 (1994);
[8] R. Barbieri, L. Hall and A. Strumia, Nucl. Phys. B **455**, 219 (1995);
[9] J. Hisano, D. Nomura and T. Yanagida, Phys. Lett. B **437**, 351 (1998);
[10] M. Raidal et al., Eur. Phys. J. C **57**: 13 (2008);
[11] P. Paradisi, Nucl. Phys. B (Proc. Suppl.) **248-250**, 8 (2014);
[12] A. Crivellin *et al.* J. High Ene. Phys. **05** 117 (2017). https://doi.org/10.1007/JHEP05(2017)117
[13] A. Crivellin *et al.* arXiv:1611.03409v1 [hep-ph];
[14] J. Albrecht *et al.*, arXiv:1311.5278v2 [hep-ex];
[15] A. de Gouvea and N. Saoulidou, Annu. Rev. Nucl. Part. Sci. **60**, 513 (2010).
[16] W. J. Marciano, T. Mori and J. M. Roney, Annu. Rev. Nucl. Part. Sci. **58**, 315 (2008);
[17] R. H. Bernstein and P. S. Copper, Phys. Rep. **532** 27 (2013);
[18] S. Mihara, J.P. Miller, P. Paradisi, and G. Piredda, Annu. Rev. Nucl. Part. Sci. **63**, 531 (2013);
[19] L. Calibbi and G. Signorelli, Nuovo Cimento **41**, 71 (2018);
[20] G. Cavoto, A. Papa *et al.*, Eur. Phys. J. C **78**, 37 (2018);
[21] R. Abela, F. Foroughi, D. Renker *The Future of Muon Physics* (Springer, Berlin, Heidelberg, New York 1991) 240;
[22] T. Prokscha *et al.*, Nucl. Instr. and Meth. A **595**, 317 (2008);
[23] R. Abramishvili *et al.* (COMET Collaboration), COMET-TDR-2016.pdf
[24] L. Bartoszek *et al.* (Mu2e Collaboration), https://arxiv.org/pdf/1501.05241.pdf
[25] P. R. Kettle, contribution to Future Muon Sources 2015, University of Hudderseld, United Kingdom; A. Knecht, contribution to NUFACT2017, Uppsala, Sweden;
[26] F. Berg *et al.*, Phys. Rev. Accel. Beams **19** no.2 024701 (2016);
[27] A. M. Baldini *et al.* (MEG Collaboration), MEG Proposal (2002);
[28] J. Adam *et al.* (MEG Collaboration), Eur. Phys. J. C **73** 2365 (2013);
[29] A. M. Baldini *et al.* (MEG Collaboration), Eur. Phys. J. C **76** 434 (2016).
[30] J. Adam *et al.* (MEG Collaboration), Phy. Rev. Lett. **110**, 201801 (2013);
[31] J. Adam *et al.* (MEG Collaboration), Phy. Rev. Lett. **107**, 171801 (2011) ;
[32] J. Adam *et al.* (MEG Collaboration), Nucl. Phys. B **834**, 1 (2010);
[33] M. L. Brooks *et al.* (MEGA Collaboration), Phys. Rev. Lett. **83** 1521 (1999);
[34] A.M. Baldini *et al.* (MEGII Collaboration), arXiv:1301.7225v2 [physics.ins-det];
[35] A.M. Baldini *et al.* (MEGII Collaboration), arXiv:1801.04688 [physics.ins-det];
[36] A. Blondel *et al.* (Mu3e Collaboration), arXiv:1301.611v1[physics.ins-det];
[37] U. Bellgardt *et al.* (SINDRUM Collaboration), Nucl.Phys. B **299** 1 (1988);
[38] W. Bertl *et al.* (SINDRUM II Collaboration), Eur. Phys. J. C **47** 337 (2006);
[39] S. Mihara *et al.*, J. Phys. Conf. Ser. **308**, 012009 (2011);
[40] W. Ootani *et al.* PoS(PhotoDet 2012)035;
[41] T. Iwamoto *et al.* JINST **9**, C09037 (2014);
[42] M. Hildebrandt, Nucl. Instr. Meth. Phys. Res. Sect. A **623**, 111 (2010);
[43] A. M. Baldini *et al.* JINST **11**, P07011 (2016);
[44] G. Chiarello *et al.* JINST **12**, C03056 (2017);
[45] M. De Gerone *et al.*, IEEE Trans. Nucl. Sci. **59** 379 (2012);
[46] M. De Gerone *et al.* JINST **9** C02035 (2014);
[47] A. Papa, IL NUOVO CIMENTO B **122** 627 (2007);
[48] J. Adam *et al.* (MEG Collaboration), Nucl. Instr. Meth. Phys. Res., Sect. A **641** 19 (2011);
[49] A. Baldini *et al.*, Nucl. Instr. Meth. Phys. Res., Sect. A **565** 589 (2006);
[50] A. Papa *et al.* Nucl. Phys. B **248-250** (Proc. Suppl.) 121 (2014);

[51] A. Papa *et al.*, JINST **9** C05066 (2014);

[52] A. Papa, G. Cavoto, E. Ripiccini, Nucl. Instr. Meth. Phys. Res., Sect. A **718** 580 (2013);

[53] A. Papa *et al.*, Nucl. Instr. Meth. Phys. Res., Sect. A **787** 130 (2015);

[54] A. Papa *et al.*, Nucl. Instr. Meth. Phys. Res., Sect. A **824** 128 (2016);

[55] G. Rutar *et al.*, Nucl. Instr. Meth. Phys. Res., Sect. A **824** 575 (2016);

[56] S. Ritt, R. Dinapoli and U. Hartmann, Nucl. Instr. Meth. Phys. Res., Sect. A **623**, 486 (2010);

[57] S.Ritt *The WaveDAQ system for the MEGII upgrade* contribution to 13th Pisa Meeting on Advanced Detectors (2015);

[58] L. Galli *et al.* JINST **8**, P01008 (2013);

[59] L. Galli *et al.* JINST **9**, P04022 (2014);

[60] M. Francesconi *et al.*, IEEE doi: 10.1109/MOCAST.2016.7495118

[61] I. Peric, Nucl. Instr. Meth. Phys. Res., Sect. A **582** 876 (2007);

[62] I. Peric *et al.*, Nucl. Instr. Meth. Phys. Res., Sect. A **731** 131 (2013);

[63] N. Berger *et al.*, Nucl. Instr. Meth. Phys. Res., Sect. A **732** 61 (2013);

[64] S. Bachmann *et al.*, JINST **9**, C01011 (2014);

[65] S. Bravar *et al.*, JINST **12**, C07011 (2017);

[66] H. Augustin *et al.*, Nucl. Instr. Meth. Phys. Res., Sect. A **845** 194 (2017);

[67] S. Ritt, http://www.psi.ch/drs/evaluation-board;

[68] A. Damyanova and A. Bravar, Nucl. Instr. Meth. Phys. Res., Sect. A **845** 475 (2017);

[69] T. Harion *et al.*, JINST **9**, C02003 (2014);

[70] T. Harion *et al.*, IEEE doi:10.1109/NSSMIC.2012.6551231;

[71] H. P. Eckert *The Mu3e Tile Detector*, Thesis dissertation: *https : //www.psi.ch/mu3e/ThesesEN/DissertationEckert.pdf*

[72] H. Chen *et al.*, JINST **12**, C01043 (2017);

Effective-field theories for charged lepton flavour violation

Giovanni Marco Pruna[1]

[1] INFN, Laboratori Nazionali di Frascati, Via E. Fermi 40, 00044 Frascati (Rome), Italy

Abstract. These proceedings review the status of present and future bounds on muonic lepton flavour violating transitions in the context of an effective-field theory defined below the electroweak scale. A specific focus is set on the phenomenology of $\mu \to e\gamma$, $\mu \to 3e$ transitions and coherent $\mu \to e$ nuclear conversion in the light of current and future experiments. Once the experimental limits are recast into bounds at higher scales, it is shown that the interplay between the various experiments is crucial to cover all corners of the parameter space.

1 Introduction

Lepton flavour violation (LFV) is strongly suppressed in the framework of the Standard Model (SM) of particle physics. Any observation of such phenomenon in the charged lepton sector would indicate unambiguously the presence of new physics (NP). Therefore, various experimental plans have taken place worldwide to confirm the vanishing prediction of the SM, and investigations are scheduled for the future to explore deeper regions of the NP parameter space.

Remarkable limits have been established with complementary tests of muonic LFV processes by the MEG [1, 2] and SINDRUM [3, 4] collaborations:

$$\mathrm{Br}(\mu^+ \to e^+\gamma) \leq 4.2 \times 10^{-13}\,, \tag{1}$$

$$\mathrm{Br}(\mu^+ \to e^+e^-e^+) \leq 1.0 \times 10^{-12}\,, \tag{2}$$

$$\mathrm{Br}^{\mathrm{Au}}_{\mu\to e} \equiv \frac{\Gamma(\mu^- \,\mathrm{Au} \to e^- \,\mathrm{Au})}{\Gamma^{\mathrm{capt}}_{\mathrm{Au}}} \leq 7 \times 10^{-13}\,. \tag{3}$$

Various experiments are already planned to improve these values by orders of magnitude: the MEG II upgrade [5] with an expected sensitivity of $\mathrm{Br}(\mu \to e\gamma) \sim 5 \times 10^{-14}$, the Mu3e experiment [6] with an improvement up to four orders of magnitude with respect to SINDRUM, and Mu2e at FNAL and COMET at J-PARC [7–9] aiming to improve the sensitivity by four orders of magnitude compared with SINDRUM II.

Consequently, these experimental efforts have to be supported both by an accurate theoretical interpretation of possible signals (or absence of signals) in terms of viable NP parameter space and a precise determination of the fundamental backgrounds[1].

These proceedings explore the possibility to give a model-independent phenomenological interpretation of muonic LFV transitions by adopting an effective-field-theory (EFT) description of NP interactions.

[1] Recent progress has been made in the precise estimation of the $\mu \to e\gamma$ [10, 11] and $\mu \to 3e$ [12, 13] fundamental background.

The EFT approach applied to LFV transitions has a long tradition: in the context of neutrino oscillations the first papers were published decades ago [14, 15], while the first complete dimension-six parameterisation at low energy for charged LFV appeared in [16]. On the other hand, the first systematic treatments of charged LFV in the context of SM EFT [17, 18] were published only a couple of years ago [19–24].

This note summarises the main results obtained in [25, 26], where the EFT parameterisation introduced by [16] was adopted to recast the current and future experimental limits on muonic LFV transitions in terms of bounds on the NP parameter space at the electroweak (EW) energy scale by exploiting a systematic renormalisation-group-equation (RGE) analysis.

2 Parameterisation

An effective Lagrangian for the $\mu \to e$ transitions valid below some scale Λ with $m_W \geq \Lambda \gg m_b$ is considered. It consists of all the operators invariant under $U(1)_{\mathrm{QED}} \times SU(3)_{\mathrm{QCD}}$ and contains all the SM fermion fields (except for the top quark) and the QED and QCD gauge fields:

$$\begin{aligned}\mathcal{L}_{\mathrm{eff}} &= \mathcal{L}_{\mathrm{QED}} + \mathcal{L}_{\mathrm{QCD}} \\ &+ \frac{1}{\Lambda^2}\Big\{ C^D_L O^D_L \\ &+ \sum_{f=q,\ell} \left(C^{V\,LL}_{ff} O^{V\,LL}_{ff} + C^{V\,LR}_{ff} O^{V\,LR}_{ff} + C^{S\,LL}_{ff} O^{S\,LL}_{ff} \right) \\ &+ \sum_{h=q,\tau} \left(C^{T\,LL}_{hh} O^{T\,LL}_{hh} + C^{S\,LR}_{hh} O^{S\,LR}_{hh} \right) + L \leftrightarrow R \Big\},\end{aligned} \tag{4}$$

plus the Hermitian conjugate components wherever required, and the explicit form of the operators given with obvious notation by

$$O_L^D = e\, m_\mu \left(\bar{e}\sigma^{\mu\nu} P_L \mu\right) F_{\mu\nu}, \qquad (5)$$

$$O_{ff}^{V\,LL} = \left(\bar{e}\gamma^\mu P_L \mu\right)\left(\bar{f}\gamma_\mu P_L f\right), \qquad (6)$$

$$O_{ff}^{V\,LR} = \left(\bar{e}\gamma^\mu P_L \mu\right)\left(\bar{f}\gamma_\mu P_R f\right), \qquad (7)$$

$$O_{ff}^{S\,LL} = \left(\bar{e} P_L \mu\right)\left(\bar{f} P_L f\right), \qquad (8)$$

$$O_{hh}^{S\,LR} = \left(\bar{e} P_L \mu\right)\left(\bar{h} P_R h\right), \qquad (9)$$

$$O_{hh}^{T\,LL} = \left(\bar{e}\sigma_{\mu\nu} P_L \mu\right)\left(\bar{h}\sigma^{\mu\nu} P_L h\right), \qquad (10)$$

with $P_{L/R} = \left(\mathbb{I} \mp \gamma^5\right)/2$. In the equations above, f represents any fermion below the scale m_W, and $h \in \{u, d, c, s, b, \tau\}$.

In the scenario where NP physics is realised at a scale $\Lambda < m_W$, NP gives rise to the interactions described in $\mathcal{L}_{\rm eff}$. If BSM physics is beyond the EW scale, $SU(2)$-invariant higher-dimensional operators are generated in the SMEFT. Then, the higher-dimensional operators in $\mathcal{L}_{\rm eff}$ stem from the matching of the SMEFT to our theory, as performed at the tree level in [27].

3 Observables

Given that the expressions for the rates of the processes $\mu^+ \to e^+\gamma$, $\mu^+ \to e^+e^-e^+$, and coherent muon-to-electron conversion in muonic atoms $\mu^- N \to e^- N$ will be exploited in the forthcoming section, we present them in terms of the effective coefficients (in the limit $m_e \ll m_\mu$).

For $\mu^+ \to e^+\gamma$, at the tree level, the branching ratio is

$$\mathrm{Br}(\mu \to e\gamma) = \frac{\alpha_e m_\mu^5}{\Lambda^4 \Gamma_\mu}\left(\left|C_L^D\right|^2 + \left|C_R^D\right|^2\right), \qquad (11)$$

where Γ_μ is the width of the muon.

The branching ratio of $\mu \to 3e$ expressed in terms of effective coefficients is

$$\begin{aligned}\mathrm{Br}(\mu \to 3e) = &\\ &\frac{\alpha_e^2 m_\mu^5}{12\pi\Lambda^4\Gamma_\mu}\left(\left|C_L^D\right|^2 + \left|C_R^D\right|^2\right)\left(8 \log\left[\frac{m_\mu}{m_e}\right] - 11\right)\\ &+ \frac{m_\mu^5}{3(16\pi)^3\Lambda^4\Gamma_\mu}\Big(\left|C_{ee}^{S\,LL}\right|^2 + 16\left|C_{ee}^{V\,LL}\right|^2 + 8\left|C_{ee}^{V\,LR}\right|^2\\ &+ \left|C_{ee}^{S\,RR}\right|^2 + 16\left|C_{ee}^{V\,RR}\right|^2 + 8\left|C_{ee}^{V\,RL}\right|^2\Big) + X_\gamma, \end{aligned} \qquad (12)$$

where the interference term with the dipole operator is given by

$$\begin{aligned} X_\gamma = -\frac{\alpha_e m_\mu^5}{3(4\pi)^2\Lambda^4\Gamma_\mu}\Big(&\mathrm{Re}[C_L^D\left(C_{ee}^{V\,RL} + 2C_{ee}^{V\,RR}\right)^*] +\\ &\mathrm{Re}[C_R^D\left(2C_{ee}^{V\,LL} + C_{ee}^{V\,LR}\right)^*]\Big). \end{aligned} \qquad (13)$$

Once relativistic and finite nuclear size effects are taken into account for heavy nuclei [28–30], the transition amplitudes for the coherent $\mu \to e$ transition in nuclei exhibit different sensitivities with respect to the atomic number involved. Consequently, different target atoms provide different limits on the coefficients of the involved class of operators. The SINDRUM collaboration has presented limits for gold, titanium and lead [4, 31, 32], but the upcoming experiments mostly concentrate on aluminium.

For this process, the Lagrangian $\mathcal{L}_{\rm eff}$ as given in Eq. (4) is not directly applicable. Instead, a Lagrangian at the nucleon level containing proton and neutron fields is required. This Lagrangian is obtained in two steps: integrating out heavy quarks and matching at a scale of $\mu_n = 1$ GeV to an effective nuclear Lagrangian. Following [33], the transition rate $\Gamma_{\mu\to e}^N = \Gamma(\mu^- N \to e^- N)$ is

$$\begin{aligned}\Gamma_{\mu\to e}^N = \frac{m_\mu^5}{4\Lambda^4}\Big|4\left(G_F m_\mu m_p \tilde{C}_{(p)}^{SL} S_N^{(p)} + \tilde{C}_{(p)}^{VR}\, V_N^{(p)} + p \to n\right)&\\ + e\, C_L^D\, D_N\Big|^2 + L \leftrightarrow R,&\end{aligned} \qquad (14)$$

where p and n denote the proton and the neutron, respectively. The effective couplings in Eq. (14) can be expressed in terms of our Wilson coefficients as

$$\tilde{C}_{(p/n)}^{VR} = \sum_{q=u,d,s}\left(C_{qq}^{V\,RL} + C_{qq}^{V\,RR}\right) f_{Vp/n}^{(q)}, \qquad (15)$$

$$\begin{aligned}\tilde{C}_{(p/n)}^{SL} = &\sum_{q=u,d,s}\frac{\left(C_{qq}^{S\,LL} + C_{qq}^{S\,LR}\right)}{m_\mu m_q G_F} f_{Sp/n}^{(q)}\\ &- \frac{1}{12\pi}\sum_{q=c,b}\frac{C_{qq}^{S\,LL} + C_{qq}^{S\,LR}}{G_F\, m_\mu m_q} f_{Gp/n},\end{aligned} \qquad (16)$$

with analogous relations for $L \leftrightarrow R$.

The nucleon form factors can be recast from [34, 35]. A next-to-leading-order computation of the nuclear Wilson coefficients appeared recently in the literature [36]; however, the inclusion of this result does not change any qualitative conclusion of the present study.

The branching ratio used in the following section is defined as the transition rate in Eq. (14), divided by the capture rate of the considered atom. For the latter, the values taken from [37] are adopted.

4 Connecting different energy scales

The operators in $\mathcal{L}_{\rm eff}$ carry the phenomenological information obtained from the experimental constraints. At the tree level, however, a direct interpretation would result in bounds on the Wilson coefficients evaluated at the experimental scales, *i.e.* $C(\lambda = m_\mu)$ or $C(\lambda = \mu_n)$.

A more meaningful procedure consists of extracting limits on the Wilson coefficients at scales that are different/higher than the phenomenological scale. If the NP is not far above the EW scale, the interconnection among scales is realised by the QED and QCD RGE of the effective operators, with the coefficients evolved between the experimental scale and the NP scale, *i.e.* $m_\mu, \mu_N \leq \Lambda \leq m_W$.

Under the RGE, the various operators in $\mathcal{L}_{\rm eff}$ mix among each other. To encode the leading mixing effects, at least the one-loop anomalous dimensions for all the operators must be considered. Moreover, the dipole operator plays a prominent role in all $\mu \to e$ transitions, and two-loop effects of direct mixing into C_L^D and C_R^D that are at the leading order must be considered (also to ensure a regularisation-scheme-independent result [38, 39]).

Including only a subset of two-loop leading-order contributions, while phenomenologically useful, is not a self-consistent procedure and should not be understood as a replacement of a genuine two-loop precise calculation, but rather as a qualitative indication of leading-order effects.

The evolution of the full set of operators is described by the following equations, where

$$\dot{C} \equiv (4\pi)\mu \frac{\mathrm{d}}{\mathrm{d}\mu} C \tag{17}$$

is understood, and the remaining notation is clear.

In the BMHV scheme [40, 41], the coefficient of the dipole operator runs according to

$$\begin{aligned}
\dot{C}_L^D = {} & 16\,\alpha_e\, Q_l^2 C_L^D - \frac{Q_l}{(4\pi)}\frac{m_e}{m_\mu} C_{ee}^{S\,LL} - \frac{Q_l}{(4\pi)} C_{\mu\mu}^{S\,LL} \\
& + \sum_h \frac{8Q_h}{(4\pi)}\frac{m_h}{m_\mu} N_{c,h}\, C_{hh}^{T\,LL}\, \Theta(\mu - m_h) \\
& - \frac{\alpha_e Q_l^3}{(4\pi)^2}\left(\frac{116}{9} C_{ee}^{V\,RR} + \frac{116}{9} C_{\mu\mu}^{V\,RR}\right. \\
& \left. - \frac{122}{9} C_{\mu\mu}^{V\,RL} - \left(\frac{50}{9} + 8\frac{m_e}{m_\mu}\right) C_{ee}^{V\,RL}\right) \\
& - \sum_h \frac{\alpha_e}{(4\pi)^2}\left(6Q_h^2 Q_l + \frac{4Q_h Q_l^2}{9}\right) N_{c,h}\, C_{hh}^{V\,RR}\, \Theta(\mu - m_h) \\
& - \sum_h \frac{\alpha_e}{(4\pi)^2}\left(-6Q_h^2 Q_l + \frac{4Q_h Q_l^2}{9}\right) N_{c,h}\, C_{hh}^{V\,RL}\, \Theta(\mu - m_h) \\
& - \sum_h \frac{\alpha_e}{(4\pi)^2} 4Q_h^2 Q_l N_{c,h} \frac{m_h}{m_\mu} C_{hh}^{S\,LR}\, \Theta(\mu - m_h)\,.
\end{aligned} \tag{18}$$

The running of the whole set of vector operators is given by the following two equations:

$$\begin{aligned}
\dot{C}_{ff}^{V\,RR} = {} & \frac{4\,\alpha_e}{3} Q_f \left(2Q_l \sum_{\ell=e,\mu} C_{\ell\ell}^{V\,RR} + Q_l C_{\tau\tau}^{V\,RR} + Q_l \sum_l C_{ll}^{V\,RL}\right. \\
& \left. + N_c \sum_q Q_q \left(C_{qq}^{V\,RR} + C_{qq}^{V\,RL}\right) + 9 Q_l C_{ff}^{V\,RR}\right),
\end{aligned} \tag{19}$$

$$\begin{aligned}
\dot{C}_{ff}^{V\,RL} = {} & \frac{4\,\alpha_e}{3} Q_f \left(2Q_l \sum_{\ell=e,\mu} C_{\ell\ell}^{V\,RR} + Q_l C_{\tau\tau}^{V\,RR} + Q_l \sum_l C_{ll}^{V\,RL}\right. \\
& \left. + N_c \sum_q Q_q \left(C_{qq}^{V\,RR} + C_{qq}^{V\,RL}\right) - 9 Q_l C_{ff}^{V\,RL}\right).
\end{aligned} \tag{20}$$

The running of the leptonic scalar and tensorial operators is summarised by the following equations:

$$\dot{C}_{\ell\ell}^{S\,LL} = 12\,\alpha_e\, Q_l^2 C_{\ell\ell}^{S\,LL} \qquad \text{for } \ell \in \{e, \mu\}, \tag{21}$$

$$\dot{C}_{\tau\tau}^{S\,LL} = -12\,\alpha_e\, Q_l^2 \left(C_{\tau\tau}^{S\,LL} + 8C_{\tau\tau}^{T\,LL}\right), \tag{22}$$

$$\dot{C}_{\tau\tau}^{S\,LR} = -12\,\alpha_e\, Q_l^2 C_{\tau\tau}^{S\,LR}, \tag{23}$$

$$\dot{C}_{\tau\tau}^{T\,LL} = -2\,\alpha_e\, Q_l^2 \left(C_{\tau\tau}^{S\,LL} - 2C_{\tau\tau}^{T\,LL}\right). \tag{24}$$

The running of the scalar and tensorial quark operators is given by

$$\begin{aligned}
\dot{C}_{qq}^{S\,LL} = {} & \left(-6\,\alpha_e\left(Q_l^2 + Q_q^2\right) - 6C_F\,\alpha_s\right) C_{qq}^{S\,LL} \\
& - 96\,\alpha_e\, Q_l Q_q C_{qq}^{T\,LL},
\end{aligned} \tag{25}$$

$$\dot{C}_{qq}^{S\,LR} = \left(-6\,\alpha_e\left(Q_l^2 + Q_q^2\right) - 6C_F\,\alpha_s\right) C_{qq}^{S\,LR}, \tag{26}$$

$$\begin{aligned}
\dot{C}_{qq}^{T\,LL} = {} & -2\,\alpha_e\, Q_l Q_q C_{qq}^{S\,LL} \\
& + \left(2\,\alpha_e\left(Q_l^2 + Q_q^2\right) + 2C_F\,\alpha_s\right) C_{qq}^{T\,LL}.
\end{aligned} \tag{27}$$

5 Phenomenological results

The effective coefficients are generated by some underlying NP theory around the EW scale. Then, the RGEs can be exploited to evolve such coefficients from the high scale m_W to the phenomenological scales μ_n and m_μ. Therefore, the predicted rates are compared with the experimental limits. This procedure will shape the constraints of various Wilson coefficients at the NP scale.

RGE mixing can generate important effects at the phenomenological scale even for vanishing Wilson coefficients at the high scale. This allows one to place bounds on coefficients that would be unconstrained if loop effects are not taken into account.

Considering the experimental limits listed in Section 1, the final aim is to compare the exploratory power of current and future $\mu \to e\gamma$, $\mu \to 3e$ and $\mu \to e$ conversion experiments for specific effective interactions at the NP scale.

First, we consider the coefficient for which the MEG experiment and future updates will deliver the best performances: $C_{\mu\mu}^{S\,LL}$. In Figure 1, the current and future branching ratio for $\mu \to e\gamma$ and $\mu \to 3e$ experiments are compared with future $\mu N \to eN$ prospects.

Here, the horizontal dashed-red line indicates that a small limit $\mathrm{Br}(\mu \to e\gamma) \lesssim 10^{-12}$ displays an equivalent exploratory power as the future Mu3e limit $\mathrm{Br}(\mu \to 3e) < 5 \times 10^{-15}$. Instead, for muon conversion experiments (vertical dashed-red line), a limit of $\mathrm{Br}(\mu N \to eN) < 10^{-15}$ would be required to perform as good as $\mu \to e\gamma$ experiments. In other words, the future MEG II experiment will place the strongest limit on $C_{\mu\mu}^{S\,LL}$ unless the COMET or Mu2e experiments could improve their expected limit to reach at least $\mathrm{Br}(\mu N \to eN) < 5 \times 10^{-17}$. Curiously, one must notice that experiments exploring the dipole interaction at lower scales will deliver the best future performance in testing a four-fermion interaction defined at higher scales. This goes against one of the most deeply embedded prejudices in the theoretical and experimental LFV community.

Now, we considered scenarios where two Wilson coefficients are non-vanishing at the EW scale. For this scope, the allowed parameter space is plotted in light of current and future experimental limits for all three processes. For clarity's sake, they are displayed on a pseudo-logarithmic scale.

In Figure 2, the allowed regions in the $C_{ee}^{S\,LL} - C_{ee}^{V\,RR}$ plane are plotted (with coefficients defined at the EW

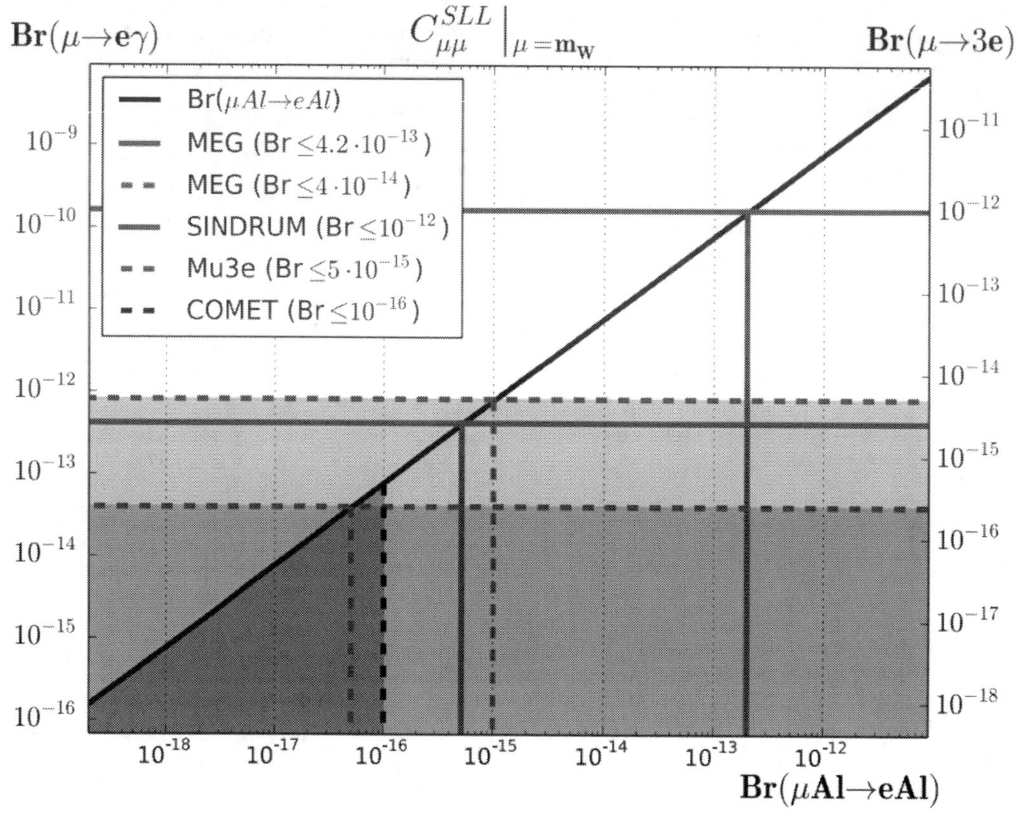

Figure 1. Br($\mu \to e\gamma$) (Br($\mu \to 3e$)) plotted on the left (right) y-axis against Br($\mu N \to eN$) for a fixed value of $C^{S\,LL}_{\mu\mu}$ given at the scale $\mu = m_W$. Current and future experimental limits are displayed.

scale). Current (solid lines) and future (dashed lines) limits on $\mu \to e\gamma$ (green) and $\mu \to e$ conversion (blue) are less stringent than those coming from $\mu \to 3e$ (red). This indicates that $\mu \to 3e$ experiments are and will be the most sensitive to these Wilson coefficients. Moreover, $\mu \to 3e$ experiments do not provide any blind region of the parameter space for this specific choice of coefficients. This occurs because these operators produce $\mu \to 3e$ already at the tree level, while they give rise to the other processes only via mixing effects.

Figure 3 shows an analogous plot for the Wilson coefficients $C^{V\,RL}_{\tau\tau}$ and $C^{V\,RR}_{bb}$. Here, $\mu \to e$ conversion experiments display a superior capability to probe vectorial four-fermion operators generated at the phenomenological scale by mixing effects. Even the future Mu3e experiment will perform just around the current $\mu \to e$ conversion limit established by the SINDRUM II collaboration more than a decade ago. This shows that future Mu2e and COMET experiments will represent the best opportunity to test four-fermion vectorial interactions containing *any type of heavy fermion*, including τ leptons. Again, this result challenges the biased opinion that coherent LFV conversion in nuclei can test only contact interactions with valence quarks.

Furthermore, the plot displays an interesting complementarity among various experiments: assuming that the underlying theory produces a cancellation both in $\mu \to e$ conversion and $\mu \to 3e$, then $\mu \to e\gamma$ experiments will provide a complementary limit, ultimately circumscribing the allowed region of the parameter space.

Figure 4 shows the allowed regions in the $C^D_L - C^{S\,LR}_{bb}$ plane. For this choice of coefficients, one may immediately notice that $\mu \to 3e$ experiments are less constraining than the other two options. On the other hand, the COMET and Mu2e experiments will indeed set the best limits on each uncorrelated Wilson coefficient. However, there is a big portion of the parameter space where $\mu \to e$ conversion experiments are completely blind. To cover this region, results from MEG II will be required.

With Figure 5, we conclude our overview by considering the coefficients C^D_L and $C^{V\,RR}_{ee}$. In this case, it is well established that the current limit on the C^D_L coefficient comes from the MEG experiment, and the future limit will be set by Mu2e and COMET. Focusing on the $C^{V\,RR}_{ee}$ coefficient, even with intuitive arguments one must realise that $\mu \to 3e$

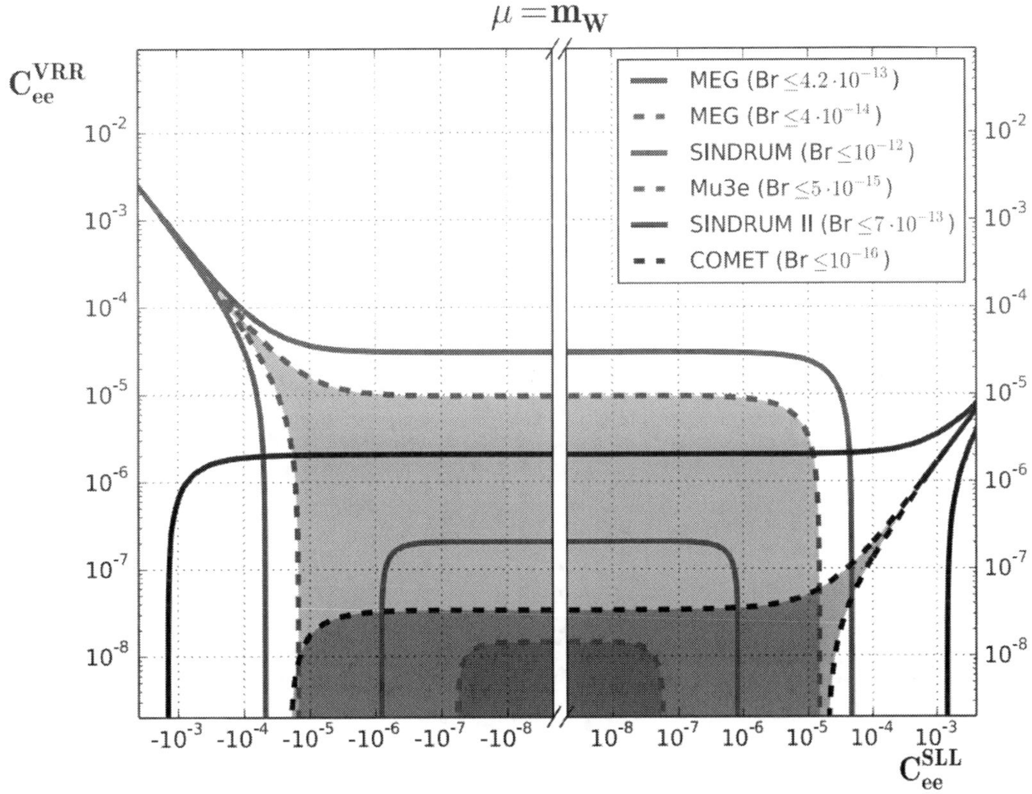

Figure 2. Allowed regions in the $C_{ee}^{SLL} - C_{ee}^{VRR}$ plane from $\mu \to e\gamma$ (green), $\mu \to 3e$ (red) and $\mu \to e$ conversion (blue) for current (straight) and future (dashed) experimental limits.

experiments (SINDRUM in the past and Mu3e in the future) give the most significant limits. However, in the corners of the parameter space where potential cancellations might occur, an interesting interplay between the observables implies that all of the future experimental limits are useful to ensure that no blind spots in parameter space remain unexplored.

Of course, our choice of combinations for the free parameters is far from being exhaustive. However, the presented analysis carries two main messages:

- many opinions that are deeply entrenched in the LFV theoretical and experimental community are undermined by a systematic RGE treatment of effective coefficients;
- the interplay between the various experiments is crucial to cover all corners of the NP parameter space, especially where cancellations can result in blind spots for one or even two specific experiments.

6 Conclusion

In these proceedings, an RGE-improved analysis of the three $\mu \to e$ processes $\mu \to e\gamma$, $\mu \to 3e$ and $\mu \to e$ conversion in nuclei in the context of EFT was briefly reviewed.

The complete set of dimension-six operators giving rise to point-like muonic LFV interactions invariant under $U(1)_{QED} \times SU(3)_{QCD}$ was introduced.

The relevant observables for muonic LFV transitions and the formulae for their transition rates were recollected.

The complete one-loop RGEs for an EFT with dimension-six operators invariant under $U(1)_{QED} \times SU(3)_{QCD}$ were presented. Furthermore, the leading two-loop QED effects for the mixing of vector operators into the dipole operators were included.

Then, the resulting bounds on the Wilson coefficients (given at the scale m_W) were calculated. Afterwards, some benchmark scenario was considered to stress both the complementarity of the three $\mu \to e$ processes and the inaccuracy of some opinions that are widely held in the LFV community. Concerning the former, the capability of covering regions of parameter space that would be blind spots for a single process was clearly pointed out. Con-

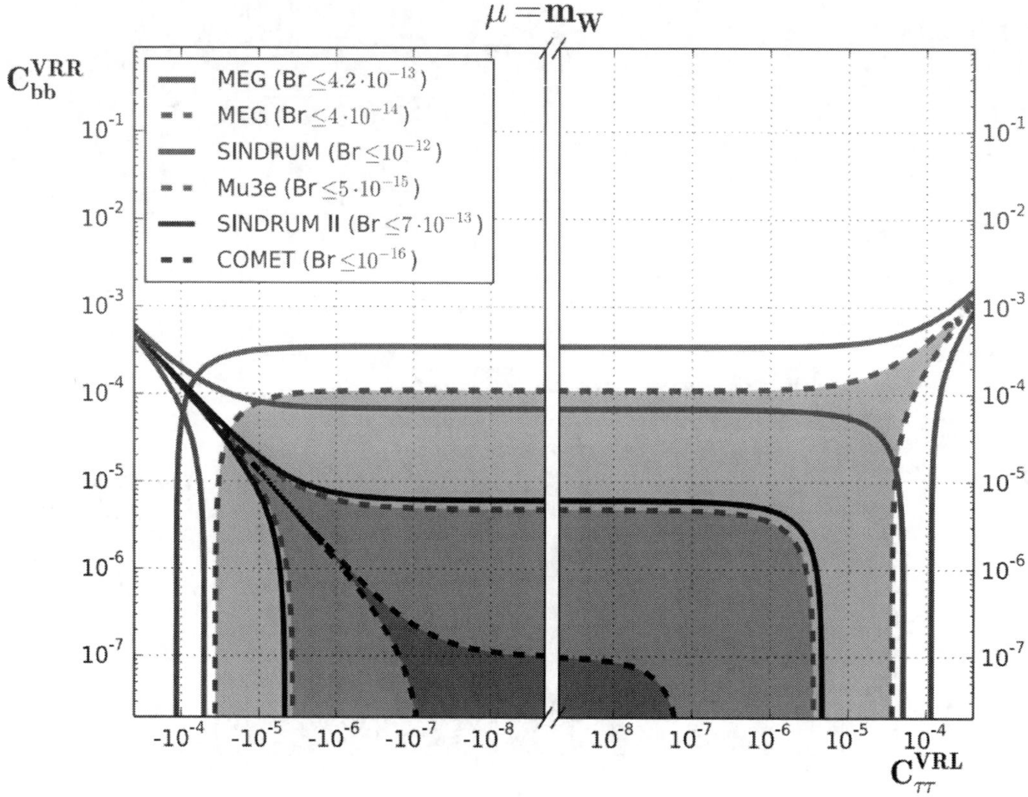

Figure 3. Allowed regions in the $C_{\tau\tau}^{VRL} - C_{bb}^{VRR}$ plane from $\mu \to e\gamma$ (green), $\mu \to 3e$ (red) and $\mu \to e$ conversion (blue) for current (straight) and future (dashed) experimental limits.

cerning the latter, the potential of $\mu \to e\gamma$ experiments to explore scalar four-fermion interactions and the impact from coherent $\mu \to e$ conversion on vectorial four-fermion interactions with heavy fermions was clearly displayed.

The present analysis is far from being exhaustive; indeed, intriguing new ideas to explore the effective parameter space appeared recently in the literature [42–46], and their RGE-improved analysis should be included in the present treatment.

Finally, even if a systematic study was never performed in this direction, we claim that future developments of the μe-scattering (MUonE) experiment [47, 48] could increase our current knowledge of muon LFV transitions and a dedicated EFT analysis should be performed in a future study.

References

[1] J. Adam et al. (MEG), Phys. Rev. Lett. **110**, 201801 (2013), 1303.0754

[2] A. M. Baldini et al. (MEG), Eur. Phys. J. **C76**, 434 (2016), 1605.05081

[3] U. Bellgardt et al. (SINDRUM), Nucl. Phys. **B299**, 1 (1988)

[4] W. H. Bertl et al. (SINDRUM II), Eur. Phys. J. **C47**, 337 (2006)

[5] A. M. Baldini et al. (2013), 1301.7225

[6] A. Blondel et al. (2013), 1301.6113

[7] R. M. Carey et al. (Mu2e), FERMILAB-PROPOSAL-0973 (2008)

[8] R. K. Kutschke, *The Mu2e Experiment at Fermilab*, in *Proceedings, 31st International Conference on Physics in collisions (PIC 2011): Vancouver, Canada, August 28-September 1, 2011* (2011), 1112.0242

[9] Y. G. Cui et al. (COMET), KEK-2009-10 (2009)

[10] M. Fael, L. Mercolli, M. Passera, JHEP **07**, 153 (2015), 1506.03416

[11] G. M. Pruna, A. Signer, Y. Ulrich, Phys. Lett. **B772**, 452 (2017), 1705.03782

[12] M. Fael, C. Greub, JHEP **01**, 084 (2017), 1611.03726

[13] G. M. Pruna, A. Signer, Y. Ulrich, Phys. Lett. **B765**, 280 (2017), 1611.03617

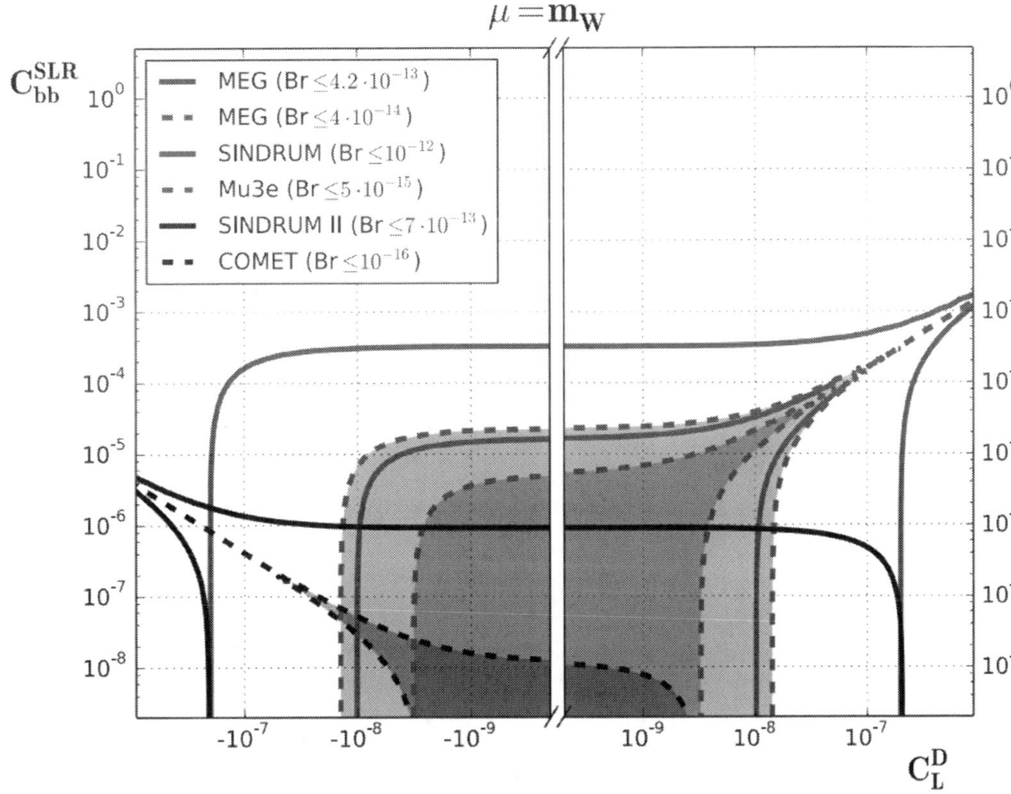

Figure 4. Allowed regions in the $C_L^D - C_{bb}^{SLR}$ plane from $\mu \to e\gamma$ (green), $\mu \to 3e$ (red) and $\mu \to e$ conversion (blue) for current (straight) and future (dashed) experimental limits.

[14] S. T. Petcov, Sov. J. Nucl. Phys. **25**, 340 (1977), [Erratum: Yad. Fiz.25,1336(1977)]
[15] P. Minkowski, Phys. Lett. **67B**, 421 (1977)
[16] Y. Kuno, Y. Okada, Rev. Mod. Phys. **73**, 151 (2001), hep-ph/9909265
[17] W. Buchmuller, D. Wyler, Nucl. Phys. **B268**, 621 (1986)
[18] B. Grzadkowski, M. Iskrzynski, M. Misiak, J. Rosiek, JHEP **10**, 085 (2010), 1008.4884
[19] A. Crivellin, S. Najjari, J. Rosiek, JHEP **04**, 167 (2014), 1312.0634
[20] E. E. Jenkins, A. V. Manohar, M. Trott, JHEP **10**, 087 (2013), 1308.2627
[21] E. E. Jenkins, A. V. Manohar, M. Trott, JHEP **01**, 035 (2014), 1310.4838
[22] R. Alonso, E. E. Jenkins, A. V. Manohar, M. Trott, JHEP **04**, 159 (2014), 1312.2014
[23] G. M. Pruna, A. Signer, JHEP **10**, 014 (2014), 1408.3565
[24] G. M. Pruna, A. Signer, EPJ Web Conf. **118**, 01031 (2016), 1511.04421
[25] A. Crivellin, S. Davidson, G. M. Pruna, A. Signer, *Complementarity in lepton-flavour violating muon decay experiments*, in *18th International Workshop on Neutrino Factories and Future Neutrino Facilities Search (NuFact16) Quy Nhon, Vietnam, August 21-27, 2016* (2016), 1611.03409
[26] A. Crivellin, S. Davidson, G. M. Pruna, A. Signer, JHEP **05**, 117 (2017), 1702.03020
[27] S. Davidson, Eur. Phys. J. **C76**, 370 (2016), 1601.07166
[28] O. U. Shanker, Phys. Rev. **D20**, 1608 (1979)
[29] A. Czarnecki, W. J. Marciano, K. Melnikov, AIP Conf. Proc. **435**, 409 (1998), hep-ph/9801218
[30] R. Kitano, M. Koike, Y. Okada, Phys. Rev. **D66**, 096002 (2002), [Erratum: Phys. Rev.D76,059902(2007)], hep-ph/0203110
[31] C. Dohmen et al. (SINDRUM II), Phys. Lett. **B317**, 631 (1993)
[32] W. Honecker et al. (SINDRUM II), Phys. Rev. Lett. **76**, 200 (1996)
[33] V. Cirigliano, R. Kitano, Y. Okada, P. Tuzon, Phys. Rev. **D80**, 013002 (2009), 0904.0957

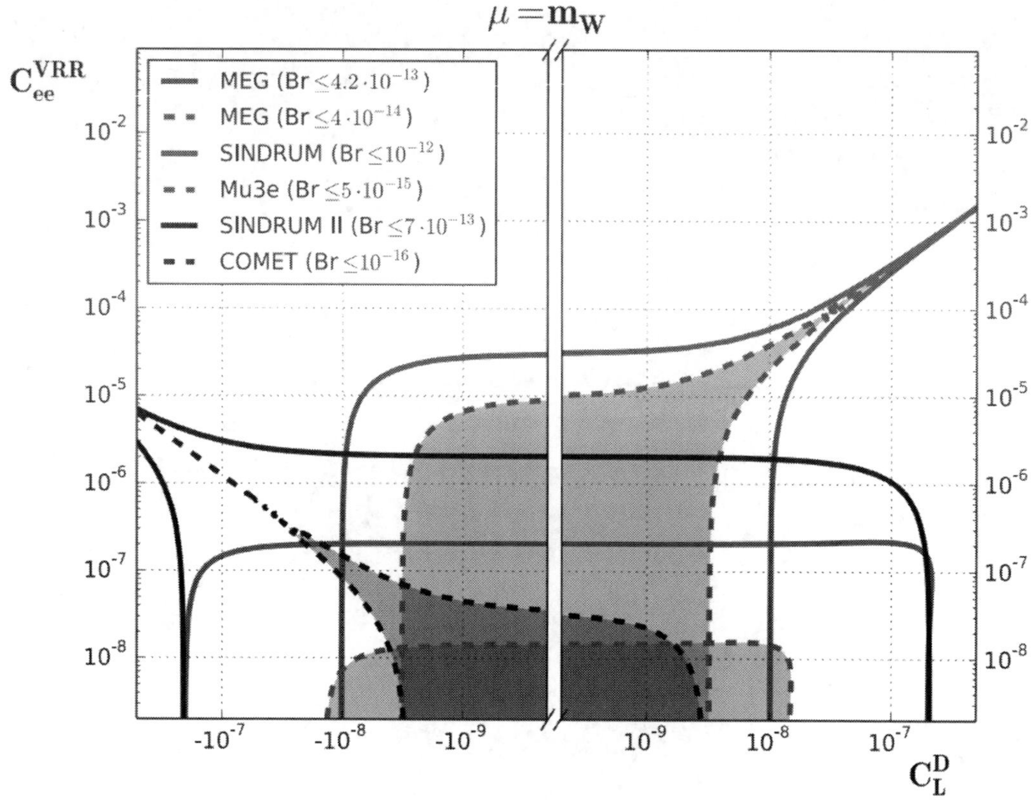

Figure 5. Allowed regions in the $C_L^D - C_{ee}^{VRR}$ plane from $\mu \to e\gamma$ (green), $\mu \to 3e$ (red) and $\mu \to e$ conversion (blue) for current (straight) and future (dashed) experimental limits.

[34] M. Hoferichter, J. Ruiz de Elvira, B. Kubis, U. G. Meißner, Phys. Rev. Lett. **115**, 092301 (2015), 1506.04142

[35] P. Junnarkar, A. Walker-Loud, Phys. Rev. **D87**, 114510 (2013), 1301.1114

[36] A. Bartolotta, M. J. Ramsey-Musolf (2017), 1710.02129

[37] T. Suzuki, D. F. Measday, J. P. Roalsvig, Phys. Rev. **C35**, 2212 (1987)

[38] M. Misiak, Phys. Lett. **B321**, 113 (1994), hep-ph/9309236

[39] M. Ciuchini, E. Franco, L. Reina, L. Silvestrini, Nucl. Phys. **B421**, 41 (1994), hep-ph/9311357

[40] G. 't Hooft, M. J. G. Veltman, Nucl. Phys. **B44**, 189 (1972)

[41] P. Breitenlohner, D. Maison, Commun. Math. Phys. **52**, 11 (1977)

[42] M. Koike, Y. Kuno, J. Sato, M. Yamanaka, Phys. Rev. Lett. **105**, 121601 (2010), 1003.1578

[43] Y. Uesaka, Y. Kuno, J. Sato, T. Sato, M. Yamanaka, Phys. Rev. **D93**, 076006 (2016), 1603.01522

[44] Y. Uesaka, Y. Kuno, J. Sato, T. Sato, M. Yamanaka (2017), 1711.08979

[45] V. Cirigliano, S. Davidson, Y. Kuno, Phys. Lett. **B771**, 242 (2017), 1703.02057

[46] S. Davidson, Y. Kuno, A. Saporta (2017), 1710.06787

[47] C. M. Carloni Calame, M. Passera, L. Trentadue, G. Venanzoni, Phys. Lett. **B746**, 325 (2015), 1504.02228

[48] G. Abbiendi et al., Eur. Phys. J. **C77**, 139 (2017), 1609.08987

Status of the PADME experiment and review of dark photon searches

M. Raggi[1]

[1] Sapienza Università di Roma ed INFN, Sezione di Roma
P.le A. Moro 2, I-00185 Roma (Italy).
e-mail: mauro.raggi@roma1.infn.it

Abstract. While accelerator particle physics has traditionally focused on exploring dark matter through high-energy experiments, testing dark-sectors hypothesis requires innovative low energy experiments that use high-intensity beams and high-sensitivity detectors. In this scenario attractive opportunities are offered to low energy machines and flavour experiments. In this paper we will focus our attention on the Dark Photon (DP) scenario, reviewing the current status of searches and new opportunities with particular attention to the PADME experiment at Laboratori Nazionali di Frascati.

1 Introduction

The existing experimental program to search for dark matter is focused primarily on weakly-interacting massive particles (WIMPs). At the same time, given the importance of dark matter, there are strong motivations to explore a broader set of dark matter candidates. Recent null results obtained in Dark Matter candidates searches at the Large Hadron Collider, together with the increasing sensitivity reached in direct detection without any evidence of WIMPs, have revitalised the idea that dark matter can have sub-GeV mass. In addition recent anomalies in the magnetic momentum of the muon, in the decay of ^8Be, and proton radius can be solved introducing light new states[1]. A simple possibility is that dark matter interacts through a new force that is similar in structure to the known forces, but couples only indirectly to ordinary matter through new particles called "mediators".

2 Dark Sector Panorama

A dark sector is a collection of particles that are not directly charged under the Standard Model (SM) forces. Such particles are assumed to possess gravitational interactions, and may also interact with familiar matter through several "portal" interactions that are constrained by the symmetries of the Standard Model. The dark sector typically include one or more mediator particles coupled to the SM via the portal interaction. The portal relevant for dark sector-SM interactions depends on the mediator spin and parity: it can be a scalar, a pseudo-scalar, a fermion, or a vector.

We will focus on the vector portal, but we will briefly comment on the other possible mediators. If the mediator is a scalar, it can interact via the Higgs portal and can be probed in various ways, including exotic Higgs decays at high-energy colliders such as the LHC. If the mediator is a pseudoscalar a valuable candidate is the pseudo Nambu-Goldstone boson of the Peccei -Quinn U(1) symmetry: the axion. A possible existence of a sterile neutrino N may lead to the addition of Yukawa term in the lagrangian requiring a fermionic mediator.

The most general interaction of an electrically neutral vector particle A with the Standard Model fermions can be written in the form of a U(1) interaction. Its mediator will be a massive "photon like" particle which couples to SM with a strength ϵe. In the present review we focus mostly on the vector portal as this scenario is able produce signatures accessible in flavour and low energy experiments.

3 Dark Photon Models

We will call dark photon model every model that predicts the existence of a new neutral vector particle A' which has a non-vanishing coupling to the standard fermions. The A' could be the mediator between the SM and the dark sector. The origin of the coupling of A' to the fermion fields could arise in various models. Since almost any extension of the Standard Model introduces new symmetries and gauge groups, the wide range of possibilities go from maximally universal models to the ones including a single type of fermions or even a single lepton generation[1].

The production of A' in collider experiments relies on the DP couplings to SM particles, primarily electrons and quarks. In the simplest scenario the couplings arise from the "kinetic mixing" interaction, which mixes the A' boson of the non-SM "dark" gauge group $U(1)_D$ with the SM photon:

$$\mathcal{L} = \frac{1}{2}\epsilon F^{\mu\nu}F'_{\mu\nu} \tag{1}$$

where ϵ is a dimensionless parameter while $F^{\mu\nu}$ is the tensor of the SM $U(1)_{em}$ and $F'_{\mu\nu}$ of the dark $U(1)_D$ respectively.

Due to the mixing with the standard model photon, the DP can be produced in electromagnetic processes similar to those in the SM. The most considered are: emission of A' in electron bremsstrahlung (A'-strahlung), $e^+e^- \to \gamma A'$ annihilation, and mesons decays with photons in the final state ($\pi^0, \eta \to \gamma A'$).

The DP decay modes depend on its mass and on the hypothesis on the dark sector particles mass spectrum. The most commonly used are to consider the DP the lightest state in the dark sector and that it decays to SM particles, "visible decays". Depending on the A' mass di-lepton or hadronic decays are allowed. In more general models the existence of dark fermions χ with mass $m_\chi < m_{A'}/2$, coupling to DP with a strength α_D, allows the $A' \to \chi\bar{\chi}$ decay channel. In natural scenario, $\alpha_D >> \alpha\epsilon^2$ since there is no necessity to suppress the interactions in the hidden sector, the dominant decay channel would be $A' \to \chi\bar{\chi}$, "invisible decay.".

The two scenarios, so called "visible" and "invisible", produce very different experimental signatures. In the visible case a narrow resonance might be observed in the di-lepton or, in general, in the di-particle invariant mass spectrum, while in the invisible case the DP could manifest itself through the search for missing mass or missing energy. A detailed review of the DP photon physics and experimental bounds can be found in [1] [2].

3.1 Visible dark photon search status

In the case of the visible dark photon model with universal coupling to SM fermions, the present experimental situation is summarised in Fig. 1. Experiments searching for bumps in di-lepton invariant mass distributions ruled out ϵ values greater than 10^{-3} in the 10 MeV-10 GeV mass range. NA48/2 at CERN [3], A1 at Mainz [4], and BaBar at SLAC [5] provide the strongest experimental bounds. These experiments employ different dark photon production mechanisms, including meson decays (NA48/2), electron bremsstrahlung (A1), and $e^+ e^-$ to two photon annihilation (BaBar). The region of $\epsilon < 10^{-4}$ is out of reach for bump hunt experiments due to standard model background. Beam dump experiments, such as E137[6] at SLAC profit of massive absorbers, beam dumps, to kill the SM background and detect the decay product of long living A' traversing the dump without interacting. There are also theoretical constraints derived from the measurement of the anomalous magnetic moment of the electron and muon a_e, a_μ. All the existing constraints, in the hypothesis of universal coupling to fermions, rule out the possibility that the visible dark photon model can explain the observed anomaly in the magnetic moment of the muon.

3.2 Invisible dark photon search status

The parameter space is significantly less constrained in the case of A' decaying to dark-sector ("invisible") final states, see Fig.2. In this case, the only bounds come from BaBar[7], using $e^+e^- \to \gamma A'$, NA64 at CERN[8] using missing energy. The bounds from the a_e measurement also applies, since it relies on virtual dark photon contribution and as such is insensitive to the A' decay mode. Recently the Babar and NA64 measurements covered the parameter region favoured by the $(g-2)_\mu$ anomaly explanation, disfavouring DP as a candidate solution. Additional constraints can be extracted by looking for the scattering dark fermions (χ), obtained from DP decays, in beam dump experiments. A summary of such a kind of bounds is presented in [1].

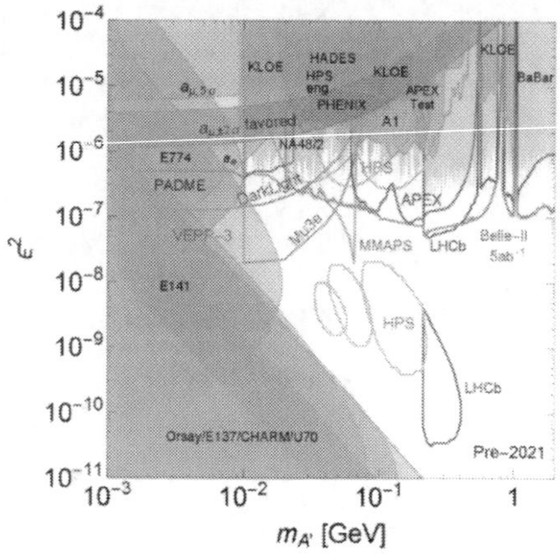

Figure 1. Status of DP visible decay searches [1]. Filled area are already excluded while lines represent projections for future dedicated searches.

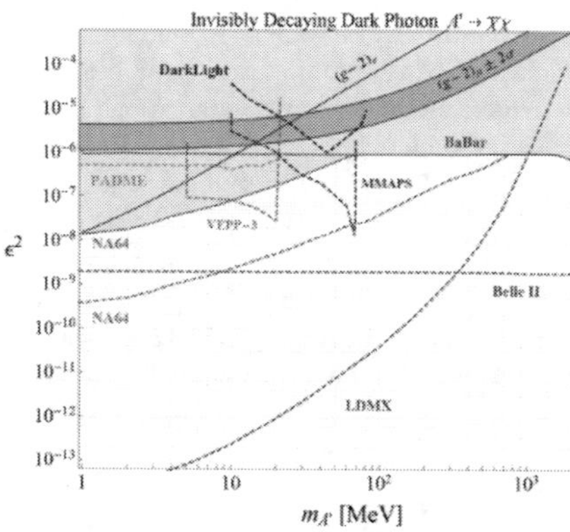

Figure 2. Status of DP invisible decay searches [1]. Filled area are already excluded while dashed lines represent projections for future dedicated searches.

4 Opportunities in Kaon decays

Kaon physics experiments have a long tradition in precision measurements. Huge samples of Kaon decays have been collected during the last 2 decades allowing NA48/2 to set already a strong bound in the visible DP parameter space. New possibilities can still be explored with extended mass reach or different production and decay mechanisms.

Kaon radiative decays are a natural place where A' can be produced. Since the interaction is mediated by mixing with the photon, the natural place to look for A' are the decays of kaons with photon or lepton pairs in the final state. We will consider two important processes:

- $K^+ \to \pi^+\gamma^*$ mixing with $K^+ \to \pi^+ A'$
- $K^+ \to \ell^+\nu\gamma^*$ mixing with $K^+ \to \ell^+\nu A'$

Modern kaon experiments, in particular NA62[9] and KOTO[10], are able to explore kaon decays with the very high luminosity and precision detectors needed to search for new light weakly coupled particle.

4.1 The $K^+ \to \pi^+ A'$ decay

An A' mixing to virtual photon can be produced in the decay $K^+ \to \pi^+ X$. In the visible dark photon scenario the A' will promptly decay in standard model leptons producing, depending on its mass, pairs of electrons or muons. The invariant mass spectrum of e^+e^- or $\mu^+\mu^-$, in $K^+ \to \pi^+e^+e^-$ an $K^+ \to \pi^+\mu^+\mu^-$ decays respectively, may reveal the presence of a resonance via the bump hunt search. The recent null result by NA48/2 in bump hunt search in the di-muon invariant mass of $K^+ \to \pi^+\mu^+\mu^-$ decay[11] even if is not competitive with the best limits in the kinetic mixing scenario, might have interesting implications in case of non universal flavour couplings. More stringent limits are in within reach of NA62 for the invisible decay scenario: $K^+ \to \pi^+ A'$ with $A' \to \chi\bar{\chi}$. The new physics signature $K^+ \to \pi^+\ missing-energy$ can only be realised in the SM via the strongly suppressed FCNC process $K^+ \to \pi^+\nu\bar{\nu}$, and therefore even a relatively modest excess of events may lead to a clear discovery. NA62 experiment, currently taking data at CERN SPS, aiming at collecting 100 events of $K^+ \to \pi^+\nu\bar{\nu}$ with very low background contamination, has the possibility to improve strongly the current bound coming from E787-E949 (See Fig.5). Invisibly decaying DP will manifest itself as an anomaly in the Missing Mass distribution of the background in the NA62 data sample.

4.2 The $K^+ \to \ell^+\nu A'$ decay

Rare radiative leptonic decays might reveal signatures of particles that couple predominantly to leptons, avoiding part of the constraints shown in Fig 1. In the $K^+ \to \mu^+\nu e^+e^-$ and $K^+ \to \mu^+\nu\mu^+\mu^-$ processes, kaons are a source of tagged muons, and the A' can be produced even in absence of coupling to quarks. Models have been proposed with light scalars only coupling to muons[12] thus circumventing almost all present bounds. In this scenario an interesting role might be played by $K^+ \to \mu^+\nu\mu^+\mu^-$ final state, in which both production, A'-strahlung from a μ leg, and the A' decay, $A' \to \mu^+\mu^-$, are allowed even if DP only couples to muons. NA48/2 collaboration is currently exploring the di-electron decay mode [13].

5 The PADME experiment at LNF

The PADME experiment [14][15] at the DAΦNE LINAC of the INFN Laboratori Nazionali di Frascati (LNF), has been designed to search for the dark photon by using a positron beam hitting a low Z target. The A' can be observed by searching for a narrow unexpected peak in the spectrum of the missing mass measured in single photon final states. The measurement requires the precise determination of the 4-momentum of the recoil photon and of the impinging positron.

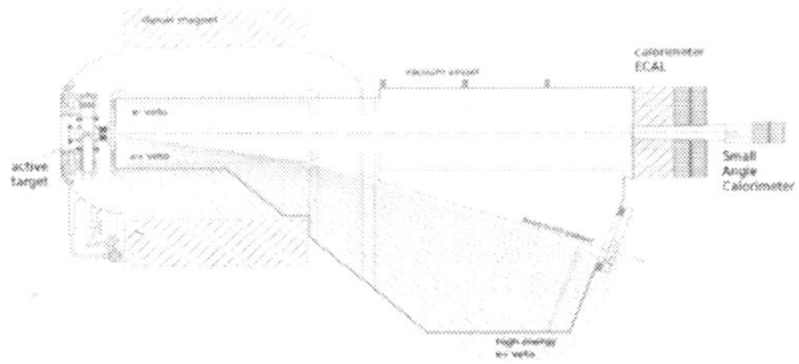

Figure 3. Layout of the PADME detector.

The collaboration aims to complete the construction of the experiment in early 2018 and to collect ~10^{13} positrons on target by the end of 2018. The data taking is scheduled to start by April 2018. PADME experiment at LNF-INFN, shown in Fig.3 will use 550 MeV positron impinging on 100μm thick active target made of polycrystalline diamond. The recoil photons from $e^+e^- \to \gamma A'$ annihilation process will be detected by a quasi-cylindrical BGO crystal calorimeter made of 616 20x20x230 mm^3 crystals, located 3.3 m downstream the target, providing very precise energy and position measurements. The non-interacted beam positrons, will be deflected outside the acceptance of the calorimeter by a 1 m long dipole magnet. Three different sets of plastic scintillator bars, two of which located inside the magnet gap, will serve to detect the charged particles, thus providing an efficient veto for for the positron bremsstrahlung background. In addition, a very fast small angle calorimeter will be placed along the beam axis, to help with the suppression of the three photon annihilation background. The entire setup will be located in vacuum to minimise the possible beam-residual gas interactions.

5.1 The PADME detector status

The construction of the PADME experiment is expected to be completed in early 2018 and the preparation of the different detectors proceeds according to the schedule. In Fig.4 some pictures of the prototypes used to asses the detector performance are collected.

The 100 μm active diamond target with graphite strips readout[16] was one of the most challenging parts of the PADME experiment. After several test on the BTF beam line in Frascati [17] the detector construction has been successfully completed and the in vacuum support system tested.

The PADME dipole magnet, an MBP-S dipole of the SPS transfer lines, providing 0.55 T magnetic field over a large (200x500x1000 mm^3) gap, has been characterised at Laboratori Nazionali di Frascati and the mechanical support for the integration is in construction.

Figure 4. Prototypes of the PADME detector.

The prototypes of the electromagnetic calorimeters and charged veto detector systems have been finalised and tested using electrons and positrons from the LNF Beam Test Facility [18]. Encouraging performance, in agreement with previous L3 results obtained with BGO crystals, have been achieved with the calorimeter prototype[19]. All the BGO crystals and plastic scintillator bars needed for the full detector construction are currently under test to be ready for final assembly.

The PADME experiment readout is based on the CAEN V1742 digitiser board (1-5 Gs/s and 12bit ADC) with optical link readout[20]. About 1000 channels system including the readout of the electromagnetic calorimeter, SAC, and charged particle veto sub-detectors is already available and under test using the prototype data acquisition software. A prototypes of the Level 0 trigger distribution system has been successfully tested achieving the required synchronisation performance. The running of the experiment will start in April 2018 with the commissioning of the PADME beam line.

5.2 The PADME experiment sensitivity

The PADME experiment sensitivity estimate is based on a GEANT4 simulation extrapolated to 10^{13} positrons on target (POT). This number of particles can be obtained by running PADME for 2 years at 50% efficiency with 5000 e^+ per 40 ns bunch at a repetition rate of 49 Hz. Recent upgrade of the DAΦNE LINAC gun allows to extend the bunch length up to 200 ns, when the LINAC is operated in dedicated mode. In this conditions a data sample of 1×10^{13} POT can be obtained in 6 month while in 2 years a sample of 4×10^{13} POT is within reach. The bounds obtained for A' decaying into invisible particles are shown in Fig. 5. The red line shows the expected bound for 1×10^{13} POT while the red dashed the one for 4×10^{13} POT. The favoured $(g\text{-}2)_\mu$ region can be explored in a model independent way (the only hypothesis is that A' couples to leptons) up to masses of 23.7 MeV [15] . Exploiting the capability of detecting photon and charged particles, the PADME experiment will be able to explore other type of dark sectors mediators such as Axion Like Particles, protophobic X boson[21], and Dark Higgs.

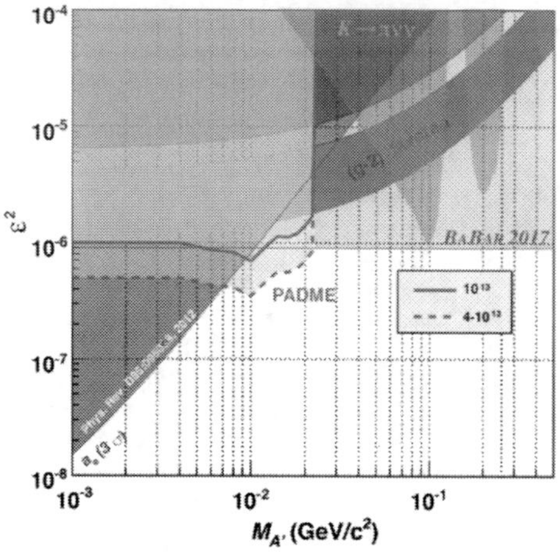

Figure 5. PADME estimated sensitivity for A' decaying into invisible particles for different values of the integrated luminosity.

6 Conclusions

Dark sectors models offer a simple and phenomenological rich explanation for different Standard Model anomalies, together with a natural solution to the dark matter problem. The existence of new mediators, or dark matter particles, can be probed at low energy accelerators using existing precision experiments or with new dedicated detectors able to explore the intensity frontier. In the next 2-5 years, several new experimental results are expected to enlighten our knowledge on the nature of dark matter.

References

[1] M. Battaglieri *et al.*, arXiv:1707.04591 [hep-ph].

[2] M. Raggi and V. Kozhuharov, Riv. Nuovo Cim. **38**, no. 10, 449 (2015).

[3] J. R. Batley *et al.* [NA48/2 Collaboration], Phys. Lett. B **746**, 178 (2015)

[4] H. Merkel *et al.*, Phys. Rev. Lett. **112**, no. 22, 221802 (2014)

[5] J. P. Lees *et al.* [BaBar Collaboration], Phys. Rev. Lett. **113**, no. 20, 201801 (2014)

[6] J. D. Bjorken *et al.*, Phys. Rev. D **38**, 3375 (1988).

[7] J. P. Lees *et al.* [BaBar Collaboration], Phys. Rev. Lett. **119**, no. 13, 131804 (2017)

[8] D. Banerjee *et al.* [NA64 Collaboration], Phys. Rev. Lett. **118**, no. 1, 011802 (2017)

[9] E. Cortina Gil *et al.* [NA62 Collaboration], JINST **12**, no. 05, P05025 (2017)

[10] T. Yamanaka [KOTO Collaboration], PTEP **2012**, 02B006 (2012). doi:10.1093/ptep/pts057

[11] J. R. Batley *et al.* [NA48/2 Collaboration], Phys. Lett. B **769**, 67 (2017)

[12] C. Y. Chen, M. Pospelov and Y. M. Zhong, Phys. Rev. D **95**, no. 11, 115005 (2017)

[13] R. Marchevski *et al.* [NA48/2 Collaboration], PoS ICHEP **2016**, 636 (2016).

[14] M. Raggi and V. Kozhuharov, Adv. High Energy Phys. **2014**, 959802 (2014)

[15] M. Raggi, V. Kozhuharov and P. Valente, EPJ Web Conf. **96**, 01025 (2015)

[16] M. De Feudis *et al.*, Nuovo Cim. C **39**, no. 1, 254 (2016). doi:10.1393/ncc/i2016-16254-5

[17] R. Assiro *et al.*, arXiv:1709.07081 [physics.ins-det].

[18] P. Valente *et al.*, arXiv:1603.05651 [physics.acc-ph].

[19] M. Raggi *et al.*, Nucl. Instrum. Meth. A **862**, 31 (2017)

[20] E. Leonardi, M. Raggi and P. Valente, J. Phys. Conf. Ser. **898**, no. 3, 032024 (2017).

[21] J. L. Feng, B. Fornal, I. Galon, S. Gardner, J. Smolinsky, T. M. P. Tait and P. Tanedo, Phys. Rev. D **95**, no. 3, 035017 (2017)

Hadronic vacuum polarization and $e^+e^- \to \mu^+\mu^-$ cross section

Reanalysis with new precise data for σ_h with 4π final states included

Vladimir Sauli

[1] Institute of Nuclear Physics Rez near Prague, CAS, Czech Republic

Abstract. The interference effect between leptonic radiative corrections and hadronic polarization functions is calculated via optical theorem for μ−pair production in vicinity of narrow resonances. Within seven most dominant exclusive channels of the production cross section $\sigma_h(e^+e^- \to hadrons)$ one achieves high acuracy which is necessary for the comparison with experiments. The result is compared with KLOE and KLOE2 experiments for $\mu^-\mu^+$ and $\mu^-\mu^+\gamma$ productions at ϕ and ω/ρ meson energy.

1 Introduction

Comparisons between theory and experiment are used to test Standard Theory for decades. For an accurate measurement the studies require consistent account of leptonic as well as hadronic virtual corrections. The hadronic contribution to photon vacuum polarization function plays particularly important role, since it is the main source of uncertainties in theoretical calculation of muon anomalous magnetic moment a_μ. The last precise measurement of a_μ, together with the last decades data for electro-hadron production, leads to an evidence of tension between Standard Theory and experiments [1, 2]. Similar confrontation of the theoretical technique with the experimental accuracy is offered by long time known [3, 4] interference effect between leptonic and hadronic vacuum polarization functions in close vicinity of narrow resonances: ω and ϕ as well as heavier quarkonia the $J/\Psi, \Psi$ and Υ's. There, the hadronic vacuum polarization is enhanced several orders of magnitude above remaining hadronic background. In practice, the effect is explored in the so called B-factories like BABAR, BELLE or BES or more earlier in Frascati with detectors and accelerator tuned at the ϕ meson mass. Most recent precise measurements of muon production $e^+e^- \to \mu^+\mu^-$ by SND, CMD-2,3 and the KLOE/KLOE2 detectors allow to another test of the Standard theoretical assumptions. With incomes of many new precise data for hadroproduction cross section σ_h we reanalyzed the calculation for [5] within the new data used. Having extracted hadron polarization function, we provide independent comparison of theory and recent KLOE2 [6] measurement of the fine structure coupling constant.

2 $\sigma_{\mu\mu}$ for KLOE 2004

In the first part I will present the comparison between calculated cross section $\sigma(e^+e^- \to \mu^+\mu^-)$ and the high precision measurement obtained by KLOE detector [5]. The method of extraction of hadronic polarization function Π_h will be briefly mentioned. As we shall see the recent knowledge of σ_h allows theoretical determination of $\sigma_{\mu\mu}$ with theoretical error comparable to KLOE σ_{stat} for $\sigma_{\mu\mu}$. Thus the error of theory due to the uncertainty in σ_h becomes 10 times smaller then the experimental error of 2004 KLOE measurement. The result presented here represents a precise prediction for a future more accurate experiments.

The details of theory of $e^+e^- \to \mu^+\mu^-$ for KLOE detector can be found in [7]. The integral cross section formula is proportional to the square of the fine structure constant $\alpha(s)$, which reads

$$\alpha(s) = \frac{\alpha}{1 - \Pi(s)}, \tag{1}$$

with $\alpha = \alpha(0) = 1/137.0359991390$ and the polarization function $\Pi(s) = \Pi_l(s) + \Pi_h(s)$ is made of the leptonic l and the hadronic h part.

Purely QED contributions, represented by Π_l, are well known from perturbation theory and relevant terms are listed in [7], however Π_h which is partially known in the spacelike domain [8] is not directly calculable from the equation of motions in the Minkowski space. Instead of this, with ceratin assumptions, it is obtained via Unitarity relation within the help of of many other experiments $e^+e^- \to hadrons^*$, where the stars $*$ means that the final state photons should be included as well, while in opposite, the initial ones should be subtracted. In fact, the evaluation of Π_h relies on numerical evaluation of the following singular integral [9, 10]:

$$\Pi_h(s) = \frac{s}{4\pi^2\alpha} \int_{m_\pi^2}^{\infty} d\omega \frac{\sigma_h(\omega)\left[\frac{\alpha}{\alpha(\omega)}\right]^2}{\omega - s + i\epsilon}. \tag{2}$$

Recall, the use of experimental data straightforwardly would lead to a large numerical noise and lost of required accuracy. I found the usual way (for the method of clusters see [11, 12]) produces cusps and spikes. They cause the error stemming from such interpolation, averaging and integration procedure is hard to estimate due to the presence of Principal value integration (recall for clarity, such numerical problem is avoided in the case of a_μ evaluation as the integral is regular).

Therefore, instead of direct use of experimental data, the fit (including errors) is made for each combinations of measurements. I used the fine selection method of the data, which is based on the following simple criterion:

$$\sigma^2_{syst} + \sigma^2_{stat} < \epsilon^2_h, \tag{3}$$

where on the left side there is sum of statistical and systematic error of the data points and on the right side a suited choice of error function evaluated at data points is employed. The data satisfying the inequality (3) are used to establish the fits, wherein the experimental errors are replaced by the inflated error function (IEF) ϵ_h. While all data points not satisfying the rule (3) are not used for making a fits.

We stress here, that with the choice of a given IEF ϵ_h, the condition (3) does not automatically ensure the existence of a good global fit satisfying $\chi^2 < 1$. It actually happens in cases when one combine various experiments with the error underestimated by the experimental group, which usually happens when the systematic error is not completely known (not named explicitly, the older threshold data extracted by ISR method are a typical example). Impossibility of minimizing χ^2 such that $\chi^2 < 1$ (note $\chi^2 \simeq 1$ is valid only for non-inflated error) indicates the badness, or rather say the incompatibility of the data. In this case we are either force to inflate further the IEF (by changing the prescription for ϵ_h) or we discard the problematic data set from the fine selection.

There is an obvious price to pay and one must to perform cross section fits explicitly. To this point, well established interpolating fits to the existing experimental data for σ_h at each exclusive channel have been found during several last years. Using these, the large number of generated quasidata points makes systematic error from principal value integration in (2) immaterial and the error of theory for $\sigma_{\mu\mu}$ is almost solely due to the propagation of "inflated" error ϵ_h. More explicitly, the systematic error due to the integration procedure has been minimized with the relative precision smaller then 0.005/40 for the muon production cross section (the absolute value is approximately hundred times smaller then the experimental error [5]).

The functional form of the error function ϵ_h can be taken arbitrary, however the choice, which does not reduce n.d.f. too much and is simultaneously simple enough, is preferred. For this purpose we have used the following IEF

$$\epsilon_h(s) = c_l\sqrt{\sigma_h(s)}\ \ , \epsilon_h(s) = c_s\sigma_h(s)\,, \tag{4}$$

where the left equation in (4) is used for $\sigma_h(s)$ larger the $1nb$, while the right one is used for small $\sigma_h(s) < 1nb$. It is enough to take constant values for parameters $c_{l,s}$, noting that values $c_l = 0.8nb^{1/2}$, $c_s = 1/3$ in Eq. (4) were most recently used in almost all the hadronic channels.

Recent measurements are taken completely into account and the complete list of selected experimental data and their fits will be presented in an updated version of [7]. Without specifications of channels, the choice (4) fully accept the last experiments, eg. CMD-3, BES-III, KLOE, also most CMD-2 measurements as well as large-s BABAR data are fully taken, it cuts partially some data from SND, CMD-2, it also cuts on resonance data from BABAR, while in practice we can freely discard the data from and old experiments completely (CMD,DM,NA7,OLYA,TOF).

In order to evaluate Π_h, the main exclusive channels:$\eta\gamma$,$\pi\gamma$, $\pi\pi$, K^+K^-, K_LK_S and $\pi\pi\pi$ as well as 4π have been considered. Contributions from final states, like $KK\pi\pi$ and the one with higher multiplicity were estimated and neglected as their amounts is fairly below the systematic error of $\sigma_{\mu\mu}$ cross section. The effect of well established vector charmonia and bottomonia was found to be nonnegligible and has been included through σ_h by using their BW forms with PDG experimentally determined values.

The result is shown in the Fig. 1, where the old analyses (set of green lines) is compared with the new one (violet line with crosses). The new analysis includes in addition the data from KLOE, BES-III and BABAR for KK channels, which have no practical effect to the final curve. However, more importantly, it newly includes $\eta\gamma$, $\pi\gamma$ and 4π channels of σ_h, which have been neglected in the previous analyses of ϕ- meson study. The green solid line states for the old result shows the effect of neglecting the three mentioned channels. The band between $\pm\epsilon$ (dashed and dot dashed line) reflects the propagation of inflated error ϵ_h into the muon pair cross section and was obtained with $c_l = 1nb^{1/2}$, $c_s = 1$ in the previous analyses. The error band for a newest analyses is even more tight and not shown in the figure. The KLOE data points are represented by triangles, noting the statistical deviations roughly correspond with the size of the triangle. For interest, we also show the systematic deviation which stems from one order decrement of the number of integration points in Eq. (2), these points are labeled by red crosses with no line.

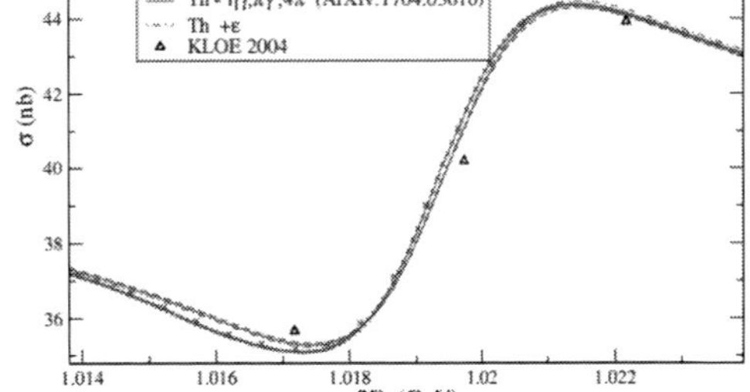

Figure 1. Muon pair cross section. The comparison between theory and experiment as described in text.

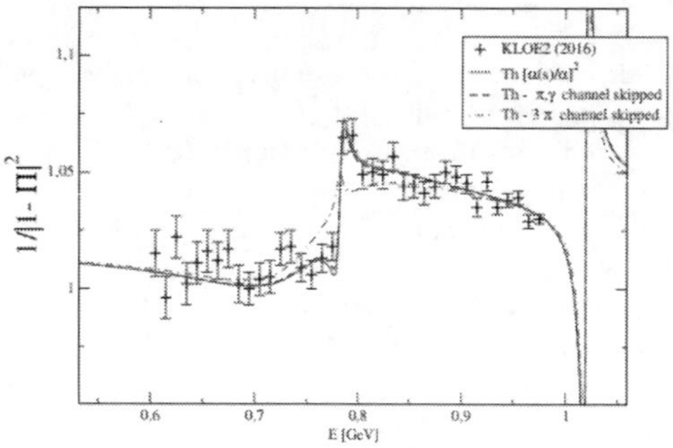

Figure 2. Square of fine structure constant. To see the portion of 3π and π,γ contributions we also show the results with these channels subtracted from σ_h.

To conclude the first part of the talk, the observed ϕ-meson interference effect in $\mu\mu$ spectrum is roughly reproduced by the Standard Theory dispersion relation. Remind that the detector measurement provided three points with very small statistical error ($\sigma_{stat} = 0.1nb$), unhappily the total error was governed by systematic error due to the luminosity and detection uncertainties ($\delta_{syst} = 1.2\%$). Small $1.7\sigma_{tot}$ difference from SM prediction does not represent large tension between theory and experiment. Lowering the systematic error would be not only experimental challenge for precise experimental facilities like KLOE2, CMD3, but also for the Standard Theory.

3 α_{QED} at KLOE2

As a bonus we get also similar interference effect in the ω energy region. This, somehow smaller, $3nb$ sized zig-zag structure has been measured by radiative return method only very recently by the KLOE2 collaboration [6]. In this case the experiment is in complete agreement with the theory, noting the relative errors are much larger in the ρ/ω region. The comparison with KLOE2 experiment and the way the ρ/ω peak is pronounced in the QED running coupling is shown in Fig. 2. More precise comparison remains challenging for both the Standard Theory and new generation experiments as well.

References

[1] G. Bennet et al., Phys. Rev. Lett. 89, 101804 (2002).
[2] K. Olive et al., Rewiev of Particle Physics, The Muon Anomalous Magnetic Moment by A. Hoecker and B. Marciano, Chin. Phys. C 38, 090001 (2014).
[3] J. E. Augustin, et al.; Phys. Rev. Lett. 30, 462 (1973).
[4] F. A. Berends and G.J. Komen, Nuc. Phys. B 115, 114-140 (1976).
[5] F. Ambrosino, et al.; Phys. Lett. B 608, 199-205 (2005).
[6] KLOE2 Collab. and F . Jegerlehner, Phys. Lett. B 767, 485–492 (2017).
[7] V. Sauli, ArXiv:1704.01887.
[8] T. Goecke, C.S. Fischer, R. Williams, Phys. Lett. B704, 211 (2011).
[9] N. Cabibbo and R. Gatto, Phys. Rev. 224, N.5,1577-1595 (1961).
[10] S. Eidelman, F. Jegerlehner, Z. Phys. C67, 585-602 (1995).
[11] M. Davier, A. Hoecker, B. Malaescu, Z. Zhang, Eur. Phys. J. C71, 1515 (2011).
[12] M. Benayoun et al.; Hadronic contributions to the muon anomalous magnetic moment Workshop. $(g-2)\mu$: Quo vadis? Workshop. Mini proceedings, arXiv:1407.4021.

The muon $g-2$ for low-mass pseudoscalar Higgs in the general 2HDM*

Adriano Cherchiglia[1], Dominik Stöckinger[2], and Hyejung Stöckinger-Kim[2]

[1] Universidade Federal do ABC - Centro de Ciências Naturais e Humanas, Santo André - Brazil
[2] Institut für Kern- und Teilchenphysik, TU Dresden, 01069 Dresden, Germany

Abstract. The two-Higgs doublet model is a simple and attractive extension of the Standard Model. It provides a possibility to explain the large deviation between theory and experiment in the muon $g-2$ in an interesting parameter region: light pseudoscalar Higgs A, large Yukawa coupling to τ-leptons, and general, non-type II Yukawa couplings are preferred. This parameter region is explored, experimental limits on the relevant Yukawa couplings are obtained, and the maximum possible contributions to the muon $g-2$ are discussed.

1 Introduction

So far the experiments at the LHC have not identified clear evidence for physics beyond the Standard Model (SM). This is regularly taken as an indication that new physics particles, if they exist, must be rather heavy. The persisting discrepancy between measurement and SM theory in the muon anomalous magnetic moment a_μ, however, cannot be explained by arbitrarily heavy new particles. Currently it is given by

$$a_\mu^{\text{Exp-SM}} = \begin{cases} (26.8 \pm 7.6) \times 10^{-10}[1], \\ (28.1 \pm 7.3) \times 10^{-10}[2], \\ (31.3 \pm 7.7) \times 10^{-10}[3]. \end{cases} \quad (1)$$

This situation has motivated extensive studies of a_μ in cases with high-mass new physics. E.g. Ref. [4] has studied the largest possible a_μ contributions in the MSSM and identified the largest possible SUSY mass scale for which the current a_μ deviation could be fully explained. Ref. [5] considered precision computations in the MSSM, which are particularly important in cases with a partially heavy spectrum of supersymmetric particles, and all these results were implemented in the program GM2Calc [6].

In the remainder of these proceedings we focus on an alternative idea, to consider models which can provide significant contributions to a_μ only for very light new particles, which however are nevertheless still viable. Specifically we focus on the two-Higgs doublet model (2HDM), the model realizing a non-minimal scalar sector breaking electroweak symmetry in the simplest possible way. It has been known for a long time that it can give rise to significant contributions to a_μ only if one of the extra Higgs bosons, specifically the pseudoscalar A-boson, is fairly light, lighter than the observed SM-like Higgs boson with mass of 125 GeV. Recently it has been stressed that the parameter space relevant for this is still viable, see particularly Refs. [7, 8].

In Ref. [9] the 2HDM prediction for a_μ has been derived at the full two-loop level (which corresponds to leading order), and in reference [10], experimental constraints are exploited to obtain the possible allowed ranges of all contributions and thus identify the parameter space of the general 2HDM, which can explain a_μ. Here we review these results.

2 Two-Higgs doublet model

We consider the general 2HDM, with only few restrictions on the parameters. We take the Higgs potential to be [11, 12]

$$\begin{aligned} V(\phi_1, \phi_2) =& m_{11}^2 \phi_1^\dagger \phi_1 + m_{22}^2 \phi_2^\dagger \phi_2 - m_{12}^2 \left(\phi_1^\dagger \phi_2 + \phi_2^\dagger \phi_1\right) \\ &+ \frac{\lambda_1}{2} \left(\phi_1^\dagger \phi_1\right)^2 + \frac{\lambda_2}{2} \left(\phi_2^\dagger \phi_2\right)^2 + \lambda_3 \phi_1^\dagger \phi_1 \phi_2^\dagger \phi_2 \\ &+ \lambda_4 \phi_1^\dagger \phi_2 \phi_2^\dagger \phi_1 + \frac{\lambda_5}{2} \left[\left(\phi_1^\dagger \phi_2\right)^2 + \left(\phi_2^\dagger \phi_1\right)^2\right], \end{aligned} \quad (2)$$

where we only neglect the so-called $\lambda_{6,7}$ parameters, which we checked to have negligible influence on our later results.

The potential parameters determine the physical masses $M_{h,H,A,H^\pm}$, where h corresponds to the SM-like Higgs boson, H is assumed to be heavier, A is a CP-odd scalar, and $H^\pm$ is a charged Higgs boson. Further, the potential determines the ratio between the two vacuum expectation values, $\tan\beta$, and the mixing angle $\eta \equiv \frac{\pi}{2} - (\beta - \alpha)$, which LHC-data forces to be small.

The Yukawa couplings are assumed to be "aligned" in the sense of Ref. [13]. I.e. we do not assume one of the usual type I,II,X,Y models but only require that the couplings of both Higgs doublets to fermions are proportional to each other, to prevent violations of FCNC bounds. The

*Presented at Workshop on Flavour Changing and Conserving Processes (FCCP2017), September 2017

Yukawa couplings between the scalar S and fermion f are then determined by the factors Y_f^S given for small η by

$$Y_f^h = 1 + \eta\zeta_f,\ Y_f^H = -\zeta_f + \eta,$$
$$Y_f^A = -\Theta_f^A\zeta_f,\ \Theta_{d,l}^A = 1,\ \Theta_u^A = -1. \tag{3}$$

The units are such that the respective SM-coupling would correspond to $Y_f^h = 1$. The couplings depend on the fermion type, $f \in \{u, d, l\}$.

In contrast to the usual type I,II,X,Y models, the Yukawa couplings are determined by free parameters $\zeta_{u,d,l}$ for each fermion type, but they are independent of $\tan\beta$. $\tan\beta$ is a parameter whose role is confined to the Higgs sector.

3 Muon $g-2$ in the 2HDM at two-loop order

In the 2HDM the muon $g-2$ receives contributions from one- and two-loop diagrams. The one-loop diagrams, however, are suppressed by two additional powers of the muon mass and thus very small (except at Higgs masses below around 20 GeV). The parametrically leading diagrams are of two-loop order. In this sense Ref. [9] has completed the full leading-order computation of a_μ in the 2HDM; see Ref. [14] for a previous advanced calculation and references to earlier works.

The full result, including one- and two-loop contributions can be written as

$$a_\mu^{\text{2HDM},2} = a_\mu^{\text{2HDM},1} + a_\mu^{\text{B}} + a_\mu^{\text{F}} + a_\mu^{\Delta r\text{-shift}}. \tag{4}$$

We now briefly discuss all contributions.

- $a_\mu^{\text{2HDM},1}$ are the 2HDM one-loop contributions (which are not contained in the SM prediction). They arise from one-loop diagrams with a single Higgs exchange and are suppressed by two powers of the muon mass divided by the Higgs mass. We will later consider the case of light CP-odd scalar mass M_A. In this case the one-loop contributions are negative and make it more difficult to explain the deviation (1).
- a_μ^{B} are the bosonic two-loop contributions, i.e. the contributions arising from diagrams without closed fermion loop. They contain the diagrams with three vertices on the muon line, computed for the first time in Ref. [9], but also Barr-Zee-type diagrams with charged Higgs loop, computed in Ref. [14] and Ref. [9] using different methods. They turn out to be rather small and dominated by diagrams which involve the triple Higgs coupling $C_{HH^+H^-}$ between H and two charged Higgs bosons.
- a_μ^{F} are the two-loop contributions with closed fermion loop. These are dominated by the 3rd generation fermions, which have the largest Yukawa couplings. Among the common type I,II,X,Y models, only the type X (or lepton-specific) model is capable of providing significant contributions [7]; in this case the τ-loop contributions are proportional to $\zeta_l = -\tan\beta$ and dominate by far, but top- and bottom-loop contributions are suppressed as $1/\tan\beta$. In the more general aligned model, the top-loop can also contribute significantly (see later and Ref. [8]).
- $a_\mu^{\Delta r\text{-shift}}$ is a shift required because the SM contribution is parametrized in terms of the muon decay constant which receives additional 2HDM contributions. This contribution is below 2×10^{-12} and will hence be neglected in the following.

Useful numerical approximations for the contributions are [10], using $\hat{x}_S \equiv M_S/100\,\text{GeV}$ and setting $M_{H^\pm} = M_H$.

$$a_\mu^{\text{2HDM},1} \simeq \left(\frac{\zeta_l}{100}\right)^2 \left\{\frac{-3-0.5\ln(\hat{x}_A)}{\hat{x}_A^2}\right\} \times 10^{-10}, \tag{5}$$

$$a_\mu^{\text{F}} \simeq \left[\left(\frac{\zeta_l}{100}\right)^2 \left\{\frac{8+4\hat{x}_A^2+2\ln(\hat{x}_A)}{\hat{x}_A^2}\right\} + \left(\frac{-\zeta_l\zeta_u}{100}\right)\left\{54-14\ln(\hat{x}_A)-15\ln(\hat{x}_H)\right\}\right] \times 10^{-10} \tag{6}$$

$$|a_\mu^{\text{B}}| \simeq \rho|C_{HH^+H^-}/\text{GeV}|\,|\zeta_l| \times 10^{-15} \tag{7}$$

The terms in the second line are mainly due to two-loop diagrams with τ-loop and top-loop, respectively, and in the result for a_μ^{B}, ρ is typically in the range $1\ldots6$ for $M_H, M_{H^\pm} = 150\ldots300$ GeV and the triple Higgs coupling $C_{HH^+H^-}$ is limited by perturbativity to values up to around 1000 GeV.

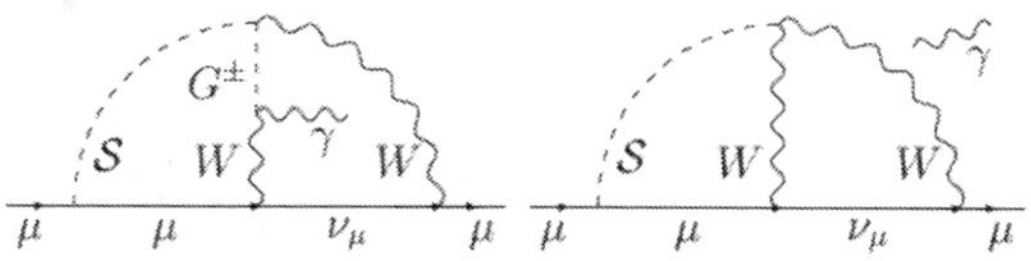

Figure 1. Bosonic two-loop diagrams with three vertices on the fermion line. From [9].

4 Muon $g-2$ contributions constrained by experiment

In this section we provide an overview of the three single largest contributions to a_μ in the 2HDM and describe how they are constrained by current experiments.

- τ-loop contribution: the most important contribution is given by the Barr-Zee-like two-loop diagram with a τ-loop generating a γ–γ–A interaction. The diagram has a behaviour proportional to ζ_l^2/M_A^2, see the approximation (6). Hence it is of interest to determine in general the constraints on the lepton Yukawa parameter ζ_l.

  Such constraints have already been studied in the literature, particularly in Refs. [7, 15], for the type X model, where $\zeta_l = -\tan\beta$. The constraints found there arise from τ-decays, which can be influenced by additional 2HDM-contributions, and by universality of $Z \to ll$ decays. Fig. 2(left) shows a scatter plot of allowed parameter points in the 2HDM in the M_A–ζ_l plane, for fixed $M_H = M_{H^\pm} = 200$ GeV and $\zeta_u = 0$, where consistency with τ-decays and $Z \to ll$ was checked as described in

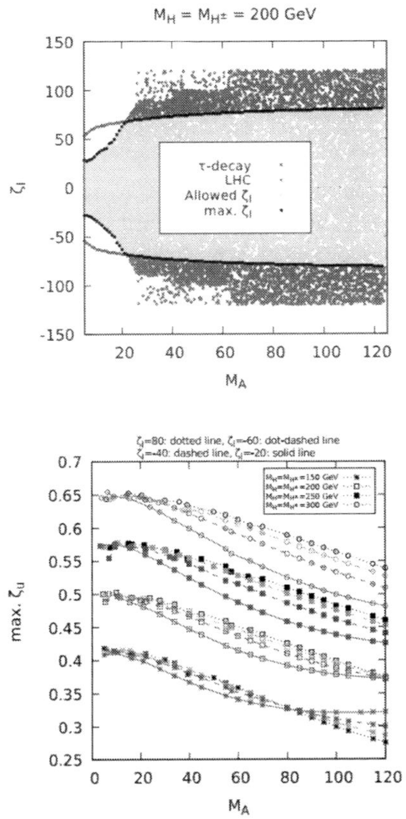

Figure 2. *Top:* Scatter plot of the possible parameter points in the 2HDM in the M_A–ζ_l plane, for fixed $M_H = M_{H^\pm} = 200$ GeV and $\zeta_u = 0$. Yellow points correspond to the points allowed both by collider constraints (labelled by "LHC") and τ-physics constraints, red and blue corresponds to the points allowed by one of these constraints, ignoring the other one. The maximum and minumum boundary of ζ_l is depicted as black points. The result for the allowed ζ_l values corresponds to the maximum allowed values of $\tan\beta$ in the 2HDM type X, studied in Ref. [15]. *Bottom:* Maximum values of the top-Yukawa parameter ζ_u allowed by B-physics (see text), as a function of M_A.

Ref. [15] and consistency with Higgs measurements was checked using 2HDMC and HiggsBounds [16, 17]. The scatter plot shows that the τ-decay and $Z \to ll$ constraints carry over from the type X model to the more general model without change, and that there is a small region at very small M_A, which turns out to be excluded by LEP measurements. We refer to Ref. [10] for further details.

The exclusion bound depends slightly on the chosen values of M_H and $M_{H^\pm}$, but generally the maximum allowed value of $|\zeta_l|$ varies between around 40 (for smaller Higgs masses) and 100 (for larger Higgs masses).

- top-loop contribution: the next important contribution is given by the Barr-Zee like two-loop diagram with a top-loop. It depends on $\zeta_u\zeta_l$ and on M_A and M_H, and hence the constraints on ζ_u are now crucial.

  ζ_u is constrained by B-physics and by direct Higgs searches. It is clear that the corresponding Feynman diagrams depend on the same parameters ζ_u, ζ_l, M_A as the desired a_μ-contribution. Using the results of Ref. [18] we found that the most constraining B-physics observables are $b \to s\gamma$ and $B_s \to \mu\mu$. Implementing them and checking consistency with data, we obtain the plot of Fig. 2(right), showing the maximum allowed values of ζ_u, as a function of M_A, for various choices of ζ_l and the other Higgs masses. The maximum allowed values are between $\zeta_u < 0.3$ and $\zeta_u < 0.7$.

  The constraints from LHC Higgs searches place additional limits on ζ_u, which have a more intricate parameter dependence. We provide an example of the LHC constraints for $M_A = 80$ GeV and $\zeta_l = -40$ and $M_H = M_{H^\pm} \in [200, 300]$ GeV in Fig. 3(left). The colours show parameter points in the ζ_u–$C_{HH^+H^-}$-plane which successively fulfil bounds from S, T, U parameters, HiggsBounds, HiggsSignals, and tree-level stability, unitarity and perturbativity (checked by 2HDMC [17]). The decisive constraints are perturbativity, which provides an upper limit on the triple Higgs coupling $C_{HH^+H^-}$, and the LHC constraint on the $H \to \tau\tau$ decay. The latter constraint limits ζ_u as a function of the triple Higgs coupling, because Higgs production is governed by ζ_u, while the branching fraction for the decay to $\tau\tau$ can be suppressed by large triple Higgs coupling. In the example of Fig. 3(left), the LHC constraints on ζ_u are weaker than the ones from B-physics. This is true in a significant part of the parameter space, hence we will not discuss these LHC constraints further here and refer to Ref. [10] for details.

- bosonic contributions: in the promising parameter region with small M_A, the bosonic contributions are dominated by the Barr-Zee-like diagram with charged Higgs loop and scalar H exchange. It is proportional to the triple Higgs coupling, which appears in Fig. 3(left). Inserting the upper limit on the triple Higgs coupling we obtain a possible range for a_μ^{B} in agreement with the approximation Eq. (7).

Combining all the results on upper limits on ζ_l, ζ_u, and on the bosonic contributions, we can obtain the maximum possible 2HDM two-loop contribution to a_μ, as a function of M_A, for fixed values of the other Higgs masses. Fig. 3(right) shows the resulting maximum contribution for $M_H = M_{H^\pm} = 150, 200, 250, 300$ GeV. For each M_A and $M_H, M_{H^\pm}$, we first maximize ζ_l. This provides the lower (dotted) curves, which correspond to the τ-loop alone, equivalent to the result in the type X model. Then, for the given M_A and ζ_l we maximize ζ_u, given B-physics and LHC constraints. This provides the middle (dashed) curves, corresponding to the maximum full fermionic two-loop contribution in the general 2HDM. Adding the maximum bosonic two-loop contribution then yields the upper (continuous) curves.

The figure shows that the 2HDM is well capable of providing large contributions to a_μ, even larger than the deviation (1). However, the necessary parameter region is quite specific: the CP-odd Higgs boson mass must be smaller than M_Z, and ζ_u and ζ_l must both be close to their respective maximum values. In particular going beyond the type X model, i.e. having ζ_u of order 1 instead of order $1/|\zeta_l|$, increases the maximum a_μ almost by a factor two.

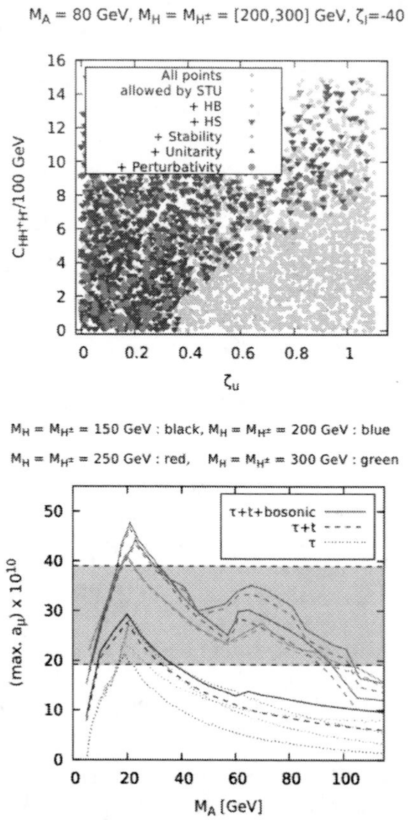

Figure 3. *top:* Allowed ranges of ζ_u and the triple Higgs coupling $C_{HH^+H^-}$, given certain constraints, see legend and text. The constraints are applied successively. *Bottom:* Maximum values of the two-loop contribution to $g-2$ in the general 2HDM for several values of $M_H = M_{H^\pm}$, given the constraints discussed in the text (from [10]). For each mass choice, the lowest line (dotted) corresponds to the maximum contribution from the τ-loop alone, equivalent to the type X model. The middle lines (dashed) correspond to adding the maximum possible top-loop contribution, and the top lines (continuous) to the overall maximum contribution, including bosonic two-loop contributions.

References

[1] M. Davier, A. Hoecker, B. Malaescu and Z. Zhang, Eur. Phys. J. C **71** (2011) 1515 [Erratum-ibid. C **72** (2012) 1874] [arXiv:1010.4180 [hep-ph]]; arXiv:1706.09436 [hep-ph].

[2] A. Keshavarzi, T. Teubner, talks at "g-2 theory initiative," June 2017, Fermilab, and at "PhiPsi 2017", June 2017, Mainz.

[3] F. Jegerlehner, arXiv:1705.00263 [hep-ph].

[4] M. Bach, J. h. Park, D. Stöckinger and H. Stöckinger-Kim, JHEP **1510** (2015) 026 doi:10.1007/JHEP10(2015)026 [arXiv:1504.05500 [hep-ph]].

[5] H. G. Fargnoli, C. Gnendiger, S. Paßehr, D. Stöckinger and H. Stöckinger-Kim, Phys. Lett. B **726** (2013) 717 [arXiv:1309.0980 [hep-ph]]. H. G. Fargnoli, C. Gnendiger, S. Paßehr, D. Stöckinger and H. Stöckinger-Kim, JHEP **1402** (2014) 070 [arXiv:1311.1775 [hep-ph]].

[6] P. Athron *et al.*, Eur. Phys. J. C **76** (2016) no.2, 62 doi:10.1140/epjc/s10052-015-3870-2 [arXiv:1510.08071 [hep-ph]].

[7] A. Broggio, E. J. Chun, M. Passera, K. M. Patel and S. K. Vempati, JHEP **1411** (2014) 058 [arXiv:1409.3199 [hep-ph]].

[8] T. Han, S. K. Kang and J. Sayre, JHEP **1602**, 097 (2016) [arXiv:1511.05162 [hep-ph]].

[9] A. Cherchiglia, P. Kneschke, D. Stöckinger and H. Stöckinger-Kim, JHEP **1701** (2017) 007 doi:10.1007/JHEP01(2017)007 [arXiv:1607.06292 [hep-ph]].

[10] A. Cherchiglia, D. Stöckinger and H. Stöckinger-Kim, arXiv:1711.11567 [hep-ph].

[11] G. C. Branco, P. M. Ferreira, L. Lavoura, M. N. Rebelo, M. Sher and J. P. Silva, Phys. Rept. **516**, 1 (2012) [arXiv:1106.0034 [hep-ph]].

[12] J. F. Gunion and H. E. Haber, Phys. Rev. D **67**, 075019 (2003) [hep-ph/0207010].

[13] A. Pich and P. Tuzon, Phys. Rev. D **80**, 091702 (2009) [arXiv:0908.1554 [hep-ph]].

[14] V. Ilisie, JHEP **1504** (2015) 077 [arXiv:1502.04199 [hep-ph]].

[15] E. J. Chun and J. Kim, JHEP **1607** (2016) 110 doi:10.1007/JHEP07(2016)110 [arXiv:1605.06298 [hep-ph]].

[16] P. Bechtle, O. Brein, S. Heinemeyer, O. Stål, T. Stefaniak, G. Weiglein and K. E. Williams, Eur. Phys. J. C **74** (2014) no.3, 2693 doi:10.1140/epjc/s10052-013-2693-2 [arXiv:1311.0055 [hep-ph]].

[17] D. Eriksson, J. Rathsman and O. Stal, Comput. Phys. Commun. **181**, 189 (2010) [arXiv:0902.0851 [hep-ph]]; Comput. Phys. Commun. **181**, 833 (2010).

[18] T. Enomoto and R. Watanabe, JHEP **1605** (2016) 002 doi:10.1007/JHEP05(2016)002 [arXiv:1511.05066 [hep-ph]].

Radiative Corrections for a Precision Determination of the Fine Structure Constant *

Luca Trentadue [1,2,a]

[1] Dipartimento di Matematica Fisica e Informatica, Università di Parma, 43123 Parma
[2] INFN, Sezione di Milano Bicocca, 20126 Milano, Italy

Abstract. We discuss the implications of a new proposed approach to determine a_μ^{HLO} and α_{QED} by using space-like kinematics.

1 Introduction

This talk is dedicated to the memory of Lev Nikolaevich Lipatov (1940-2017).

Why physicists carry such complex, lenghty and cumbersome calculations. Marcus Tullius Cicero states [1]: "Historia magistra vitae (est)". Let us, therefore, recall a few examples. Let us start with Tychonis (Tyco) Brahe who collected for decades a huge amount of sky observations and of astronomical data on the positions of the planets with no telescopes at naked eye. Johannes Kepler did analyse the Brahe's data and did publish in 1609 the book "Astronomia Nova" [2]. We all know what was the result of these observations and of their interpretation: the three laws of planetary motion. We still have, as well, the pages of Kepler's log book. These pages and the density of Kepler's mathematical work do speak by themselves of the amount of work that Kepler did to collect and process the data of Brahe's observations. Sir Isaac Newton was great and skilled not only in the production of ideas but also in many elaborate calculations. To come to this workshop Stefano Laporta will present us the result of his work that lasted several years. Although Stefano has accustomed us to the complexity and accuracy of his elaborate computational techniques, the results of this last one are really something extraordinary. We could go on to list many other examples of physicists who, for years, have engaged themselves in complex and challenging calculations crucial for understanding aspects of primary importance in physics. Considering Quantum Electrodynamics in a renowned article of more than sixty years ago [4], Freeman J. Dyson reached the conclusion that all the expansions in quantum electrodynamics power series would have been divergent after the renormalization of charge and mass. These divergences, even if they did not compromise the accuracy of the calculations, represented an important, unresolved, matter of principle. This work is, nowadays, well present in the scientific literature since, despite the fact that more than sixty years have passed still receives a considerable attention. It has received 22 citations in 2016 and 15 in 2017 to date, September 4 2017. However, despite the aforementioned matters of principle, we continue to carry out complex calculations in Quantum Electrodynamics. Silvan S. Schweber in his book : "QED and the men who made it: Dyson, Feynman, Schwinger and Tomonaga"[5] quotes an interview to Dyson in which he is stating:" I always felt was a miracle that electrons actually behaved the way the theory said". And later: "Truth to me means agreeing with the experiments,...For a theory to be true it has to describe accurately what really happens in the experiments". Moreover " The nature of a future theory is not a profitable subject for theoretical speculations. [A] future theory will be built first upon the results of future experiments" [5]. Despite the serious doubts of P. A. M. Dirac [6] about the whole renormalization procedure we continue, stubbornly, to evaluate α_{QED} with extremely accurate measurements [7]. Concerning the theoretical evaluations the determination of α_{QED} strongly depends on the accurate evaluation of the radiative corrections. Radiative corrections have started to play a prominent role only when the field has become assessed and mature and only when also the collection of experimental data has become abundant and accurate. These are the conditions to define a solid theoretical and experimental basis for further developments. The case we are considering in this talk requires a well founded approach both experimentally and theoretically.

*Work done in collaboration with G. Abbiendi, C. Carloni Calame, U. Marconi, C. Matteuzzi, G. Montagna, O. Nicrosini, M. Passera, F. Piccinini, R. Tenchini and G. Venanzoni.

[a] e-mail: luca.trentadue@unipr.it

2 a_μ^{HLO} calculation

The leading-order hadronic contribution to the muon g-2 is given by the well-known formula

$$a_\mu^{\rm HLO} = \frac{\alpha}{\pi^2}\int_0^\infty \frac{ds}{s}\, K(s)\, {\rm Im}\Pi_{\rm had}(s+i\epsilon), \qquad (1)$$

where $\Pi_{\rm had}(s)$ is the hadronic part of the photon vacuum polarization, $\epsilon > 0$, and

$$K(s) = \int_0^1 dx \frac{x^2(1-x)}{x^2+(1-x)(s/m_\mu^2)} \qquad (2)$$

is a positive kernel function with m_μ the muon mass.
As the total cross section for hadron production in low-energy e^+e^- annihilations is related to the imaginary part of $\Pi_{\rm had}(s)$ via the optical theorem, the dispersion integral in eq. (1) is computed integrating experimental time-like ($s > 0$) data up to a certain value of s. The high-energy tail of the integral can be calculated by using perturbative QCD.
Alternatively, if we exchange the x and s integrations in eq. (1) we obtain:

$$a_\mu^{\rm HLO} = \frac{\alpha}{\pi}\int_0^1 dx\,(x-1)\,\overline{\Pi}_{\rm had}[t(x)]\,, \qquad (3)$$

where $\overline{\Pi}_{\rm had}(t) = \Pi_{\rm had}(t) - \Pi_{\rm had}(0)$ and

$$t(x) = \frac{x^2 m_\mu^2}{x-1} < 0 \qquad (4)$$

is a space-like squared four-momentum. If we invert eq. (4), we get $x = (1-\beta)\,(t/2m_\mu^2)$, with $\beta = (1-4m_\mu^2/t)^{1/2}$, and from eq. (3) we obtain

$$a_\mu^{\rm HLO} = \frac{\alpha}{\pi}\int_{-\infty}^0 \overline{\Pi}_{\rm had}(t)\left(\frac{\beta-1}{\beta+1}\right)^2 \frac{dt}{t\beta}. \qquad (5)$$

Equation (5) has been used for lattice QCD calculations of $a_\mu^{\rm HLO}$ [8]; while the results are not yet competitive with those obtained with the dispersive approach via time-like data, their errors are expected to decrease significantly in the next few years [8].

The effective fine-structure constant at squared momentum transfer q^2 is defined above and is $\Delta\alpha(q^2) = -{\rm Re}\overline{\Pi}(q^2)$. The purely leptonic part, $\Delta\alpha_{\rm lep}(q^2)$, can be calculated order-by-order in perturbation theory – it is known up to three loops in QED and up to four loops in specific q^2 limits. As ${\rm Im}\overline{\Pi}(q^2) = 0$ for negative q^2, eq. (3) can be rewritten in the form

$$a_\mu^{\rm HLO} = \frac{\alpha}{\pi}\int_0^1 dx\,(1-x)\,\Delta\alpha_{\rm had}[t(x)]\,. \qquad (6)$$

We are going to proceed differently, we are proposing [9] to calculate eq. (6) by measurements of the effective electromagnetic coupling in the space-like region. The hadronic contribution to the running of α in the space-like region, $\Delta\alpha_{\rm had}(t)$, can be extracted by comparing Bhabha scattering data theory predictions. However the Bhabha cross section receives contributions from t- and s-channel photon exchange amplitudes.

Always within the space-like approach, an alternative possibility has been investigated [10]. It consists in using a fixed target $\mu \to e$ scattering process and to analyse the elastic muon distribution in the forward region. A new experiment has to be devised to measure the running of the fine-structure constant in the space-like region by scattering high-energy muons on atomic electrons of a low-Z target through the process $\mu e \to \mu e$. The differential cross section of this process, measured as a function of the squared momentum transfer $t = q^2 < 0$, provides direct sensitivity to the leading-order hadronic contribution to the muon anomaly $a_\mu^{\rm HLO}$. By using a muon beam of 150 GeV, with an average rate of $\sim 1.3 \times 10^7$ muon/s, currently available at the CERN North Area, a statistical uncertainty of $\sim 0.3\%$ can be achieved on $a_\mu^{\rm HLO}$ after two years of data taking. As the Bhabha process also this direct measurement of $a_\mu^{\rm HLO}$ will provide an independent determination, potentially competitive with the time-like dispersive approach, to consolidate, with a firmer interpretation of the measurements of the future muon g-2 experiments, the theoretical prediction for the muon g-2 in the Standard Model.
In this workshop there will be the talks of Umberto Marconi, Marina Marinkovic, Pierpaolo Mastrolia and Fulvio Piccinini examining varius implications of this idea.
A series of preliminary considerations on the detector have been discussed [10].
We consider a possible setup to measure the following observables:

- direction and momentum of the incident muon;
- directions of the outgoing electron and muon.

The CERN muon beam M2, used at 150 GeV, has the characteristics needed for such a measurement. The beam intensity appears to be adequate to provide the required event yield. The beam time structure allows to tag the incident muon while keeping low the background related to incoming particles (*e.g.* electrons). The electrons contamination is very small. The beam provides both positive and negative muons, which we plan to use.

The target consists of atomic electrons. To reach the required statistics, the target must consist of an adequate amount of material to give a sufficient number of electron scattering centres. The target has to be made of a low-Z material to minimize the impact of multiple scattering and the background due to bremsstrahlung and pair production processes. A promising idea for the detector is to use 20 layers of Be (or C) coupled to Si planes, spaced by intermediate air gaps, located at a relative distance of one meter from each other.
The arrangement provides both a distributed target with low-Z and the tracking system. As downstream particle identifiers we plan to use a calorimeter for the electrons and a muon system for the muons (a filter plus active planes). This particle identifier system is required to solve the muon-electron ambiguity for electron scattering angles around (2–3) mrad. The preliminary studies of such an ap-

paratus, performed by using GEANT4, indicate an angular resolution for the outgoing particles of ~ 0.02 mrad.

The detector acceptance must cover the region of the signal, with the electron emitted at extremely forward angles and high energies, as well as the normalization region, where the electron has much lower energy (around 1 GeV) and an emission angle of some tens of mrad. The boosted kinematics of the process allows the detector to cover almost 100% of the acceptance.

The incoming muons have to be tagged and their direction and momentum precisely measured. The angle of the scattered electron and muon are correlated[10]. This constraint is extremely important to select elastic scattering events, rejecting background events from radiative or inelastic processes and to minimize systematic effects in the determination of t. Note that for scattering angles of (2–3) mrad there can be an ambiguity between the outgoing electron and muon, as their angles and momenta are similar. To associate them correctly it is necessary to identify the two particles by means of downstream dedicated detectors (calorimeter and muon detectors).
In order to perform the planned measurement to the required precision in addition to a dedicated detector an extremely accurate evaluation of the QED scattering amplitude is necessary.
A series of systematic uncertainties should be taken into account.

Significant contributions of the hadronic vacuum polarization to the $\mu e \to \mu e$ differential cross section are essentially restricted to electron scattering angles below 10 mrad, corresponding to electron energies above 10 GeV. The net effect of these contributions is to increase the cross section by a few per mille: a precise determination of a_μ^{HLO} requires not only high statistics, but also a high systematic accuracy, as the final goal of the experiment is equivalent to a determination of the differential cross section with ~10 ppm systematic uncertainty at the peak of the integrand function.

Such an accuracy can be achieved if the efficiency is kept highly uniform over the entire q^2 range, including the normalization region, and over all the detector components. This motivates the choice of a purely angular measurement: an acceptance of tens of mrad can be covered with a single sensor of modern silicon detectors, positioned at a distance of about one meter from the target. It has to be stressed that particle identification (electromagnetic calorimeter and muon filter) is necessary to solve the electron-muon ambiguity in the region below 5 mrad. The wrong assignment probability can be measured with the data by using the rate of muon-muon and electron-electron events.

Another requirement for reaching very high accuracy is to measure all the relevant contributions to systematic uncertainties from the data themselves. An important effect, which distinguishes the normalization from the signal region, is multiple scattering, as the electron energy in this region is as low as 1 GeV. In addition, multiple scattering in general causes acoplanarity, while two-body events are planar, within resolution. These facts allow multiple scattering effects to be modelled and measured by using data.

In experiments dedicated to high-precision measurements, several systematic effects can be explored within the experiment itself. In this respect the proposed modularity of the apparatus will help. A test with a single module could provide a proof-of-concept of the proposed methods.

From the theoretical point of view, the control of the systematic uncertainties requires the development of high-precision Monte Carlo tools, including the relevant radiative corrections to reach the needed theoretical precision. To this aim, QED radiative corrections at leading-logarithmic level resummed at all orders of perturbation theory and matched to the exact $O(\alpha)$ correction are mandatory in order to reach a theoretical precision at the level of $O(10^{-5})$ on the differential cross section. Moreover, by using the ratio of the cross sections in the signal and normalization regions, we expect that the theoretical uncertainties will be further reduced to the level of $O(10^{-5})$, due to partial cancellation of common radiative corrections. Work is in progress to extend to $\mu e \to \mu e$ scattering and to quantify the actual accuracy on the computation of the ratio of cross sections by means of dedicated Monte Carlo simulations. Any further improvement in the theoretical accuracy would require the matching of QED resummation of leading and NNLO corrections with exact two-loop corrections, which are not yet available at present for the $\mu e \to \mu e$ process but are within reach.

All these requirements set an unprecedented standard of accuracy both theoretically and experimentally.

The my knowledge the closest example of a comparable accuracy for a process within QED with a multiple particle final state is represented by the evaluation of the Bhabha cross-section in the small angle limit for the determination of the LEP luminosity.
I would like to remember my personal experience. Together with Nikolay Merenkov, Victor Fadin, Eduard Kuraev, and Lev Lipatov (Andrei Arbuzov joined us later), in 1992 we started to work to reach the highest possible precision in the evaluation of the Bhabha cross section in the small angle limit. In a series of papers [11] we developed a comprehensive method to sistematically take into account the various contributions needed to assess the aimed, in those days unprecedented, precision. At the end the precision reached was estimated to be of the O(10^{-3}) [12]. The work on the Bhabha continued even in the following years and a number of further contributions were calculated and have been added [13].

In order to reach a more accurate determination of the fundamental parameters as α_{QED} and a_μ^{HLO} experimental and theoretical improvements, tests, calculations, have to be worked out. Even if the goal may look challenging and difficult history of the past experiences encourages us not to give up and to go forward.

3 Acknowledgments

I would like to thank the organizers for inviting me to this lively workshop.

References

[1] M.T. Cicero, De Oratore, II, 36.

[2] T. Brahe, Astronomia Nova seu physica coelestis, tradita commentariis de motibus stellae Martis ex observationibus G.V. Tychonis Brahe, Plurium annorum pertinaci studio elaborata Pragae mathematico Joanne Keplero.

[3] S. Laporta, Phys. Lett. B772 (2017) 232.

[4] F. J. Dyson, Phys. Rev. 85 (1952) 631.

[5] S. S. Schweber, "QED and the men who made it: Dyson, Feynman, Schwinger and Tomonaga", Princeton University Press, 1994.

[6] D.W. Duke and J.F. Owens Editors, Perturbative quantum chromodynamics (Tallahassee, 1981). New York : American Institute of Physics, 1981 477 p. , p. 129.

[7] R. Bouchendira et al., Phys. Rev. Lett. 106, 080801 (2011)

[8] M. Della Morte, B. Jager, A. Juttner, H. Wittig, JHEP **1203** (2012) 055.

[9] C. Carloni Calame, M. Passera, G. Venanzoni, L. Trentadue, Phys. Lett. B **746** (2015) 325-329.

[10] G. Abbiendi, C.M. Carloni Calame, U. Marconi, C. Matteuzzi, G. Montagna, O. Nicrosini M. Passera, F. Piccinini, R. Tenchini, L. Trentadue, G. Venanzoni, Eur. Phys. J. C **77** (2017) no.3, 139 doi:10.1140/epjc/s10052-017-4633-z [arXiv:1609.08987 [hep-ex]].

[11] N. P. Merenkov, A. B. Arbuzov, V. S. Fadin, E. A. Kuraev, L. N. Lipatov and L. Trentadue, Acta Phys. Polon. B **28** (1997) 491. A. B. Arbuzov, V. S. Fadin, E. A. Kuraev, L. N. Lipatov, N. P. Merenkov and L. Trentadue, Nucl. Phys. Proc. Suppl. **51C** (1996) 154 A. B. Arbuzov, V. S. Fadin, E. A. Kuraev, L. N. Lipatov, N. P. Merenkov and L. G. Trentadue, Phys. Lett. B **394** (1997) 218 A. B. Arbuzov, V. S. Fadin, E. A. Kuraev, L. N. Lipatov, N. P. Merenkov and L. Trentadue, Nucl. Phys. B **485** (1997) 457 A. Arbuzov, V. S. Fadin, E. Kuraev, L. Lipatov, N. Merenkov and L. Trentadue, Phys. Part. Nucl. **27** (1996) 510 V. S. Fadin, E. A. Kuraev, L. Lipatov, N. P. Merenkov and L. Trentadue, Phys. Atom. Nucl. **56** (1993) 1537 [Yad. Fiz. **56N11** (1993) 145]. JINR-E2-92-577.
See also: A. B. Arbuzov, E. A. Kuraev, N. P. Merenkov and L. Trentadue, Phys. Atom. Nucl. **60** (1997) 591 [Yad. Fiz. **60N4** (1997) 673]; J. Exp. Theor. Phys. **81** (1995) 638 [Zh. Eksp. Teor. Fiz. **108** (1995) 1164]; Nucl. Phys. B **474** (1996) 271.

[12] A. B. Arbuzov, V. S. Fadin, E. A. Kuraev, L. N. Lipatov, N. P. Merenkov and L. G. Trentadue, Phys. Lett. B **394** (1997) 218; Nucl. Phys. B **485** (1997) 457.

[13] See for example (only a partial list): A. Penin and N. Zerf, Phys. Lett. B 760 (2016) 816; Z. Bern et al., Phys. Rev. D63,053007 (2001); E. Glover et al.,Phys. Lett. B 516 (2001) 33; R. Bonciani et al., Nucl. Phys. B 716 (2005) 280; R. Bonciani and A. Ferroglia, Phys. Rev. D72, 056004 (2005); S. Actis et al., Eur.Phys.J. C66 (2010) 585.

AUTHOR INDEX

AUTHOR INDEX

AUTHOR INDEX

EDP Sciences – Web of Conferences
17, Avenue du Hoggar
Parc d'Activité de Courtabouf
BP 112
F-91944 Les Ulis Cedex A
France

ISBN 978-1-5108-6390-3

9 781510 863903

CURRAN ASSOCIATES INC.
proceedings.com

Twelfth International Conference on Advanced Geographic Information Systems, Applications, and Services (GEOProcessing 2020)

Held at DigitalWorld 2020

Valencia, Spain
21 - 25 November 2020

IARIA